Materials Horizons: From Nature to Nanomaterials

Series Editor

Vijay Kumar Thakur, School of Aerospace, Transport and Manufacturing, Cranfield University, Cranfield, UK

Materials are an indispensable part of human civilization since the inception of life on earth. With the passage of time, innumerable new materials have been explored as well as developed and the search for new innovative materials continues briskly. Keeping in mind the immense perspectives of various classes of materials, this series aims at providing a comprehensive collection of works across the breadth of materials research at cutting-edge interface of materials science with physics, chemistry, biology and engineering.

This series covers a galaxy of materials ranging from natural materials to nanomaterials. Some of the topics include but not limited to: biological materials, biomimetic materials, ceramics, composites, coatings, functional materials, glasses, inorganic materials, inorganic-organic hybrids, metals, membranes, magnetic materials, manufacturing of materials, nanomaterials, organic materials and pigments to name a few. The series provides most timely and comprehensive information on advanced synthesis, processing, characterization, manufacturing and applications in a broad range of interdisciplinary fields in science, engineering and technology.

This series accepts both authored and edited works, including textbooks, monographs, reference works, and professional books. The books in this series will provide a deep insight into the state-of-art of Materials Horizons and serve students, academic, government and industrial scientists involved in all aspects of materials research.

More information about this series at http://www.springer.com/series/16122

Amit Kumar Nayak · Md Saquib Hasnain
Editors

Biomedical Composites

Perspectives and Applications

Springer

Editors
Amit Kumar Nayak
Department of Pharmaceutics
Seemanta Institute of Pharmaceutical Sciences
Mayurbhanj, Odisha, India

Md Saquib Hasnain
Department of Pharmacy
Palamau Institute of Pharmacy
Chianki, Daltonganj, Jharkhand, India

ISSN 2524-5384 ISSN 2524-5392 (electronic)
Materials Horizons: From Nature to Nanomaterials
ISBN 978-981-33-4752-6 ISBN 978-981-33-4753-3 (eBook)
https://doi.org/10.1007/978-981-33-4753-3

This Springer imprint is published by the registered company Springer Nature Singapore Pte Ltd.
The registered company address is: 152 Beach Road, #21-01/04 Gateway East, Singapore 189721, Singapore

Preface

Composites designed for the uses in various biomedical applications are usually well-known as "biomedical composites". There are numerous prospective biocomposites, which can be classified into various overlapping categories. The most important applications of biomedical composites include drug delivery, gene delivery, cancer therapy, antimicrobial, tissue regeneration, wound healing, dentistry, prosthesis, bioimaging, biosensors, *etc.*

The current book entitled *Biomedical Composites: Perspectives and Applications* covers the recent innovations in the developments of various biomedical composites and their applications in different biomedical uses like drug delivery, gene delivery, cancer therapy, antimicrobial, tissue regeneration, wound healing, dentistry, prosthesis, bioimaging, biosensors, *etc.* The current book is a collection of 9 chapters presenting different key topics along with emphasis on recent advances in the biomedical field by academicians and researchers across the world. A concise account on the contents of each chapter has been described to provide a glimpse of the book to the readers.

The Chapter entitled "**Biomedical Applications of Nanosilicate Composites**" describes different uses of nanosilicate composites in tissue engineering and regenerative medicine, dentistry, drug delivery, wound healing, *etc.* In addition, a brief introduction of biomedical composites has been presented.

The Chapter entitled "**Polysaccharide-Based Composites for Biomedical Applications**" presents the most relevant and more recent advances in the development of polysaccharide-based composites and their uses in biomedical fields like drug delivery, biosensors, wound dressing, tissue engineering, *etc.*

The Chapter entitled "**Biomedical Nanocomposites**" offers the most recent significant researches and developments on various nanocomposites for their potential uses in drug delivery, gene delivery, antimicrobial, tissue regeneration, wound healing, dentistry, prosthesis, bioimaging, and biosensors. In addition, nanocomposites and biomedical nanocomposites have been defined with a brief introductory understanding.

The Chapter entitled "**Antimicrobial Nanocomposites**" presents a comprehensive review on the nanocomposites as antimicrobial agents in food packaging, wound dressings, water treatment, anticancer, and dental applications.

The Chapter entitled "**Potential Application of Silver Nanocomposites for Antimicrobial Activity**" highlights the pathogenic strains responsible for the microbial infections, the limitations of the conventional treatment strategies, emergence of silver nanocomposites as the new treatment strategy. Various synthesis methods of the silver nanocomposites and their applications in the area of antimicrobial activity and its ultimate mechanism of action behind it are discussed. In addition, the challenges and future perspectives to be met in order for the silver nanocomposites as excellent antimicrobial agents in the upcoming period of commercial market have been addressed.

The Chapter entitled "**Cellulose and Chitin Nanofibers: Potential Applications on Wound Healing**" discusses the uses of cellulose, chitin, and their derivatives to prepare cellulose nanofibers and chitin nanofibers for wound healing applications.

The Chapter entitled "**Bionanocomposites for In Situ Drug Delivery in Cancer Therapy: Early and Late Evaluations**" highlights the recent advances in the development of bionanocomposites for *in situ* drug delivery, their efficacy and impact on cancer therapy, how to select the right polymer, and how to study polymeric-drug interactions, *in silico*.

The Chapter entitled "**Tumor Microenvironment and Intracellular Signal-Activated Nanocomposites for Anticancer Drug Delivery**" summarizes the recent advances in various tumor microenvironments and intracellular signal-activated nanocomposites for anticancer drug delivery. Future perspectives of design consideration were also discussed in detail.

The Chapter entitled "**Nanocomposites for Cancer Targeted Drug Delivery Therapeutics**" deals with the materials that can be combined to form the most interesting nanocomposites will be summarized by making a classification depending on the nature of one of the constituent materials. Then, polymer-based nanocomposites, clay-based nanocomposites, metal-based nanocomposites, silica-based nanocomposites, magnetic-based nanocomposites, and carbon-based nanocomposites will be described in relation to their application in nanomedicine for cancer therapy.

We would like to convey our sincere thanks to all the authors of the chapters for providing timely and valuable contributions. We thank the publisher—**Springer Nature**. We specially thank **Dr. Vijay Kumar Thakur** (Series Editor, **Materials Horizons: From Nature to Nanomaterials, Springer Nature**), **Swati Meherishi**, **Sushmitha Shanmuga Sundaram**, **Vindhya H Pillai**, and **Ashok Kumar** for their invaluable support in the organization of the editing process right through the beginning to the finishing point of this book. We gratefully acknowledge the permissions to reproduce copyright materials from various sources. Finally, we would like to thank our family members, all respected teachers, friends, colleagues, and dear students for their continuous encouragements, inspirations, and moral supports during the preparation of the current book. Together with our contributing authors and the publishers, we will be extremely pleased if our endeavor fulfills the needs of academicians, researchers, students, biomedical experts, pharmaceutical students, and drug delivery formulators. In a nutshell, it will also help the health professionals in academia as well as in the industries.

Mayurbhanj, India
Daltonganj, India

Amit Kumar Nayak
Md Saquib Hasnain

Contents

About the Editors

Dr. Amit Kumar Nayak (M. Pharm., Ph.D.) is currently an Associate Professor at Seemanta Institute of Pharmaceutical Sciences, India. He has earned his Ph.D. from IFTM University, Moradabad, India. He has over 12 years of research experiences in the field of pharmaceutics, especially in the development and characterization of novel biopolymeric and nanostructured drug delivery systems. He has authored over 123 research and review publications in various high impact peer-reviewed journals and 82 book chapters. He has edited/authored 9 international books. He has earned highly impressive publishing and cited record in Scopus (h-index: 35) and in Google Scholar (h-index: 40, i10-Index: 110). He has received University Foundation Day Research Award-2019 by Biju Patnaik University of Technology, Odisha. Dr. Nayak is life member of Association of Pharmaceutical Teachers of India (APTI) and a Registered Pharmacist.

Dr. Md Saquib Hasnain (M. Pharm., Ph.D.) has over 9 years of research experience in the field of drug delivery and pharmaceutical formulation analyses, especially systematic development and characterization of diverse nanostructured drug delivery systems, controlled release drug delivery systems, bioenhanced drug delivery systems, nanomaterials, and nanocomposites employing Quality by Design approaches and many more. He has authored over 45 publications in various high impact peer-reviewed journals, 80 book chapters, and 10 books to his credit. He is also serving as the reviewer of several prestigious journals. He has an impressive publishing and cited records in Google Scholar (h-index: 18, i10-Index: 35). He has also participated and presented his research work at over ten conferences in India, and abroad. He was also the member of scientific societies i.e., Royal Society of Chemistry, Great Britain, International Association of Environmental and Analytical Chemistry, Switzerland, and Swiss Chemical Society, Switzerland.

Biomedical Applications of Nanosilicate Composites

Ashwini Kumar and Awanish Kumar

1 Biomedical Composites: Definition and Applications

The term "composite" used either by general material scientists or by biomedical engineers can in general be defined as a material, not existing naturally, composed of two or more chemically and physically distinct materials that are arranged in a manner that the new property of the composite is completely different from the individual components. Generally, a composite has a continuous bulk phase called matrix and a non-continuous phase known as reinforcement. The bulk phase or matrix generally has lower Young's modulus and higher elasticity while the reinforcement has high load capacity, physical, and mechanical characteristics. In biomedical engineering, the matrix of a composite is generally a biopolymer while an inorganic bioactive material is considered for reinforcement. These composites, that have specific applications in biomedical engineering, are termed as "Biocomposites" [1, 2]. In composites, matrix or the continuous bulk phase supports the reinforced material and improves the mechanical and physical properties needed for specific biological application. Matrix has also been reported to improve the processing parameters and characteristics of the biocomposites [1]. Biocomposites can contain polymer, ceramic, or metals that individually or together have certain bioactivity or should be bio-inert but should always be biocompatible. The term biocompatibility includes biofunctionality, bioinertness, bioactiveness, and biodegradability. Another important consideration today for choice of materials to be biomaterials is that they should avoid biofouling. Biofouling of biomaterials is one of the biggest medical threats that can result in infections that are emergency conditions [3].

In commercial terms, the global market for biomaterials is promising with greatest demand for heart valves (around 1 lakh per year globally), and orthopedic implants

A. Kumar · A. Kumar (✉)
Department of Biotechnology, National Institute of Technology, Raipur 492010, Chhattisgarh, India
e-mail: awanik.bt@nitrr.ac.in

A. K. Nayak et al. (eds.), *Biomedical Composites*, Materials Horizons: From Nature to Nanomaterials, https://doi.org/10.1007/978-981-33-4753-3_1

(around 1 million per year globally) [3]. This figure consists of both biocomposites and metallic implants. A growing interest in natural biopolymers-based biocomposites owe to great biocompatibility of these biopolymers with the human body and are biodegradable too (if there is need). These polymers include silk, collagen, chitosan, alginate, hyaluronic acid, and starch. There are synthetic polymers such as polyethylene glycol (PEG), poly glycolic acid (PGA), poly lactic acid (PLA), poly-L-Lactic acid (PLLA), poly (lactic-co-glycolic) acid (PLGA), polycaprolactone (PCL), polyanhydrides, and poly propyl fumarate that are biocompatible but have comparatively slower biodegradation rate. These polymers have been extensively studied for various biomedical applications such as drug delivery, tissue engineering, and medical implants. Out of these PEG, PLA, PLLA, and PLGA are FDA-approved biopolymers with Generally Regarded As Safe (GRAS) status [3].

Biocomposites have found substantial applications in medical science, primarily as the biomedical implant materials. Starting with orthopedic implants, UHMWPE (ultra high molecular weight polyethylene) and PEEK (poly ether ether ketone) polymers have been the mainstay in this application to from composites. These are high molecular weight polymers exhibiting extreme resistance to chemicals, pressure, and temperature. Researchers across the globe have been trying to reinforce these polymers with certain inorganic materials for wide application purpose [3]. In case of load bearing bone implants, the traditional use of 316L stainless steel and Ti-6Al-V4 titanium alloy has witnessed an induction of composite materials such as carbon fiber reinforced PEEK or polysulfone matrices. These can be much better tailored as per the mechanical properties of the load bearing bones. The issue of biocompatibility from the carbon debris in the implants has been addressed by coating with hydroxyapatite or the titanium-based alloy. For fracture fixation surgeries, it was necessary that the bone fixation plates are slowly biodegradable so that the need for second surgery to remove the plates can be averted. In this regard, PLA was chosen as the matrix and reinforced either with carbon fiber or the calcium phosphate-glass fiber to make composites. The former inorganic filler is half biodegradable while the latter is completely resorbed gradually [1]. An Interpenetrating-Polymer Network (IPN) formed by mixing Bis-Phenol A-Glycidyl Dimethacrylate (Bis-GMA) and Triethylene Glycol Dimethacrylate (TEGDMA) displayed remarkable biomineralization characteristics and have been used as restorative composite filler in dentistry. Dental composites have been more successful as compared to bare metals and ceramics in dentistry. These composites easily match the physical and mechanical properties of the natural teeth. Dental composites have been used in the preparation of crowns and even as dental fillers. Dental crowns are usually all-ceramic composites. An important example is In-Ceram that is a composite of alumina and glass. Orthodontic archwires are another example of dental composite application. Orthodontic brackets are also formed from a composite structure composed of polyethylene matrix reinforced with HAp nanoparticles [1, 3]. Various other composites such as PCL-Polyurethane, Poly Ethylene Oxide-PCL (PEO-PCL), and Poly (ether) Urathane-Poly Dimethyl Siloxane (PEtU-PDMS) have been tested for synthetic vascular scaffolds for the purpose of arterial and venous grafting. Biocomposites of biopolymer and inorganic substances such as hydroxyapatite (HAp) and tricalcium phosphate (TCP) have been tested extensively for bone tissue engineering and dental applications. This kind of composite is said to mimic the natural bone/cartilage and imparts bone/cartilage-like characteristics to the prepared scaffold [3].

Apart from the various inorganic reinforcement materials used in various biocomposites, the use of clay minerals has recently gained momentum. These are usually also termed *biomedical clays* based on their unconventional application in biomedical field. Clays are the naturally occurring minerals that are important and major constituents of sedimentary rocks. They are primarily composed of silica (layered silicates) mostly in the form of aluminosilicates. The general chemical formula for these aluminosilicates is $(Ca, Na, H)(Al, Mg, Fe, Zn)_2(Si, Al)_4O_{10}(OH)_2xH_2O$, where x represents the amount of water present in the mineral unit. Thus, we can say that clay minerals are hydrated aluminum silicates with variable amount of magnesium, zinc, iron, and other alkali and elements. Clay minerals are broadly classified into two categories: natural and synthetic clays based on the occurrence of alternating tetrahedral SiO_2 and octahedral AlO_6 sheets with varying ratios. This ratio could be 1:1 (one octahedral layer linked to a tetrahedral one), 2:1 (two tetrahedral sheets on either side of an octahedral layer), and 2:1:1 or 2:2 (a positively charged brucite sheet between the layers that restrict the swelling ability). Common biomedical clays are grouped as kaolinite (KAO; 1:1 organization), bentonite (BEN; 2:1 organization), laponite (LAP; 2:1 organization), montmorillonite (MMT; 2:1 organization), and saponite (SAP; 2:1 organization). Most of these biomedical clays are used as nanoclays. Nanoclays are the clay minerals with at least one dimension in the nanometer range (1–100 nm). Nanoclays are either cationic or anionic depending upon their surface charge. For various biomedical applications, nanoclays have mostly been used as reinforcement in the biopolymer matrix. Because of their complete absence of toxicity, they have found their use in the research related to tissue engineering, drug delivery, and wound healing. They have also been tested as bone cement [4]. Figure 1 shows the common classification of clay minerals. ***The coming sections deal with various biomedical clays and their applications in detail.***

2 Clay Minerals and Nanosilicates

Clay minerals or clays have an interesting history of being eaten by primitive humans, Homo habilis, around 2 million years ago. This mode of nutrition was called ***Geophagy***. Even with Homo sapiens, the clays have remained in active use for more than 25,000 years ago. Moraes et al. has also reported that primitive humans used ocher clay mixed in water for wound healing and other skin ailments. Ancient Egypt called the present-day clay minerals as medicinal earth and used them for wound healing and hemostasis. Egyptians also used Nubian-earth as anti-inflammatory material [5, 6]. As stated in the section above, clays are the natural minerals that are chemically hydrous aluminum silicates with a variety of other doped cations such as magnesium, iron, and few others. Clays are considered to be environmental friendly and safe for human applications. The versatility of clay minerals for human

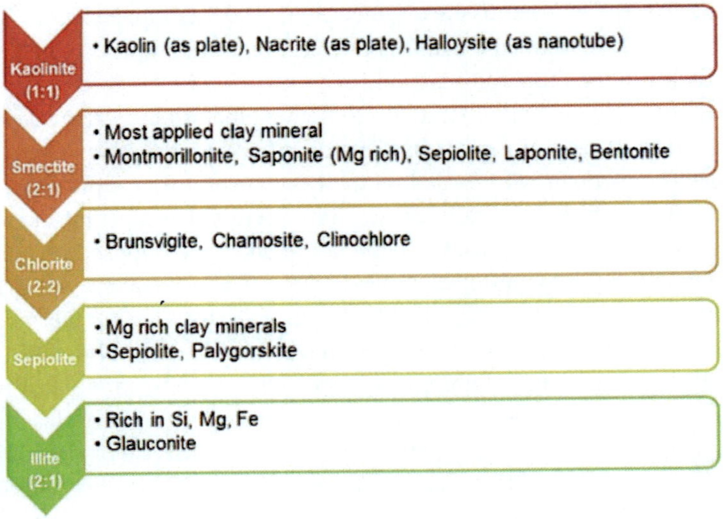

Fig. 1 Clay mineral classification (from the information provided in https://pubs.usgs.gov/of/2001/of01-041/htmldocs/clay.htm and reference # 8); The ratio mentioned shows the ratio of SiO_2 and AlO_6

application owe to their properties such as high adsorption quality, high surface area, colloidal property, thixotropy, swelling capacity, and chemical inertness. Secondly, they have very low toxicity upon oral administration (as also revealed by the geophagy culture). Among the naturally occurring clay minerals, kaolinite (kaolin, halloysite), smectites (MMT, SAP), and sepiolite are the most used for pharmacological applications. Though considered very safe on oral administration, the concern of formation of renal stone on long term consumption has also been raised. Secondly, adsorption of gastric enzymes could also take place that might disturb normal digestion process [7]. Clay minerals show different water solubility. Clays such as illite, sepiolite, and palygorskite do not swell in water while montmorillonite swells easily and forms a gel-like structure. A brilliant comprehensive review by Kotal and Bhowmik on various preparations and characteristics of nanoclays and polymer composite described various preparations, structures, and properties of such composites [8]. They have described the polymer-based nanocomposites of both natural and modified clays. Among all the clays, MMT, kaolin, and saponite have found most applications in fields such as electronics, biomedical engineering, optics, and aeronautics. The nanoscale form of clays has improved physico-mechanical characteristics such as tensile strength, modulus, and lower thermal expansion coefficient. Various methods such as solution intercalation, melt mixing, and in situ polymerization have been applied to form the polymer-based nanocomposites. Polymers with more polar groups are reported to interact well with the nanoclays [8].

To start with very basic medical applications, clay minerals have long been used in various semi-solid formulations such as gels, ointments, lotions, shampoo, and pastes. A review by Viseras et al. has mentioned various cosmetic and health-care semi-solid preparations that consist of clays such as montmorillonite, kaolin, saponite, and sepiolite [9]. Khurana et al. [10] have reviewed the pharmaceutical properties and applications of various clays. Clay minerals find their first use as excipients in pharmaceutical industry for various purposes such as enhancing color, flavor, increasing viscosity, emulsifying agent, lubricating agents, and binding agents. Apart from the application as excipients, various clays have shown to have certain protective action over pathological conditions. Clays have shown to be effective in protecting gastric ulcers by acting as antacid agents (smectite, sepiolite), anti-diarrhoeal agents (smectite, sepiolite, kaolinite), dermatological protection (kaolinite, smectite), and anti-inflammatory (kaolinite). The details of these works shall be discussed in the section specific to the application. In terms of use, Moraes et al. has reported that approximately 30 clay minerals are being used currently in various biomedical and cosmetic industries. Apart from the widespread common use of kaolinite and montmorillonite, halloysites have recently gained prominence in the field of nano-therapeutic delivery. As per the US and European pharmacopoeias, kaolinite, montmorillonite, saponite, talc, sepiolite, and palygorskite are approved for various pharmaceutical and cosmetic purposes [6]. Since the clay minerals are phyllosilicates (silicate-based minerals), these nanoclays have been described as nanosilicates by Gaharwar and team in various researches focusing multifaceted biomedical applications of clay minerals [11–13].

The wide application of Kaolinite can be seen from an extensive review by Awad et al. in 2017 [14]. Kaolinite which is chemically $Al_2Si_2O_5(OH)_4$ is one of the most abundant and inexpensive clay mineral in the earth's crust. Other members of kaolinite family are halloysite, dickite, and nacrite. Kaolinite is hydrous aluminum silicate with 1:1 ratio of alumina (octahedral surface) and silica (tetrahedral siloxane surface) arranged in dioctahedral fashion. The average platelet size of kaolinite (pseudohexagonal platelets) is reported to be <2 μm which is much lesser in size than average human RBCs. The edges of these alumina and silica layers contain oxygen atoms and hydroxyl (OH) groups. In solutions or environment with acidic range ($pH < 3.6$), the hydroxyl groups become protonated, while deprotonation happens at $pH > 3.6$. Thus, kaolinite surfaces experience two types of charges: the permanent negative charge on the tetrahedral face and the variable pH-dependent charge caused by the protonation or deprotonation of hydroxyl groups on the amphoteric sites, at the edges and the octahedral faces. Pharmaceutical grade kaolin has been in application for long as an excipient with usage such as binder, diluent, particle film coating, and disintegrant. Kaolin has been reported to be a good disintegrant in solid dosage forms due to its high porous nature and swelling ability. Kaolin has also been reported for drug delivery applications. Kaolin, in combination with other materials such as PEG and alginate has been shown to have modified dug release activity. Apart from its use as pharmaceutical excipients, the kaolin has found its commercial use as an active heamostatic agent. Because of the negative charge of kaolinites surface in the pH range of human blood, the kaolin nanoplatelets instantly coagulate the bleeding as soon as it comes in contact with the blood. Specifically, kaolin is seen to enhance the activity of coagulation factor XII that in turn activates coagulation factor XI and pre-kallikrein. Therefore, kaolin has been used in various commercially available wound dressings and hemostatic dressings such as QuikClot Combat Gauze™ (QCG), QuikClot Combat GauzeXL (QCX), QuikClot Combat Gauze TraumaPad (QCTP), and QuikClot® Interventional which are FDA approved products [7, 14]. Kaolinite has also been reported to be used in preparations and formulations concerning inflammation (Caloplast, KL kaolin), dermatological concerns (kerodex, REINOL Aquagard, CALMA mask, Lactocalamine lotion), gastrointestinal and anti-diarrhoeal agents (ASDA Tablets, Entrocalm, Boots Kaolin, Kaopectate, Kaomix suspension, Kaolin Antacil), anti-bacterial material, antiviral agents, and anti-tumor compounds [14].

Montmorillonite (MMT), which is chemically $[(Na, Ca)_{0.33}(Al, Mg)_2(Si_4O_{10})(OH)_2 \cdot nH_2O]$, is a 2:1 dioctahedral silicate mineral having two tetrahedral and one octahedral sheet. A typical MMT particle or nanoplatelet is approximately 1 nm thick and 0.2–2 μm in diameter. The layered structure and the spaces between the layers provide excellent adsorption characteristics. Depending upon the geological location, MMT is either sodium MMT or calcium MMT. It has been reported that Na-MMT shows more swelling ability than Ca-MMT, therefore is more applied for drug delivery applications. The biomedical usefulness of MMT can be seen from the fact that hardly any oral toxicity in mice was seen at a dose of 1000 mg/kg body weight. MMT has much higher swelling capacity as compared to kaolinite. MMT has been extensively evaluated as a drug carrier for novel sustained delivery formulations with drugs such as gentamicin, metronidazole, praziquantel, chlorpromazine, metformin, and insulin [7, 15]. A recent review has mentioned and referred vast literatures suggesting that MMT has been extensively used for wound healing, tissue engineering, and regenerative medicine [16]. In reports published in 2014 and 2017, researchers have reported that ingestion of Uniform Particle Size NovaSil and ACCS100 (pharmacological grade refined Ca-MMT used in adults) in aflatoxin-affected children significantly reduced the urinary metabolite of aflatoxin (AFM1) without having any secondary adverse effect [17, 18]. This further confirms the excellent biocompatibility of MMT.

Laponite is a synthetic clay belonging to the Smectite group due to its chemistry and structure. Being synthetic clay, Laponite enjoys having minimal impurities, low toxicity, and controlled nanosized structure. It is available in various grades, but Laponite XLG is the medical grade version. Laponite crystals are disc shaped with approximate diameter of 25 nm and a height of 0.92 nm with a chemical formula $Na^{+}{}_{0.7}[(Si_8Mg_{5.5}Li_{0.3})O_{20}(OH)_4]^{-0.7}$. Biocompatibility and biodegradation are two important parameters for any material to be used in biomedical applications. Laponite has been shown to degrade in acidic condition generating aqueous silica, sodium, magnesium, and lithium. Laponite XLG has been shown to be cytocompatible with human mesenchymal stem cells till a concentration of 1 mg/mL [19].

Halloysite, a natural kaolinite family member, is a 1-dimensional hollow nanotubular structure nanosilicate. Thus, it is a naturally occurring nanotube and hence referred to as halloysite nanotube (HNT). It has a chemical formula of $Al_2Si_2O_5(OH)_4.nH_2O$. The wall of the HNTs consists of 10–15 bilayered structure with an approximate spacing of 0.72 nm between the walls. HNT surface shows positive charge in the inner surface while negative charge on the outer surface on a wide pH range. Typical length of HNTs is in the range of 0.2–1.5 μm while the inner and outer diameters are in the range of 10–30 nm and 40–70 nm, respectively. Since the surfaces of the HNTs can be widely modified and HNTs are reported to be highly biocompatible, there has been a growing interest among biomedical researchers to use the natural nanotube as a drug delivery carrier. Apart from the small molecule drugs, researchers have even tested the loading and release of insulin and found that insulin is released in a sustained manner for 7 days and stability assay confirmed that the released insulin is in native state [7, 20]. The unique tubular structure of HNTs has enabled their successful application in enzyme immobilization too apart from its investigated role in drug delivery, tissue engineering, and wound healing [20]. Further insight into the characteristics and application of HNTs can be referred from reference number 20. Figure 2 is a pictorial representation of various biomedical applications of nanoclays.

Fig. 2 Various biomedical applications of nanoclays/nanosilicates

3 Nanosilicate Composites in Tissue Engineering and Regenerative Medicine

Since there is a huge gap in the number of tissues/organs available and the number of recipients waiting for the transplant, tissue engineering research is perhaps the only ray of hope for replacement of diseased or damaged tissues/organs. In the seminal paper on tissue engineering, Langer and Vacanti have clearly stated that "*Tissue engineering is an interdisciplinary field that applies the principles of engineering and the life sciences toward the development of biological substitutes that restore, maintain, or improve tissue function*" [21]. In correlation to the natural extra-cellular matrix of the tissues/organs, tissue engineering approach requires the fabrication of a suitable scaffold as the primary step. Though tissue engineering approach has been in concept since early 90s and clay minerals are used in cosmetic products for long, the application of nanoclays or nanosilicates as a component of scaffold has gained recent attention and interest. The scaffold fabrication is largely dependent upon biopolymers. A lot of importance is given to doping these polymers with suitable inorganic components to render them with better biocompatibility and functionality. Nanosilicates have gained recognition because of their compatibility with a wide range of biopolymers and their ability to be easily chemically modified. Because of their considerably good biocompatibility and bioactivity, they have been used as a dopant with biopolymers in the preparation of suitable scaffolds using either traditional way of scaffold fabrication or 3D printing. Important to mention here again, nanosilicates are disc-like 2D nanoplatelets that are approximately 1 nm in thickness and can have diameter ranging from 25 nm to over 100 nm [22]. The structure of nanosilicates is such that there are permanent negative charges on the surfaces while positive charges on the edges. The high surface area (>900 m^2/g) and effective charge distribution makes them attractive materials for chemical modification and make beautiful interaction with the biopolymers. Since there has been a lot of attention over bioink fabrication after the advent of 3D bioprinting, the remarkable cytocompatibility of nanosilicates with the mammalian cells has made these materials highly significant in recent bioink fabrication. Small amount of nanoclays or nanosilicates can significantly improve the mechanical behavior of the hydrogel network that is supposed to be used for scaffold fabrication. Another important feature of these nanosilicates is their applicability in drug delivery that is often an important consideration of tissue engineering [22]. Naumenko et al. [23] used halloysites nanotubes (HNTs) as the dopant in the chitosan-agarose-gelatin blend to create a biocompatible and mechanically superior printable bioink. Halloysite tubes have been reported to have length range of 0.3–1.5 μm while the inner and outer diameter is approximately 10–20 nm and approximately 40–70 nm, respectively. The innermost alumina surface of the HNTs is positively charged while the outermost silicate surface is negatively charged. The inner lumen of HNTs can be used as nanocontainers for loading drugs and other pharmaceutically important agents. Animal studies have shown that these HNT doped chitosan-agarose-gelatin scaffolds displayed complete biodegradation and restoration of complete blood supply in 6 weeks with significant biocompatibility [23]. Paul et al. [24] in 2016 reported the osteogenic differentiation of human mesenchymal stem cells (without any osteogenic growth factor such as bone morphogenic protein or BMPs) when seeded on nanosilicate doped GelMa (methacrylated gelatin) polymer as the scaffold polymer. There were no signs of cellular apoptosis and no evidence of inflammatory signals. Such bioactive hydrogels have a high scope of being used as viable bioink for tissue engineering applications, particularly bone tissue engineering. In support of this MSCs differentiation capacity of nanosilicates, Carrow et al. [25] analyzed the transcriptome of the MSCs upon differentiation under the influence of Laponite. Laponite XLG is a synthetic nanosilicate finding a wide range of biomedical applications recently. The intracellular dissociation of Laponite into Si, Mg, and Li is reported to be the best possible reason for the osteogenic and chondrogenic differentiation of MSCs without any specific growth factor. Carrow et al. [25] also mentioned that Si ions induced cWnt pathway leading to the chondrogenic differentiation of MSCs. Similarly, an upregulation of Col1A1 and VEGF has been shown to be triggered by magnesium ions. The effectiveness of Laponite in osteogenic and chondrogenic differentiation of MSCs was also evident from the fact that there was a significant upregulation of Col1A1 expression (osteogenic), alkaline phosphatase (ALP) expression (osteogenic), mineralization (osteogenic), glycosaminoglycan (chondrogenic), and aggrecan (chondrogenic). Nojoomi et al. [26] has reported that Laponite doped imine-functionalized cross-linkable polyethylene glycol (PEG) injectable hydrogel was quite effective as a cytocompatible biocomposites hydrogel. Balakrishnan et al. also reported the formation of composite of high-density polyethylene (HDPE) with hydroxyapatite and MMT for the purpose of fabrication of bone scaffold. They, however, concluded that doping of MMT has a favorable effect on the mechanical properties of HDPE/HA composite but there was a significant loss of bioactivity [27]. Recently, Wang et al. [28] have fabricated electrospun Laponite XLG reinforced polycaprolactone (PCL) nanofibres for bone tissue engineering application and found positive results. Recently, Carrow et al. [29] fabricated Laponite XLG doped poly(ethylene oxide terephthalate)/poly(butylene terephthalate) (PEOT/PBT) copolymer for bone tissue engineering using 3D printing. Earlier, PEOT/PBT copolymer has been shown to have calcification and bone bonding ability in vivo, while 2D nanosilicates have the property of osteogenic differentiation of human mesenchymal stem cells (hMSCs) even in the absence of osteoinductive agents or growth factors. Therefore, nanosilicate doping increased the bioactivity of the PEOT/PBT scaffold that was confirmed by observing significant upregulation of osteo-proteins and mineralization. Very recently, Nadernezhad et al. [30] successfully developed agarose-Laponite RD composite shear-thinning bioink for fabrication of scaffold for bone tissue engineering using 3D printing. Laponite was also mixed with methacrylated chitosan and methacrylated gelatin to create a suitable bioactive bioink for bone scaffold fabrication by Cebe et al. [31]. Methacrylated chitosan-Laponite composite displayed better cell adhesion, osteogenic differentiation, and mineralization as compared to the gelatin counterpart. Addition of Laopnite to human adipose-derived stem cells resulted in approximately 70-fold increase in Runx2 (osteo-specific marker), nine-fold increase in osteocalcin, and 45-fold increase in the expression of osteopontin.

Secondly, there was a significant production of collagen 1 which is a hallmark of mineralized matrix [32]. Apart from the conventional biocomposites-based scaffolds, Laponite has also shown to be a great dopant in formulation of injectable self-healing hydrogel for skeletal tissue engineering [33]. HNTs have been reported to be combined with a variety of biopolymers such as agarose, gelatin, chitosan, and gellan gum to fabricate bioinks for relevant scaffold fabrication. These HNT doped biopolymeric scaffold displayed significant cytocompatibility [34]. Ahlfeld et al. have fabricated Laponite XLG reinforced alginate-methylcellulose bioinks for tissue engineering. They successfully demonstrated that hMSCs remained viable (70–75%) for 21 days within the fabricated construct. Moreover, the shape of the construct was preserved over longer period too. Secondly, they also demonstrated that addition of Laponite resulted in sustained release of model proteins (BSA and VEGF) as compared to the construct without Laponite [35]. Very recently, Cidonio et al. fabricated VEGF loaded Laponite-GelMA bioink as a prospective scaffold candidate for bone tissue engineering. They have demonstrated significant angiogenesis through VEGF-LAP-GelMA hydrogel as compared to that from VEGF-GelMA hydrogels suggesting a promoting role of Laponite in retention of VEGF [36]. In a recent study, Cui et al. have developed photocrosslinkable methacrylated glycol chitosan reinforced with MMT as an injectable in situ polymerizing hydrogel for bone regeneration without any growth factor or other osteogenic proteins. The most important characteristic highlighted in this research was the ability of the modified chitosan-MMT blend to recruit native osteoblast cells from the injected environment to generate de novo osteogenesis [37]. As we know, 3D printing and 4D printing has gained momentum since the last few years, preparation of bioinks has become a completely separate field of research for material and biomedical scientists. Nanomaterials such as graphene, carbon nanotubes (CNTs), nanoclays, HA nanoparticles, tricalcium phosphate nanoparticles, and iron oxide nanoparticles have been doped to display favorable results. Since nanoclays or nanosilicates are much cheaper and abundant than CNTs and have much better biocompatibility, they have gained better response as reinforcement in biocomposites. These biopolymer-nanoclay composites have shown improved mechanical and physico-chemical characteristics. Till date, the nanosilicates have been used as dopant with alginate, chitosan, GelMA, and hyaluronic acid. Because of their surface charge, nanosilicates form effective chemical crosslinks with the polymer chain providing suitable characteristics. Nanosilicates can bind with multiple polymer chains simultaneously providing balanced and even stress dissipation. Overall, even small amounts of nanosilicates can be applied to significantly increase the stiffness, fracture energy, extensibility, and printability of biopolymeric hydrogel networks. The electrostatic interactions between nanosilicates and biopolymers confer the bioinks with a revocable network structure that collapses during flow though the extruder (nozzle) and then quickly reforms after printing [22].

4 Nanosilicate Composites in Dental Research

Dental restorative materials are used in dentistry to replace the lost part (dental caries or trauma, etc.) of the crown or the cavity made during root canal treatment. The restorative materials should have mechanical properties similar to the natural tooth and provide aesthetic features too. The materials should be highly biocompatible and should not be degradable. Though silver-mercury amalgam has been used for ages, the possible leakage of mercury still remains the biggest health concern. Use of dental composites has largely replaced amalgam, though is a bit expensive as compared to the latter. Composites (also called dental resin or dental adhesive) made of polymer matrix are now widely used in restorative dentistry to fill cavities (diseased or therapeutic), restore fractured teeth, and to replace missing teeth. The most popular dental composites typically consist of a polymer matrix along with a photo initiator chemical (for photo-curing), a small amount of pigment, and stabilizers along with a high percentage of inorganic filler material (mostly silica). Most common composition use monomers such as bisphenol A glyceryldimethacrylate (Bis-GMA), triethylene glycoldimethacrylate (TEGDMA), and urethane dimethacrylate (UDMA). Though composites are widely used, they face the issue of polymerization shrinkage creating a minute gap between the composite and the natural enamel. This space (microcavities) is a suitable place for bacterial adhesion and subsequent infection. Triethyleng-lycoldimetachrylate (TEGDMA) is frequently added to reduce the viscosity of Bis-GMA polymer matrix. But if the photo-curing is not proper, there is a high possibility of leakage of TEGDMA which is cytotoxic. Other challenges in conventional polymer composites are mechanical stability, biocompatibility, carries inhibition activity, and marginal leakage. Since polymerization shrinkage is a serious clinical failure, it has been reported that addition of inorganic filler materials such as nanoclays reduces the shrinkage and renders better mechanical properties to the composite matrix [38]. It has been reported that addition of nanoclays grafted filler as the adhesive resin increases the mechanical properties of the composite along with improved biocompatibility. In this regard, more than a decade ago, Atai et al. [39] used PMMA grafted MMT as the restorative filler for dental adhesive and demonstrated significantly higher mechanical properties. de Menezes and da Silva also improved the mechanical properties of the dental resin by using synthetic MMT (Dellite 67G and Viscogel B8) as the nanofillers by using only 0.2% w/w as the optimum concentration [40]. Menezes et al., in a separate work, showed that the nanoclays modified composite was much better in all aspects compared to the composites containing hydrophilic silica and organomodified silica. The results were best when the concentration of nanoclays was just 2.5% [41]. The features of nanoclays are so attractive that very recently Nikolaidis et al. used various modified MMTs as nanofillers in the dental polymeric composites [42]. Glass Ionomer Cement (GIC) are conventional filling materials in restorative dentistry. The clinical GIC consists of two mixing components viz. the liquid component (aqueous solution of polyacrylic acid stabilized with 5% tartaric acid copolymers) and the powder component that is a fluoro-aluminosilicate glass. GIC is preferred in dentistry due to good chemical bond to the tooth structure, anti-cariogenic properties, and excellent biocompatibility. But, even GIC faces the same problem of low mechanical strength and high wear chance. Fareed and Stamboulis prepared MMT linked GIC polymer and improved the mechanical properties of the GIC composite that could be even better utilized in dentistry [43].

5 Nanosilicate Composites in Drug Delivery Research

Sustained or controlled drug delivery is required for reasons such as effective dose at required site for longer time, reduced drug dose, reduced administration frequency, and overall reduced adverse effects. With this kind of delivery pattern, the drug remains at the plasma therapeutic level for a longer time. Secondly, such delivery systems can also be engineered to have a targeted delivery of the drug. Nanoparticles (or microparticles) have been used for almost last 30 years for drug delivery starting with the use of liposomes as the vehicles. Though there are commercially available nanoformulations, often the need arises to have a material that could be easily engineered for drug loading and targeting, and is extremely biocompatible, non-immunogenic, low cost, and easily available. Nanoclays fit all the above-mentioned criteria. They even have unique properties such as intercalation, swelling, and non-toxic degradation products [44].

The general preparation of clay–drug complex is usually carried out by mixing a proper volume ratio of an aqueous dispersion of clay with a suitable solution of drug. The solid phase is filtered off, washed several times with an appropriate solvent so that the physically adsorbed drug is washed off, and dried under vacuum. But the chemical reason behind the formation of nanoclay-drug complex is a simple chemistry. Most of the biomedically relevant and used nanosilicates such as kaolin, MMT, and halloysite are cationic clay, i.e., they have an overall permanent negative charge on the surface. Therefore, they are able to interact with the basic drugs. But for controlled drug release, the drugs are also intercalated between the nanosilicate layers. For example, the common anti-cancer drug doxorubicin (DOX) has been loaded in both kaolin (methoxy intercalated) and HNT with a loading efficacy of approximately 54% and 80%, respectively. It was further showed that at pH 5.5 (approximate pH of cancer tissue), cumulative release of approximately 32% DOX was observed in 30 h while the release was only about 9.5% at neutral pH. This pH responsive delivery shows to be an effective characteristic for targeted drug delivery that could minimize the adverse effects of the anti-cancer drugs [44, 45].

MMT has been one of the most investigated natural nanosilicate for biomedical applications. It is highly resistant chemically and is stable under acidic conditions too. It has an appreciable swelling capacity and is a potent detoxifier too. PLGA is an FDA approved GRAS polymer for biomedical applications. Jain and Datta developed an MMT-dexamethasone-PLGA composite for sustained release of the drug since dexamethasone (DEX) is hydrophobic with low bioavailability. Comparison with DEX-PLGA system, the introduction of MMT in the MMT-DEX-PLGA system showed no burst release with a much controlled release of the drug for over 30 h

[46]. Venlafaxine is an anti-depressant that has a short half-life of approximately 4 h. Thus, the medicine is required to be administered 2 to 3 times a day. Jain and Datta in yet another experiment reported the formulation of MMT-alginate microspheres for effective loading (~97%) of Venlafaxine while the release displayed minimal burst release with approximately 20% release in 26 h (in gastric fluid) and 22% release in 29 h (in intestinal fluid). They, therefore, concluded that the presence of MMT nanoclays prevented the loss of drug from burst release and also rendered a controlled release feature to the formulation [47]. A very comprehensive review of application of MMT in drug delivery has been reviewed by Jayrajsinh et al. [15]. Laponite, a synthetic nanoclay, has gained a huge momentum and interest in various biomedical applications. The stacked structure, charge distribution over surface, and no impurity have been the major forces behind using Laponite as a drug delivery vehicle. For example, Doxorubicin was shown to be intercalated in the LAP layers and displayed release at the acidic pH (gastric pH) while minimal release at neutral pH. In vitro anti-cancer activity demonstrated much better cytotoxicity with LAP-DOX complex as compared to the free DOX alone. Being a highly charged nanostructure, LAP is susceptible to form aggregates at high polyelectrolyte environment, such as human blood. Therefore, surface coating or encapsulation of the drug-LAP complex is essential to avoid such aggregations. The ability of LAP to solubilize water insoluble compounds has initiated an interest in LAP assisted photodynamic therapy for cancer. LAP was reported to dissolve aluminum phthalocyanine hydroxide (a photosensitizer) which could generate highly toxic singlet oxygen ion [48]. In a recent report, Becher et al. have fabricated Laponite-based nanohydrogel for drug delivery in combination with polyacrylates and sodium phosphate salts. The unique property of nanohydrogels is that they are not soluble or degradable in physiological media. It was also mentioned that soft nanohydrogels can show easy deformation that might facilitate their passage through crucial physiological barriers that results in increase in their blood circulation lifetime improving their therapeutic efficiency. These nanohydrogels very efficiently encapsulated cisplatin, cyclophosphamide, and 4-fluorouracil. It was demonstrated that IC_{50} of the drug loaded nanohydrogels was much lesser as compared to that of the free drugs. In vivo experiments also revealed that these nanohydrogels do not accumulate in any organ [49]. Supraphysiological dose of therapeutic growth factors and other therapeutic proteins that are used today results in a variety of adverse effects that could be serious. In a recent report by Cross et al. [50] Laponite nanodiscs were used for loading and delivering physiological dose of protein therapeutics using rhBMP2 and TGF-β_3 as the model drugs. This nano-system was used to differentiate human MSCs into osteoblasts (by rhBMP2) and chondrocytes (by TGF-β_3). Therefore, this LAP-based nanodisc formulation of growth factor delivery revealed the usage of physiological effect of these drugs that are being used in supraphysiological dose currently.

As mentioned in previous sections, HNTs are the natural alternative to CNTs on the structural front, but the former is much cheaper and biocompatible than CNTs. HNTs could also be seen as rolled kaolin sheets. Almost a decade ago, Vergaro et al. presented a detailed report on cytotoxicity and biocompatibility of HNTs and concluded that HNTs are highly biocompatible with very minimal cytotoxicity [51].

Lvov and team have probably done a considerable amount of work on HNTs. Three beautiful reviews by his team cited as Lvov et al. [52], Santos et al. [53], and Lvov et al. [54] concerning various detailed aspects of HNTs are worth reading.

6 Nanosilicate Composites in Wound Healing and Hemostasis

Wound healing is a complex physiological cascade that includes various hierarchical stages such as homeostasis stage (blood clotting and immune activation), inflammation (recruitment of immune cells, production of related cytokine, and growth factor), proliferation phase, tissue neoformation (reepithelialization, angiogenensis, and granulation), and subsequent remodeling of the newly generated tissue. Moreover, non-healing wounds (diabetic foot ulcers, venous leg ulcers, and pressure ulcers) and burns impose substantial morbidity and mortality, deeply affecting the quality of life. Secondly, surgical wounds are an important class of wounds that need proper care and any act of negligence could become medical emergency at times. Thirdly, battlefield wounds are serious concern that, if not provided proper treatment, could be detrimental in several ways. Severe cutaneous wounds are a major medical concern, with approximately 300 million chronic and 100 million traumatic wound patients affected worldwide. Moreover, approximately 37 million patients are affected globally by chronic wounds [55]. As mentioned in the previous section, clay minerals have been traditionally used for wound healing since ages. In modern-day practice too, clay minerals, primarily kaolin, has been an important constituent of dermatocosmetic products. Since the traditional clay mineral forms have impurities in them and because of the therapeutic effects of clay minerals over wound, they have been fabricated into better wound healing constructs for better therapeutic results. Pacelli et al. fabricated injectable hydrogel nanocomposites incorporating methacrylated gellan gum and Laponite that could release suitable drug and antibiotics for wound healing purpose [56]. Ambrogi et al. [57] have fabricated MMT-Chitosan-chlorhexidine films to inhibit biofilm formation in chronic wounds. In another report Aguzzi et al. [58] MMT-chitosan silver sulfadiazine loaded nanocomposites were fabricated for wound healing purpose. In a recent report by Peng et al. [59], tetracycline loaded Laponite nanodiscs were incorporated in PLLA solution to form a porous membrane that could be effectively used as a wound dressing. The fabricated membrane was shown to be highly effective against *Staphylococcus aureus* that was taken as the model organism. Secondly, the membrane displayed good biocompatibility in vitro. Villen et al. [55] fabricated MMT-Norfloxacin composite powder for application of various kinds of acute and chronic wounds.

The blood clotting process, medically known as hemostasis, is a physiological cascade that inhibits blood loss through the formation of a stable cellular plug known as hemostatic clot at the site of injury/bleeding. It is a complex physiological mechanism that involves the coordinated activation and action of platelets, a range of plasma proteins, and coagulation factors. These coagulation factors reach the bloodstream in an inactivated state and get activated as soon as the hemostatic cascade initiates. Overall, the hemostatic action happens via the coordinated action of three physiological mechanisms viz. vasoconstriction, platelet plug formation, and finally blood clotting. Pourshahrestani et al. [60] have published a comprehensive review regarding certain commercial hemostatic constructs that apply zeolite and nanoclays as the major hemostatic constituents. Where QuikClot granular powders, Advanced Clotting Sponge (ACS), and ACS+ are the Zeolite-based hemostatic dressings, QuikCLot Combat Gauge (QCG), QCG XL, QCG Trauma Pad, and QC Interventional are the kaolin-based commercial hemostatic products while WoundStat is a smectite-based commercial hemostatic product available in certain countries.

7 Conclusions

While a lot has already been mentioned in the chapter, conclusively we can see that natural as well as synthetic nanoclays/nanosilicates are readily available and are a cheaper alternative to various other nanoparticles available for biomedical applications. While many biomaterials find restricted applications, nanoclays have been used in perhaps all major biomedical applications ranging from tissue engineering, drug delivery, dental restoration, wound healing, and hemostatic agents because of their consistent biocompatibility, non-immunogenicity, and non-toxicity. Still, the use of nanoclays for medical application, apart from the use in hemostatic dressings, is in nascent stage. With FDA approval of various nanoclays for certain medical applications, researchers can utilize the properties of these nanoclays to construct useful products.

References

1. Iftekhar A (2009) Biomedical composites. In: Kutz M (ed) Biomedical engineering and design handbook, 2nd edn. McGraw Hill Professional, pp 339–355. ISBN 9780071704724
2. Swain SK, Pattanayak AJ, Sahoo AP (2018) Functional biopolymer composites. In: Thakur VK, Thakur MK (eds) Functional biopolymers. Springer, pp 159–182
3. Mouzakis DE (2013) Biomedical polymer composites and applications. In: Thomas S (ed) Polymer composite, 1st ed, vol 3. Wiley, pp 483–514. ISBN 9783527329809
4. Paras LP, Fernandez JAS, Vidaltamayo R (2019) Nanoclays for biomedical applications. In: Martinez LMT (ed) Handbook of ecomaterials. Springer. ISBN 978-3-319-68254-9
5. Detellier C (2018) Functional kaolinite. Chem Rec 18:1–11
6. Moraes JDD, Bertolino SRA, Cuffini SL, Ducart DF, Bretzke PE, Leonardi GR (2017) Clay minerals: properties and applications to dermocosmetic products and perspectives of natural raw materials for therapeutic purposes-a review. Int J Pharm 534(1–2):213–219
7. Massaro M, Colletti CG, Lazzara G, Riela S (2018) The use of some clay minerals as natural resources for drug carrier applications. J Funct Biomater 9(4):58
8. Kotal M, Bhowmik AK (2015) Polymer nanocomposites from modified clays: Recent advances and challenges. Prog Polym Sci 51:127–187

9. Viseras C, Aguzzi C, Cerezo P, Lopez-Galindo A (2007) Uses of clay minerals in semisolid health care and therapeutic products. App Clay Sci 36(1–3):37–50
10. Khurana IS, Kaur S, Kaur H, Khurana RK (2015) Multifaceted role of clay minerals in pharmaceuticals. Future Sci OA 1(3):FSO6
11. Cross LM, Carrow JK, Ding X, Singh KA, Gaharwar AK (2019) Sustained and prolonged delivery of protein therapeutics from two-dimensional nanosilicates. ACS Appl Mater Interfaces 11(7):6741–6750
12. Howel DW, Peak CW, Bayless KJ, Gaharwar AK (2018) 2D nanosilicates loaded with proangiogenic factors stimulate endothelial sprouting Adv Biosys 1800092
13. Carrow JK, Di Luca A, Dolatshahi-Pirouz A, Moroni L, Gaharwar AK (2019) 3D-printed bioactive scaffolds from nanosilicates and PEOT/PBT for bone tissue engineering. Regen Biomater 6(1):29–37
14. Awad ME, López-Galindo A, Setti M, El-Rahmany MM, Iborra CV (2017) Kaolinite in pharmaceutics and biomedicine. Int J Pharm 533(1):34–48
15. Jayrajsinh S, Shankar G, Agrawal YK, Bakre L (2017) Montmorillonite nanoclay as a multifaceted drug-delivery carrier: a review. J Drug Deliv Sci Tech 39:200–209
16. Viseras C, Carazo E, Borrego-Sánchez A et al (2019) Clay minerals in skin drug delivery. Clays Clay Miner 67:59–71
17. Mitchell NJ, Kumi J, Aleser M, Elmore SE, Rychlik KA, Zychowski KE et al (2014) Short-term safety and efficacy of calcium montmorillonite clay (UPSN) in children. Am J Trop Med Hyg 91(4):777–785
18. Awuor AO, Yard E, Daniel JH, Martin C, Bii C, Romoser A et al (2017) Evaluation of the efficacy, acceptability and palatability of calcium montmorillonite clay used to reduce aflatoxin B1 dietary exposure in a crossover study in Kenya. Food Addit Contam Part a Chem Anal Control Expo Risk Assess 34(1):93–102
19. Tomás H, Alves CS, Rodrigues J (2018) Laponite®: a key nanoplatform for biomedical applications? Nanomedicine 14(7):2407–2420
20. Satish S, Tharmavaram M, Rawtani D (2019) Halloysite nanotubes as a nature's boon for biomedical applications. Nanobiomedicine (Rij) 6:1–16
21. Langer R, Vacanti JP (1993) Tissue engineering. Science 260:920–926
22. Chimene D, Kaunas R, Gaharwar A (2019) Hydrogel bioink reinforcement for additive manufacturing: a focused review of emerging strategies. Adv Mater 1902026
23. Naumenko EA, Guryanov ID, Yendluri R, Lvov YM, Fakhrullin RF (2016) Clay nanotube-biopolymer composite scaffolds for tissue engineering. Nanoscale 8(13):7257–7271
24. Paul A, Manoharan V, Krafft D, Assmann A, Uquillas JA, Shin SR et al (2016) Nanoengineered biomimetic hydrogels for guiding human stem cell osteogenesis in three dimensional microenvironments. J Mater Chem B 4(20):3544–3554
25. Carrow JK, Cross LM, Reese RW, Jaiswal MK, Gregory CA, Kaunas R et al (2018) Widespread changes in transcriptome profile of human mesenchymal stem cells induced by two-dimensional nanosilicates. Proc Natl Acad Sci 115(17):E3905–E3913
26. Nojoomi A, Tamjid E, Simchi A, Bonakdar S, Stroeve P Injectable polyethylene glycol-laponite composite hydrogels as articular cartilage scaffolds with superior mechanical and rheological properties. Int J Polym Mat Polym Biomat https://doi.org/10.1080/00914037.2016.1182914.
27. Balakrishnan H, Husin MR, Wahit MU, Kadir MRA (2014) Preparation and characterization of organically modified montmorillonite-filled high density polyethylene/hydroxyapatite nanocomposites for biomedical applications. Polym Plast Technol Eng 53:790–800
28. Wang Y, Cui W, Chou J, Wen S, Sun Y, Zhang H (2018) Electrospun nanosilicates-based organic/inorganic nanofibers for potential bone tissue engineering. Colloids Surf B Biointerfaces 172:90–97
29. Carrow JK, Di Luca A, Pirouz AD, Moroni L, Gaharwar AK (2019) 3D-printed bioactive scaffolds from nanosilicates and PEOT/PBT for bone tissue engineering. Regener Biomaterials 6(1):29–37
30. Nadernezhad A, Caliscan OS, Topuz F, Afghah F, Erman B, Koc B (2019) Nanocomposite bioinks based on agarose and 2D nanosilicates with tunable flow properties and bioactivity for 3D bioprinting. ACS Appl Bio Mater 2:796–806

31. Cebe T, Ahuja N, Monte F, Awad K, Vyavhare K, Aswath P et al (2018) Novel 3D-printed methacrylated chitosan-laponite nanosilicate composite scaffolds enhance cell growth and biomineral formation in MC3T3 pre-osteoblasts. J Mater Res. https://doi.org/10.1557/jmr.2018.260
32. Gaharwar AK, Cross LM, Peak CW, Gold K, Carrow JK, Brokesh A, Singh KA (2019) 2D nanoclay for biomedical applications: regenerative medicine, therapeutic delivery, and additive manufacturing. Adv Mater 1900332
33. Talebian S, Mehrali M, Taebnia N, Pennisi CP, Kadumudi FB, Foroughi J et al (2019) Self-healing hydrogels: the next paradigm shift in tissue engineering? Adv Sci 6:1801664
34. Bonifacio MA, Gentile P, Ferreira AM, Cometa S, De Giglio E (2017) Insight into halloysite nanotubes-loaded gellan gum hydrogels for soft tissue engineering applications. Carbohydr Polym 163:280–291
35. Ahlfeld T, Cidonio G, Kilian D, Duin S, Akkineni AR, Dawson JI et al (2017) Development of a clay based bioink for 3D cell printing for skeletal application. Biofabrication 9(3):034103
36. Cidonio G, Alcala-Orozco CR, Lim KS, Glinka M, Mutreja I, Kim YH et al (2019) Osteogenic and angiogenic tissue formation in high fidelity nanocomposite Laponite-gelatin bioinks. Biofabrication 11(3):035027
37. Cui ZK, Kim S, Baljon JJ, Wu BM, Aghaloo T, Lee M (2019) Microporous methacrylated glycol chitosan-montmorillonite nanocomposite hydrogel for bone tissue engineering. Nat Commun 10:3523
38. Salernitano E, Migliaresi C (2003) Composite materials for biomedical applications: a review. J Appl Biomater Biomech 1(1):3–18
39. Atai M, Solhi L, Nodehi A, Mirabedini SM, Kasraei S, Akbari K, Babanzadeh S (2009) PMMA-grafted nanoclay as novel filler for dental adhesives. Dent Mater 25(3):339–347
40. de Menezes LR, da Silva EO (2016) The use of montmorillonite clays as reinforcing fillers for dental adhesives. Mat Res 19(1):236–242
41. Menezes LR, da Silva EO, da Silva Rocha AC, de Oliveira DCRS, Campos PRB (2018) The applicability of organomodified nanoclays as new fillers for mechanical reinforcement of dental composites. J Composite Mat 52(7):963–970
42. Nikolaidis AK, Koulaouzidou EA, Gogos C, Achilias DS (2019) Synthesis and characterization of dental nanocomposite resins filled with different clay nanoparticles. Polymers (Basel) 11(4):pii: E730
43. Fareed MA, Stamboulis A (2014) Nanoclays reinforced glass ionomer cements: dispersion and interaction of polymer grade (PG) montmorillonite with poly(acrylic acid). J Mater Sci Mater Med 25(1):91–99
44. Lazzara G, Riela S, Fakhrullin RF (2017) Clay-based drug-delivery systems: what does the future hold? Ther Deliv 8(8):633–646
45. Zhang Y, Long M, Huang P, Yang H, Chang S, Hu Y, Tang A, Mao L (2016) Emerging integrated nanoclays-facilitated drug delivery system for papillary thyroid cancer therapy. Sci Rep 6:33335
46. Jain S, Datta M (2015) Oral extended release of dexamethasone: Montmorillonite–PLGA nanocomposites as a delivery vehicle. App Clay Sci 104:182–188
47. Jain S, Datta M (2016) Montmorillonite-alginate microspheres as a delivery vehicle for oral extended release of Venlafaxine hydrochloride. J Drug Deliv Sci Tech 33:149–156
48. Tomas H, Alves CS, Rodrigues J (2017) Laponite: a key nanoplatform for biomedical applications? Nanomedicine 14(7):2407–2420
49. Becher TB, Mendonca MCP, de Farias MA, Portugal RV, de Jesus MB, Ornelas C (2018) Soft nanohydrogels based on laponite nanodiscs: a versatile drug delivery platform for theranostics and drug cocktails. ACS App Mater Interfaces 10(26):21891–21900
50. Cross LM, Carrow JK, Ding X, Singh KA, Gaharwar AK (2019) Sustained and prolonged delivery of protein therapeutics from two-dimensional nanosilicates. ACS App Mater Interfaces 11(7):6741–6750
51. Vergaro V, Abdullayev E, Lvov YM, Zeitoun A, Cingolani R, Rinaldi R, Leporatti S (2010) Cytocompatibility and uptake of halloysite clay nanotubes. Biomacromol 11(3):820–826

52. Lvov Y, Wang W, Zhang L, Fakhrullin R (2016) Halloysite clay nanotubes for loading and sustained release of functional compounds. Adv Mater 28:1227–1250
53. Santos AC, Ferreira C, Veiga F, Ribeiro AJ, Panchal A, Lvov Y, Agarwal A (2018) Halloysite clay nanotubes for life sciences applications: From drug encapsulation to bioscaffold. Adv Colloid Interface Sci 257:58–70
54. Lvov YM, Shchukin DG, Mohwald H, Price RR (2008) Halloysite clay nanotubes for controlled release of protective agents. ACS Nano 2(5):814–820
55. Villen FG, Faccendini A, Aguzzi C, Cerezo P, Bonferoni MC, Rossi S et al (2019) Montmorillonite-norfloxacin nanocomposites intended for healing of infected wounds. Int J Nanomed 5051–5060
56. Pacelli S, Paolicelli P, Moretti G, Petralito S, Giacomo SD, Vitalone A, Casadei MA (2016) Gellan gum methacrylate and laponite as an innovative nanocomposite hydrogel for biomedical applications. Eur Polym J 77:114–123
57. Ambrogi V, Pietrella D, Nocchetti M, Casagrande S, Moretti V, De Marco S, Ricci M (2017) Montmorillonite–chitosan–chlorhexidine composite films with antibiofilm activity and improved cytotoxicity for wound dressing. J Colloid Interface Sci 491:265–272
58. Aguzzi C, Sandri G, Bonferoni C, Cerezo P, Rossi S, Ferrari F et al (2014) Solid state characterization of silver sulfadiazine loaded on montmorillonite/chitosan nanocomposites for wound healing. Colloid Surf B Biointerfaces 113:152–157
59. Peng Q, Xu P, Xiao S (2018) Porous laponite/poly(L-lactic acid) membrane with controlled release of TCH and efficient antibacterial performance. Fibers Polym 19(3):477–488
60. Pourshahrestani S, Zeimaran E, Djordjevic I, Kadri NA, Towler MR (2016) Inorganic hemostats: The state-of-the-art and recent advances. Mat Sci Eng C 58:1255–1268

Polysaccharide-Based Composites for Biomedical Applications

Patrícia Alves, Filipa Gonçalves, and M. H. Gil

1 Introduction

Polysaccharides have been widely used in the most diverse applications, like packaging, water treatment, bionsensors, and more extensively in biomedical applications such as drug delivery vehicles, scaffolds, wound dressing, and biosensors, among others (Fig. 1). Natural polysaccharide-based composites offer biocompatibility, bioavailability [1], and biological recognitions, making them highly used in biomedical applications. For that reason, in this chapter, the authors gathered the most relevant and more recent advances in the development of polysaccharide-based composites and their uses in biomedical fields.

2 Natural Polysaccharides

The potential of polysaccharides in biomedical fields is explained not only by their natural sources but also due to their similarity with the extracellular matrix components (EMC) [2]. Also, the existence of distinct functional groups promotes the modification of polysaccharides and therefore their wide range of applicability in biomedical areas. Polysaccharides can have different sources: animal, plant, microorganisms, and algae. Starch, cellulose, pectins, and exudate gums are examples of the most common polysaccharides obtained from plants. Alginates and carrageenan are from algae sources while chitin, chitosan, and hyaluronic acid (HA) are derived from animal sources. Dextran, pullulan, xantham gum, bacterial cellulose are examples of

P. Alves · F. Gonçalves · M. H. Gil (✉)
CIEPQPF, Department of Chemical Engineering, University of Coimbra, 3030-790 Coimbra, Portugal
e-mail: hgil@eq.uc.pt

A. K. Nayak et al. (eds.), *Biomedical Composites*, Materials Horizons: From Nature to Nanomaterials, https://doi.org/10.1007/978-981-33-4753-3_2

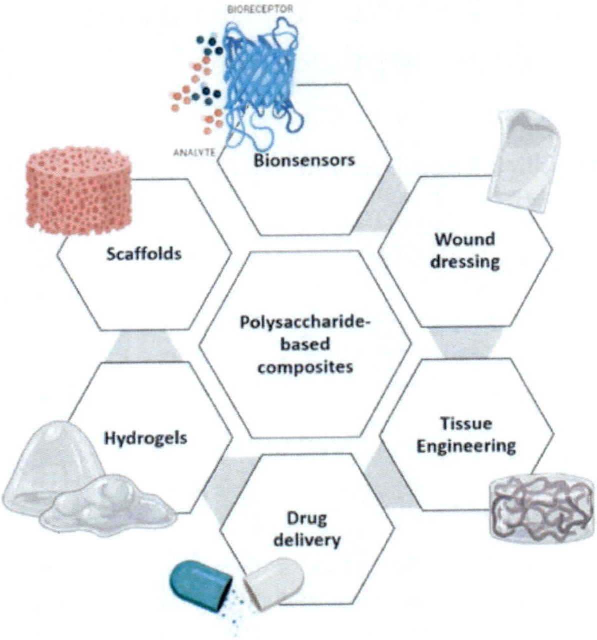

Fig. 1 Main biomedical fields of polysaccharides

polysaccharides obtained from microorganisms (Fig. 2) [1]. Among these biopolymers, cellulose, chitin, and chitosan are some of the most reported polysaccharides in biomedical field in the last decades, not only due to their properties, but also due to their high availability [3, 4].

Cellulose is the most abundant natural polysaccharide, presenting a linear-chain polymer of poly-β-(1 → 4)-D-glucosamine units, with a flat ribbon-like conformation (Fig. 3) [5]. This polysaccharide is characterized by a high amount of hydroxyl groups, three per each glucose unit. These –OH groups are extremely important since they are responsible for cellulose reactivity [2]. Although extensively used due to its high toughness, biodegradability, renewability, and availability [4], this biopolymer has some disadvantages, such as poor solubility, high hydrophilicity, low thermoplasticity, and the lack of antimicrobial properties. These drawbacks can be overcome by modifying the cellulose structure, i.e., introducing specific functional groups can overcome these weaknesses [2].

Chitin, poly (β-(1-4)-N-acetyl-d-glucosamine), is found in the exoskeleton of crustaceans (crabs, lobsters, and shrimps), insects, mollusk, and fungi (cell walls) and is the second most abundant polysaccharide in the biosphere [6]. Chitosan is the most important and used derivative of chitin and can be obtained by the deacetylation of chitin under alkaline conditions or by enzymatic hydrolysis [3]. Both biopolymers are non-toxic, biocompatible, biodegradable, and have antibacterial properties, however, chitosan presents a higher hydrophilicity and water retention capacity [4]. Also, the presence of reactive primary amines and hydroxyl groups in chitosan structure

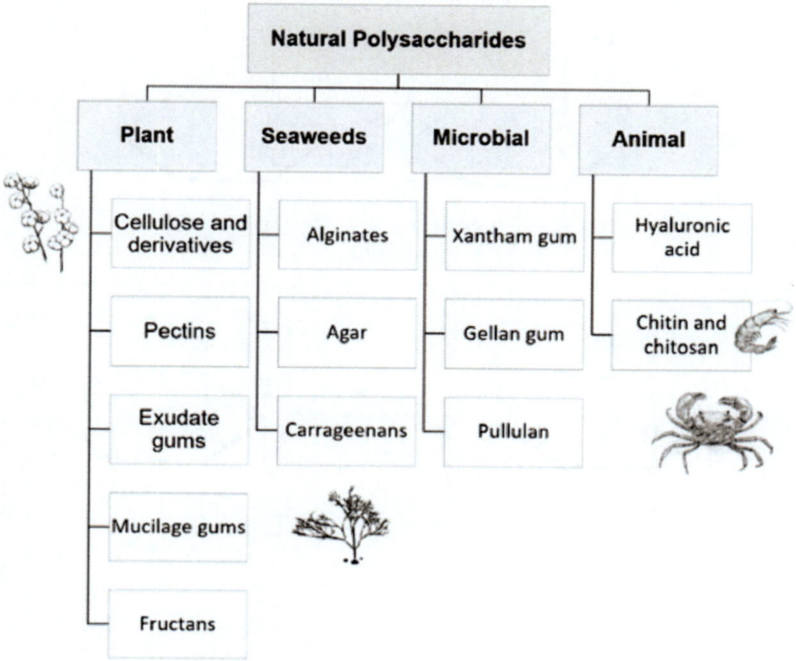

Fig. 2 Main sources of natural polysaccharides (algae, plants, animals, and microorganisms)

promotes its chemical modification [7]. The use of these natural polymers in tissue engineering and pharmaceutical fields is largely explained by their specific biological properties—antioxidant, analgesic, antitumor, and hemostatic properties—that have been recently described by several authors [6, 8, 9].

Hyaluronic acid (HA) is also a very important biopolymer, since it is one of the elements present in the extracellular matrix (ECM) with extensive usage in cosmetics for skin rejuvenescence [10]. HA is a polymer formed by repeating units of N-acetylglucosamine and D-glucuronic acid. Its large use in biomedical fields is strongly connected with its pharmacological activity—anti-inflammatory, wound healing, skin rejuvenescent, anti-aging, anticancer properties [11]—and can be used in the form of hydrogel, injectable creams and foams, scaffolds, membranes, gels, or even encapsulated in a nanocarrier. The broad range of applicability combined with their unique properties makes this polysaccharide one of the main candidate for extensive research and developments.

The previously mentioned biopolymers are highlighted and discussed in this chapter, side by side with other polysaccharides—e.g., dextran, alginic acid, cellulose derivatives—which are extensively explored for a wide range of biomedical applications (Fig. 3).

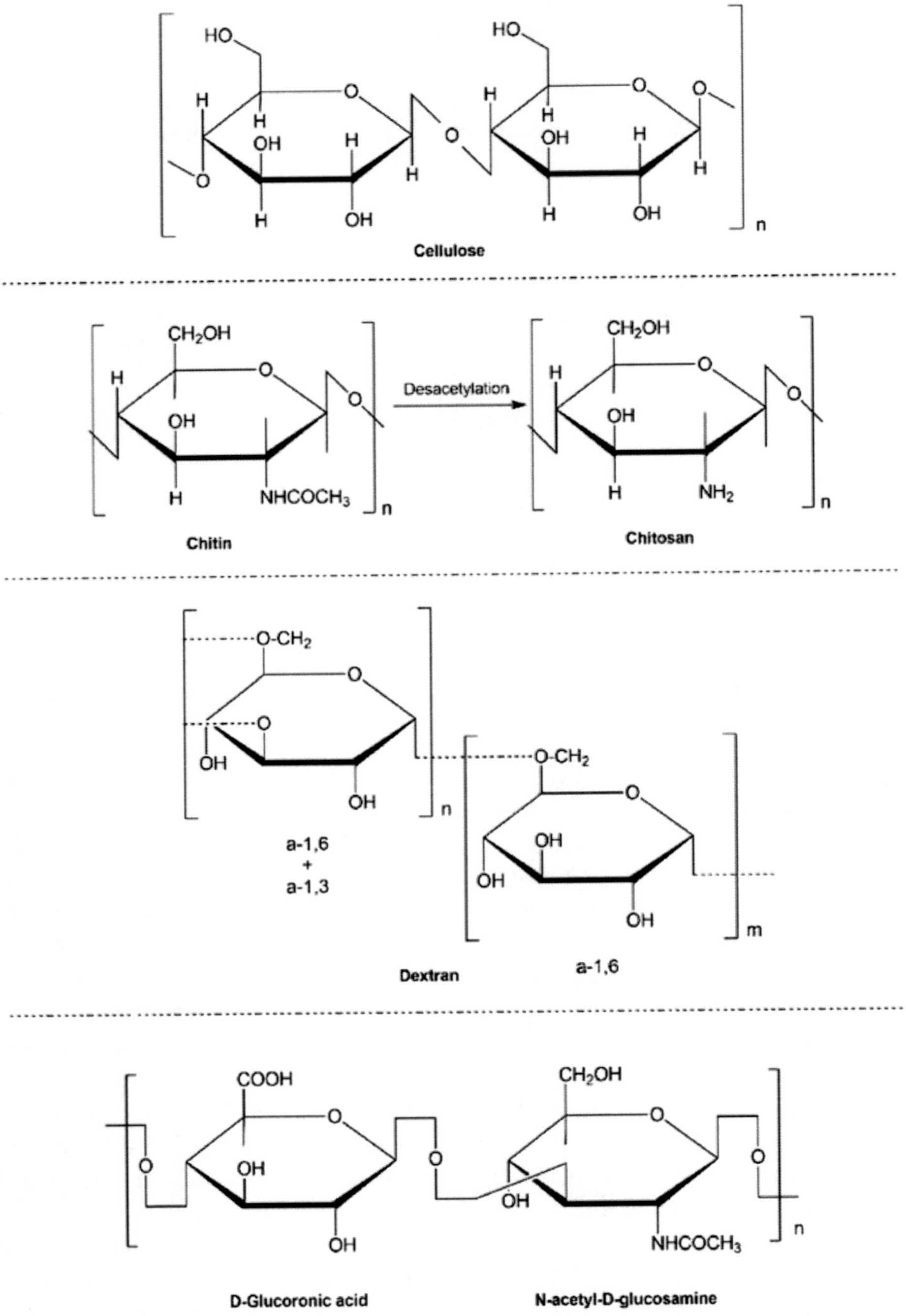

Fig. 3 Main natural polysaccharides used in biomedical fields and respective chemical structure

3 Natural Polysaccharides-Based Composites Applications

Polysaccharide-based composite materials are subject of great interest in a variety of fields, mainly in tissue engineering and drug delivery. The preparation of different polysaccharide-based composites, with different types of reinforcements, is explored in the next sections. Drug delivery, tissue engineering, hydrogels, scaffolds, bionsensors, wound dressings, are some of the most relevant applications of polysaccharide-based composites.

3.1 Drug Delivery

Several developments have been recently made in the field of delivery systems. These developments were made in order to provide therapeutic agents or natural-based active compounds to their target location for treatment of several pathologies [12, 13]. However, there are still some challenges that need to be addressed for a successful delivery of drugs to their target sites. The way that a drug is delivered highly affects the efficacy of the drug. Drug delivery systems (DDS) are pharmaceutical formulations or devices for the targeted delivery and/or controlled release of therapeutic agents. Drug delivery systems can be used either to deliver small molecules or large molecules (peptides, nucleic acids, or polymers). For this purpose, different types of delivery systems have been successfully used such as liposomes, particles, micelles, and niosomes [12]. In the last years, due to their ability to form stimuli-responsive hydrogels, polysaccharides are gaining interest as drug delivery systems. Polysaccharide-based drug delivery carriers usually present an initial burst of the drug due to their lack of mechanical properties. In order to overcome this problem, a filler is usually added into the polysaccharide matrix [1, 14], such as Fe_3O_4 [15], graphene oxide [16], or silica nanoparticles [17].

Some researchers [18] have developed a system to transport and deliver iron along the gastrointestinal tract. Iron deficiency is still a worldwide public health problem combined with the fact that the incorporation of this metal into food matrices does not completely resolve the problem if iron is not properly transported, iron-pectin beads were prepared by ionic gelation and showed to have a potential application in the food and pharmaceutical industry.

Temperature-sensitive hydrogels can be obtained from polysaccharides and applied in the drug delivery area. Almeida et al. [19] prepared thermoresponsive hydrogels from crosslinked copolymers based on dextran and chitosan to prepare Ondansetron™ delivery systems, a drug used for the treatment of nausea and vomiting, to be applied as sublingual formulations. Also, Schmitt et al. [20] produced a temperature-sensitive nanocomposite by dispersing halloysite nanotubes into a starch matrix.

One of the most common methods of drug administration is by using the oral route, since it is easy, convenient, and there is no risk of infection. But they need to pass the gastrointestinal tract, which presents a wide range of pH medium. Changes in pH may affect the efficiency of the drug or even degrade it. Therefore, pH-sensitive polysaccharide composites can be prepared in order to protect the drugs from unwanted environments especially in the stomach before reaching the intestine, where they can be absorbed. Quintana et al. [21] developed a novel strategy for probiotics controlled delivery by encapsulating bacteria using the layer-by-layer self-assembly technique (LbL) with chitosan and poly(acrylic acid) obtaining a permeable high performance drug and storage delivery system.

Another approach is to use polysaccharides to alter de drug delivery device in order to modify the surface of the original material and tune the release of the encapsulated drug. Recently, some studies reported the use of the layer-by-layer technique (LbL), using polysaccharides, such as chitosan [22, 23] and alginate [23] to control ophthalmic drug release from lens materials.

3.2 *Bionsensors*

By definition, biosensors are analytical devices applied for detection of chemical substances, by combining a biological component and a physicochemical detector (a transductor). They can be categorized either according to the biological recognition element (enzymes, DNA, whole cells and antibodies) or to the signal transduction method (electrochemical, optical, thermal, and mass-based) (Fig. 4) [24]. They can be applied in many fields, mainly in biomedical applications. In general,

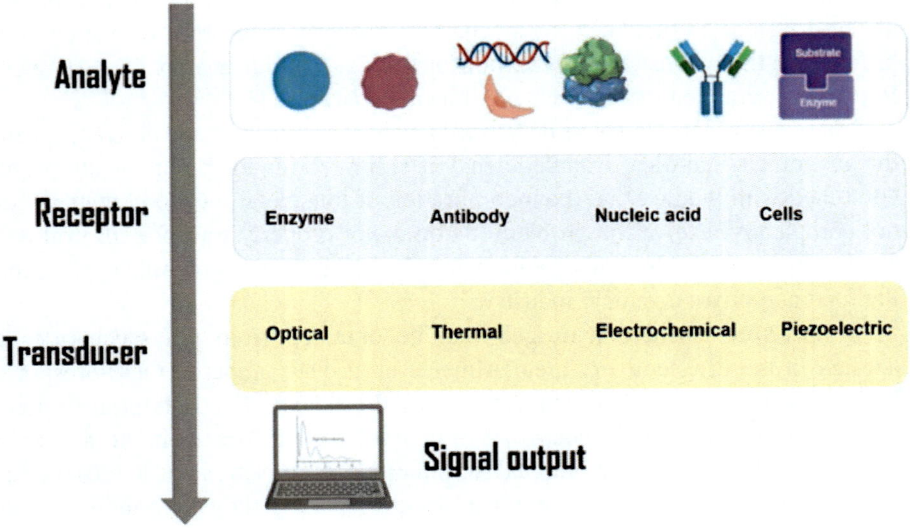

Fig. 4 Schematic representation of bionsensors actuation, divided by analyte detection and respective receptor, signal transduction method, and signal output

the sensing molecule is immobilized in a support whose properties must enhance the stability and activity of the biological compound. This matrix should have the right hydrophobicity, chemical composition, mechanical properties, and biocompatibility, considering the biosensors application as biomaterials. Additionally, it should be resistant to physiological pHs and ionic strengths. Besides, they should be able to co-immobilize more than one biological compound.

Polysaccharides are polymeric materials with excellent properties to be applied in the field of biosensors since they present the right chemical constitution, good microenvironment for the biological compound, and the right hydrophilicity. For instance, cellulose and cellulose derivatives composites have been widely used in the area of biosensors.

Gil et al. immobilized unease on cellulose derivatives for urea detection [25]. Promising results were obtained by Edwards et al. [26] who developed a composite of proteins and nano-crystalline cellulose to obtain a fluorescent biosensor for elastase. Due to the excellent properties of cellulose, namely its biocompatibility, biodegradability, and low cost, a vast amount of research work has been made with this polysaccharide and its derivatives. Ratajczaket al. [27] published a review concerning the development of cellulose paper-based biosensors for biomedical application. Abdi et al. [28] optimized the fabrication of a new cholesterol biosensor based on a composite of nanocellulose and polyaniline (PANi), with very promising results. Other researchers used a poly (4-vinylaniline) polyaniline-functionalized bacterial cellulose to develop a new flexible electrochemical biosensor [29].

Due to their chemical properties, reactivity, hydrophilicity, biocompatibility, and capacity of gel formation, chitosan and chitin are widely applied in the preparation of composites for biosensors [30–33].

Other polysaccharides are reported to be used in the preparation of biosensors. Fois et al. [34] used a starch-based hydrogel composite to develop a biosensor for glucose detection. Gautam et al. [35] characterized a composite based on PANi, starch, and a multi-wall carbon nanotube (MWCNTs) to be used in glucose and hydrogen peroxide detection.

3.3 Tissue Engineering

The incidence of bone disorder and conditions is increasing as a result of aging with obesity and few exercise. One of the most promising techniques in orthopedic surgery and tissue engineering is to repair and reconstruct bone and cartilage by autogenous tissue transplantation [36, 37]. The development of biomaterials for tissue regeneration comprises not only to create a scaffold that is biocompatible and biodegradable, but also with suitable mechanical properties and porosity that supports the differentiation of cells [38]. Sometimes is not possible to prepare a biomaterial from only one polymer with all the desired properties. Therefore, composite or hybrid materials appear in order to combine them all. Scaffolds are commonly composites produced from both synthetic and natural polymers. Among the synthetic polymers, aliphatic polyester like poly glycolic acid (PGA), poly lactic acid (PLA), poly lactic-co-glycolic acid (PLGA), polycaprolactone (PCL), are widely used in tissue engineering due to their biodegradable nature. The natural polymers commonly used are mainly polysaccharides such as chitosan (CS), alginate, agarose, and hyaluronic acid (HA), which choice is justified by their biocompatibility, renewability, and sustainability. Along with protein-based materials collagen, gelatin, and fibrin are also popular for engineering bioactive scaffolds because of their advantages in mimicking the extracellular environment [39, 40].

Jithendra et al. [39] prepared a 3D scaffold consisting of fish collagen, chitosan, and *Aloe Vera* by the freeze-drying technique. This 3D scaffold presented all the desired physical and biological properties to attract, attach, and proliferate fibroblasts, meaning that it could be used as a promising biomaterial for tissue engineering.

Electrospinning technique has been one of the most explored techniques for the preparation of polymer-based scaffolds able to sustain and improve tissue or organs regeneration. Coimbra et al. [41] prepared core-shell fibrous meshes by coaxial electrospinning, with a polycaprolactone (PCL) core and a functionalized gelatin shell, and then photocrosslinked the meshes under UV light, aiming to be used in vascular tissue regeneration. Using the same approach and for the same application, fibrous meshes were also prepared by electrospinning blends of PCL and photocrosslinkable gelatin (GelMA) [42]. It is known that the inclusion of extracellular matrix-derived materials as gelatin, in polymeric blends, increases biological activity of the scaffolds promoting cell adhesion, proliferation, and, consequently, tissue regeneration [42, 43].

Moreover, another type of hydrogels have been gaining interest in the area of tissue engineering. Injectable hydrogels have been explored as substrates into a defect, presenting several advantages. This method appears as a non-invasive approach in a way that may avoid surgery procedures and, at the same time, allows an easy and homogenous drug or cell distribution for bone tissue engineering applications [37]. Injectable hydrogels can be produced by thermal gelation, ionic interaction, physical self-assembly, photo-polymerization, and chemical crosslinking [37, 44].

Ren et al. [37] prepared a gel scaffold by the Schiff-base reaction using the linkages between oxidized alginate and carboxymethyl chitosan, which allowed to obtain a polysaccharide-based hydrogel. Hydroxyapatite nanoparticles and calcium carbonate microspheres loading tetracycline hydrochloride were blended to form a composite gel scaffold to enhance bioactive properties and mechanical strength.

Injectable beads of pullulan and dextran with hydroxyapatite were prepared by Fricain et al. [45] and showed to be safe and efficient bone engineering in the field of oral and maxillofacial surgeries.

Cheng et al. [46] prepared chitosan-cellulose nanofiber composite injectable hydrogels with self-healing properties for neural regeneration.

Also, an injectable sodium alginate/bioglass composite hydrogel containing bone marrow stem cells (BMSCs) was prepared by Zhu et al. [47] and used for subchondral bone regeneration. These authors also prepared an injectable thermosensitive alginate/agarose composite hydrogel with co-culture of BMSCs and articular chondrocytes was applied for articular cartilage regeneration.

Another interesting approach in the area of the injectable hydrogel composites was proposed by Hu et al. [48]. These authors prepared dual-structural pH-sensitive nanocarrier for growth factor delivery with the pH-sensitive acetalated β-cyclodextrin and heparin nanogel interpenetrating component for growth factor protection, which was later incorporated into a hyaluronic acid hydrogel to form a cell scaffold composite hydrogel, that maintained cell activity and supported cell growth.

3.4 *Wound Dressing*

Wound healing is a very complex process occurring after damage and is divided into four distinct steps: (a) coagulation and hemostasis; (b) inflammation and release of pro-inflammatory cytokines, (c) proliferation, resulting new tissue formation, and finally (d) remodeling where the new tissue is remodeled, defining the nature of the final scar [49]. Mainly, a wound dressing must be able to increase epidermal migration, to connect the tissue production, protect against bacterial infection, stimulate angiogenesis and connective tissue production, and also to keep a moist environment, allowing the gas exchange. Non-adhering and easy to remove after healing, non-toxic, sterile and non-allergic are also requirements for an ideal wound dressing [50]. The traditional wound dressings comprise gauze, natural and synthetic bandages, lint, and plaster. Overall these types of wound healing dressings require frequent change to avoid macerating the new tissues and its removal is usually traumatic to the patience [49]. The use of natural polysaccharides in wound dressings, besides their biocompatibility and biodegradability, is due to its similarity with biological ECM and therefore the recognition by biological systems [51]. Considering the well-known disadvantages of the traditional tissues, like the lack of a moist environment, their replacement by modern wound dressing was inevitable.

Regarding the most common polysaccharide used in wound dressing application, the use of chitosan, chitin, pullulan, starch, cellulose, and alginate can be highlighted. In recent years, several works have reported the use of polysaccharides, being chitosan one of the primary choice for wound healing—hemostasis and bacterial activity are the main reasons [52, 53]. A vast research in the preparation of wound dressings based on polysaccharides (homo and hetero), inorganic compounds, and antibiotics have been explored [54]. It is well-known that the combination of different types of polysaccharides can enhance the performance of the final wound dressing, with improved mechanical properties or antimicrobial activity [55]. Shabunin et al. [56] described the preparation via electrospinning of a two-layered wound dressing, based on a copolymer of poly(ε-caprolactam) nanofibers (CoPA) and a composite comprising chitosan and chitin nanofibrils. They showed the importance of the combination of these two layers to promote the wound healing. The preparation of cellulose-chitosan sponges, with controlled structure and morphology was also reported [52] and presented promising results as possible wound dressings for diabetic patients. Kaczmarek et al. [57] developed a polymeric blend based on chitosan and tannic acid, showing the improvement of the physiochemical properties of the final film due to the tannic acid incorporation. Other studies explored the combination between chitosan and tannic acid to obtain an enriched wound dressing [58–60]. There are several works exploring the use of polysaccharide composites with different polysaccharides but also with inorganic materials, such as the use of gold nanoparticles (AuNPs) [61–65]. These nanoparticles are characterized by a high antimicrobial activity and for being an effective antiviral agent. Tran et al. [66] prepared a composite based on AuNPs, cellulose, wool keratin, and chloroauric acid with possible application for chronic ulcerous infected wounds, or even as a controlled drug delivery. The use of metal nanoparticles, such as gold, silver, or zinc, is becoming particularly common, once these particles proved to have an outstanding antibacterial effect [67]. Along with their capability to prevent bacterial infections, they also have the ability to accelerate wound healing, and guarantee a constant moist environment. For instance, Shao et al. [68] developed a wound dressing composite with chitosan and silver sulfadiazine (AgSD) and the final composite revealed extremely good antibacterial properties.

The use of other polysaccharides, such as alginate, cellulose, and derivatives, and derivatives of chitosan had also been explored. Gao et al. [69] prepared a wound dressing based on a derivative of chitosan, carboxymethyl chitosan, and calcium alginate fibers. While carboxymethyl cellulose is characterized by its high biocompatibility, high moisture retention ability, and high antimicrobial activity, the alginate is responsible for enhancing the mechanical properties of the final wound dressing. Minocycline, a second-generation tetracycline antibiotic, was loaded in these structures to promote the antimicrobial activity of the composite. This combination resulted in a wound dressing with skin-like mechanical properties, long-term antimicrobial activity, biocompatibility, and with drug sustained-release properties.

Bacterial cellulose is also one of the main addressed topics on the preparation of improved wound dressings which is strongly related with its ability to retain water, high porosity and crystallinity, and superior mechanical properties [70, 71]. Wound dressings based on bacterial cellulose, alginate, chitosan, and copper sulphate were prepared by Wichai et al. [72], showing an enhanced antibacterial performance against gram-negative *Escherichia coli* (*E. coli*) and gram-positive *Methicillin-resistant Staphylococcus aureus* (MRSA). Recently, a composite of gelatin and bacterial cellulose was prepared using glutaraldehyde as the crosslinker [73]. This composite with a controlled morphology was loaded with ampicillin and the results suggested high antibacterial activity and biocompatibility. The construction of a wound dressing focused on a specific type of wound, a third-degree skin burn, was made through the combination of bacterial cellulose with copolymers of 3-hydroxybutyric and 4-hydroxybutyric acids [P(3HB/4HB] [74], which were also loaded with wound healing drugs and cells.

Wound dressings based on alginate composite have been widely investigated since this biopolymer presents also interesting properties for wound healing—biocompatibility, non-immunogenicity, non-toxicity, and high absorption capacity [75]. Rezvanian et al. [75] prepared a simvastatin-loaded wound dressing composite, based on alginate, pectin, and gelatin, using the solvent casting method and the obtained results proved to have the required properties to be used as wound dressings, with excellent mechanical properties. The combination of alginate with nanomaterials such as graphene oxide and polyvinyl alcohol (PVA) has also been reported by several authors [76–78].

Side by side with chitosan and cellulose, the incorporation of HA in wound dressing is also gaining attention lately. This polysaccharide is well-known for its ability to promote the wound healing with remarkable properties, and extremely used in wound dressing applications [79, 80]. Eskandarinia et al. [81] developed a wound dressing based on HA, cornstarch, and ethanolic extract of propolis, with an enhanced wound healing and good cell biocompatibility. Makvanbdi et al. [82] synthesized an injectable antibacterial thermosensitive hydrogel comprising HA, corn silk extract, and nanosilver. The use of HA with cellulose nanocrystals and gelatin is also reported in the preparation of new wound dressing with biocompatible character [83]. Other authors [84] revealed that loading an antimicrobial peptide to a composite made of alginate, HA, and collagen, resulted in a high antimicrobial activity, promoting also the re-epithelialization and angiogenesis.

The incorporation of antibacterial agents is also a procedure used to promote the functionalization of the wound dressings [85], something notorious along this section.

4 Conclusions

Polysaccharide-based composites have proved their value in the biomedical field over the past few years. The combination of these biopolymers with other type of materials—carbon nanotubes, proteins, other polysaccharides, inorganic compounds—usually results in an enhancement of the properties of the composites, in terms of biocompatibility, biodegradability, mechanical performance, and antimicrobial activity, among others. The growing use of this biopolymers in biomedical fields is not only explained by their unique properties but also by their high bioavailability, low cost, and non-toxicity.

Overall, this chapter covers the most recent developments of polysaccharide-based composites in biomedical fields, showing they are a suitable alternative to the petroleum-based polymers. By using renewable and sustainable sources, a positive outcome is expected, both economically and environmentally. Therefore, chitosan, chitin, cellulose, starch, alginate, hyaluronic acid, and several other polysaccharides will be continually used in the preparation of biocomposites to overcome the polysaccharide's main weaknesses.

References

1. Muhamad II, Lazim NAM, Selvakumaran S (2019a) Natural polysaccharide-based composites for drug delivery and biomedical applications. In: Natural polysaccharides in drug delivery and biomedical applications, pp 419–440. https://doi.org/10.1016/B978-0-12-817055-7.00018-2
2. Aravamudhan A, Ramos DM, Nada AA, Kumbar SG (2014) Natural polymers: polysaccharides and their derivatives for biomedical applications. Nat Synth Biomed Polym. https://doi.org/10.1016/B978-0-12-396983-5.00004-1
3. Jayakumar R, Menon D, Manzoor K, Nair SV, Tamura H (2010) Biomedical applications of chitin and chitosan based nanomaterials—a short review. Carbohyd Polym 82(2):227–232. https://doi.org/10.1016/j.carbpol.2010.04.074
4. Soares PIP, Echeverria C, Baptista AC, João CFC, Fernandes SN, Almeida APC, Borges JP (2017) Hybrid polysaccharide-based systems for biomedical applications. In: Hybrid polymer composite materials, pp 107–149. https://doi.org/10.1016/B978-0-08-100785-3.00004-8
5. Moon RJ, Martini A, Nairn J, Simonsen J, Youngblood J (2011) Cellulose nanomaterials review: structure, properties and nanocomposites. Chem Soc Rev 40(7):3941. https://doi.org/10.1039/c0cs00108b
6. Islam S, Bhuiyan MAR, Islam MN (2017) Chitin and chitosan: structure, properties and applications in biomedical engineering. J Polym Environ 25(3):854–866. https://doi.org/10.1007/s10924-016-0865-5
7. Thomas MS, Koshy RR, Mary SK, Thomas S, Pothan AL (2019) Starch, chitin and chitosan based composites and nanocomposites. https://doi.org/10.1007/978-3-030-03158-9
8. Adhikari HS, Yadav PN (2018) Anticancer activity of chitosan, chitosan derivatives, and their mechanism of action. Int J Biomater 2018:1–29. https://doi.org/10.1155/2018/2952085
9. Azuma K, Osaki T, Minami S, Okamoto Y (2015) Anticancer and anti-inflammatory properties of chitin and chitosan oligosaccharides. J Funct Biomater 6(1):33–49. https://doi.org/10.3390/jfb6010033
10. Vasvani S, Kulkarni P, Rawtani D (2019) Hyaluronic acid: a review on its biology, aspects of drug delivery, route of administrations and a special emphasis on its approved marketed products and recent clinical studies. Int J Biol Macromol. https://doi.org/10.1016/j.ijbiomac.2019.11.066
11. Bukhari SNA, Roswandi NL, Waqas M, Habib H, Hussain F, Khan S, Hussain Z (2018) Hyaluronic acid, a promising skin rejuvenating biomedicine: a review of recent updates and pre-clinical and clinical investigations on cosmetic and nutricosmetic effects. Int J Biol Macromol 120:1682–1695. https://doi.org/10.1016/j.ijbiomac.2018.09.188
12. Obeid MA, Al Qaraghuli MM, Alsaadi M, Alzahrani AR, Niwasabutra K, Ferro VA (2017) Delivering natural products and biotherapeutics to improve drug efficacy. Ther Deliv 8(11):947–956. https://doi.org/10.4155/tde-2017-0060
13. Patra JK, Das G, Fraceto LF, Campos EVR, del Rodriguez-Torres MP, Acosta-Torres LS, Shin H-S (2018) Nano based drug delivery systems: recent developments and future prospects. J Nanobiotechnol 16(1):71. https://doi.org/10.1186/s12951-018-0392-8
14. Wahab IF, Razak SIA (2016) Polysaccharides as composite biomaterials. Compos Renew Sustain Mater. https://doi.org/10.5772/65263
15. Hamidian H, Tavakoli T (2016) Preparation of a new Fe_3O_4/starch-g-polyester nanocomposite hydrogel and a study on swelling and drug delivery properties. Carbohyd Polym 144:140–148. https://doi.org/10.1016/j.carbpol.2016.02.048
16. Chen Y, Qi Y, Yan X, Ma H, Chen J, Liu B, Xue Q (2014) Green fabrication of porous chitosan/graphene oxide composite xerogels for drug delivery. J Appl Polym Sci 131(6):n/a–n/a. https://doi.org/10.1002/app.40006
17. Zhu M, Zhu Y, Zhang L, Shi J (2013) Preparation of chitosan/mesoporous silica nanoparticle composite hydrogels for sustained co-delivery of biomacromolecules and small chemical drugs. Sci Technol Adv Mater 14(4):045005. https://doi.org/10.1088/1468-6996/14/4/045005

18. Ghibaudo F, Gerbino E, Hugo AA, Simões MG, Alves P, Costa BFO, Simões PN (2018) Development and characterization of iron-pectin beads as a novel system for iron delivery to intestinal cells. Colloids Surf B 170:538–543. https://doi.org/10.1016/j.colsurfb.2018.06.052
19. Almeida JF, Ferreira P, Lopes A, Gil MH (2011) Photocrosslinkable biodegradable responsive hydrogels as drug delivery systems. Int J Biol Macromol 49(5):948–954. https://doi.org/10.1016/j.ijbiomac.2011.08.010
20. Schmitt H, Creton N, Prashantha K, Soulestin J, Lacrampe MF, Krawczak P (2015) Melt-blended halloysite nanotubes/wheat starch nanocomposites as drug delivery system. Polym Eng Sci 55(3):573–580. https://doi.org/10.1002/pen.23919
21. Quintana G, Simões MG, Hugo A, Alves P, Ferreira P, Gerbino E, Gómez-Zavaglia A (2017) Layer-by-layer encapsulation of Lactobacillus delbrueckii subsp. Bulgaricus using block-copolymers of poly(acrylic acid) and pluronic for safe release in gastro-intestinal conditions. J Funct Foods 35. https://doi.org/10.1016/j.jff.2017.06.007
22. Hu X, Gong X (2016) A new route to fabricate biocompatible hydrogels with controlled drug delivery behavior. J Colloid Interface Sci 470:62–70. https://doi.org/10.1016/j.jcis.2016.02.037
23. Silva D, Pinto LFV, Bozukova D, Santos LF, Serro AP, Saramago B (2016) Chitosan/alginate based multilayers to control drug release from ophthalmic lens. Colloids Surf B 147:81–89. https://doi.org/10.1016/j.colsurfb.2016.07.047
24. Mehrotra P (2016) Biosensors and their applications—a review. J Oral Biol Craniofac Res 6(2):153–159. https://doi.org/10.1016/j.jobcr.2015.12.002
25. Gil MH, Piedade A, Guthrie JT (1998) Immobilised enzymes and biosensors. In: Salamone JC (ed) Concise polymeric materials encyclopedia
26. Edwards JV, Prevost N, French A, Concha M, DeLucca A, Wu Q (2013) Nanocellulose-based biosensors: design, preparation, and activity of peptide-linked cotton cellulose nanocrystals having fluorimetric and colorimetric elastase detection sensitivity. Engineering 05(09):20–28. https://doi.org/10.4236/eng.2013.59A003
27. Ratajczak K, Stobiecka M (2020) High-performance modified cellulose paper-based biosensors for medical diagnostics and early cancer screening: a concise review. Carbohyd Polym 229:115463. https://doi.org/10.1016/j.carbpol.2019.115463
28. Abdi MM, Razalli RL, Tahir PM, Chaibakhsh N, Hassani M, Mir M (2019) Optimized fabrication of newly cholesterol biosensor based on nanocellulose. Int J Biol Macromol 126:1213–1222. https://doi.org/10.1016/j.ijbiomac.2019.01.001
29. Rebelo AR, Liu C, Schäfer K-H, Saumer M, Yang G, Liu Y (2019) Poly(4-vinylaniline)/polyaniline bilayer-functionalized bacterial cellulose for flexible electrochemical biosensors. Langmuir 35(32):10354–10366. https://doi.org/10.1021/acs.langmuir.9b01425
30. Ahmad A, Siddique JA, Setapar SHM, Lokhat D, Golandaj A, Ramjugernath D (2018) Recent advances in chitosan-based films for novel biosensor. In: Electrically conductive polymer and polymer composites, pp 137–161. https://doi.org/10.1002/9783527807918.ch7
31. Bagal-Kestwal DR, Chiang B-H (2019) Exploration of chitinous scaffold-based interfaces for glucose sensing assemblies. Polymers 11(12):1958. https://doi.org/10.3390/polym11121958
32. Jiang Y, Wu J (2019) Recent development in chitosan nanocomposites for surface-based biosensor applications. Electrophoresis 40(16–17):2084–2097. https://doi.org/10.1002/elps.201900066
33. Teepoo S, Dawan P, Barnthip N (2017) Electrospun chitosan-gelatin biopolymer composite nanofibers for horseradish peroxidase immobilization in a hydrogen peroxide biosensor. Biosensors 7(4):47. https://doi.org/10.3390/bios7040047
34. Fois M, Arrigo P, Bacciu A, Monti P, Marceddu S, Rocchitta G, Serra PA (2019) The presence of polysaccharides, glycerol, and polyethyleneimine in hydrogel enhances the performance of the glucose biosensor. Biosensors 9(3):95. https://doi.org/10.3390/bios9030095
35. Gautam V, Singh KP, Yadav VL (2018) Polyaniline/multiwall carbon nanotubes/starch nanocomposite material and hemoglobin modified carbon paste electrode for hydrogen peroxide and glucose biosensing. Int J Biol Macromol 111:1124–1132. https://doi.org/10.1016/j.ijbiomac.2018.01.094

36. Amini AR, Laurencin CT, Nukavarapu SP (2012) Bone tissue engineering: recent advances and challenges. Crit Rev Biomed Eng 40(5):363–408. https://doi.org/10.1615/CritRevBiomedEng.v40.i5.10
37. Ren B, Chen X, Du S, Ma Y, Chen H, Yuan G, Niu X (2018) Injectable polysaccharide hydrogel embedded with hydroxyapatite and calcium carbonate for drug delivery and bone tissue engineering. Int J Biol Macromol 118:1257–1266. https://doi.org/10.1016/j.ijbiomac.2018.06.200
38. Sultankulov B, Berillo D, Sultankulova K, Tokay T, Saparov A (2019) Progress in the development of chitosan-based biomaterials for tissue engineering and regenerative medicine. Biomolecules 9(9):470. https://doi.org/10.3390/biom9090470
39. Jithendra P, Rajam AM, Kalaivani T, Mandal AB, Rose C (2013) Preparation and characterization of aloe vera blended collagen-chitosan composite scaffold for tissue engineering applications. ACS Appl Mater Interfaces 5(15):7291–7298. https://doi.org/10.1021/am401637c
40. Li J, Chen G, Xu X, Abdou P, Jiang Q, Shi D, Gu Z (2019) Advances of injectable hydrogel-based scaffolds for cartilage regeneration. (May):129–140. https://doi.org/10.1093/rb/rbz022
41. Coimbra P, Santos P, Alves P, Miguel SP, Carvalho MP, de Sá KD, Ferreira P (2017) Coaxial electrospun PCL/gelatin-MA fibers as scaffolds for vascular tissue engineering. Colloids Surf B 159. https://doi.org/10.1016/j.colsurfb.2017.07.065
42. Ferreira P, Santos P, Alves P, Carvalho MP, de Sá KD, Miguel SP, Coimbra P (2017) Photocrosslinkable electrospun fiber meshes for tissue engineering applications. Eur Polymer J 97. https://doi.org/10.1016/j.eurpolymj.2017.10.018
43. Hinderer S, Layland SL, Schenke-Layland K (2016) ECM and ECM-like materials—Biomaterials for applications in regenerative medicine and cancer therapy. Adv Drug Deliv Rev 97:260–269. https://doi.org/10.1016/j.addr.2015.11.019
44. Tan H, Ramirez CM, Miljkovic N, Li H, Rubin JP, Marra KG (2009) Thermosensitive injectable hyaluronic acid hydrogel for adipose tissue engineering. Biomaterials 30(36):6844–6853. https://doi.org/10.1016/j.biomaterials.2009.08.058
45. Fricain JC, Aid R, Lanouar S, Maurel DB, Nihouannen D Le, Delmond S, Catros S (2018) Injectable polysaccharide-hydroxyapatite composite material for sinus floor augmentation. Dent Mater 1–12. https://doi.org/10.1016/j.dental.2018.03.021
46. Cheng K, Huang C, Wei Y, Hsu S (2019) Novel chitosan—cellulose nano fi ber self- healing hydrogels to correlate self-healing properties of hydrogels with neural regeneration effects. NPG Asia Mater. https://doi.org/10.1038/s41427-019-0124-z
47. Zhu Y, Kong L, Farhadi F, Xia W, Chang J, He Y, Li H (2019) Biomaterials an injectable continuous strati fi ed structurally and functionally biomimetic construct for enhancing osteochondral regeneration. Biomaterials 192(November 2018):149–158. https://doi.org/10.1016/j.biomaterials.2018.11.017
48. Hu X, Gao Z, Tan H, Wang H, Mao X, Pang J (2019) An injectable hyaluronic acid-based composite hydrogel by DA click chemistry with pH sensitive nanoparticle for biomedical application. Front Chem 7. https://doi.org/10.3389/fchem.2019.00477
49. Borda LJ, Macquhae FE, Kirsner RS (2016) Wound dressings: a comprehensive review. Curr Dermatol Rep 5(4):287–297. https://doi.org/10.1007/s13671-016-0162-5
50. Dhivya S, Padma VV, Santhini E (2015) Wound dressings—a review. BioMedicine 5(4):22. https://doi.org/10.7603/s40681-015-0022-9
51. Suarato G, Bertorelli R, Athanassiou A (2018) Borrowing from nature: Biopolymers and biocomposites as smart wound care materials. Front Bioeng Biotechnol 6(Oct):1–11. https://doi.org/10.3389/fbioe.2018.00137
52. Cheng H, Li C, Jiang Y, Wang B, Wang F, Mao Z, Sui X (2018) Facile preparation of polysaccharide-based sponges and their potential application in wound dressing. J Mater Chem B 6(4):634–640. https://doi.org/10.1039/c7tb03000b
53. Khan MA, Mujahid M (2019) A review on recent advances in chitosan based composite for hemostatic dressings. Int J Biol Macromol 124:138–147. https://doi.org/10.1016/j.ijbiomac.2018.11.045

54. Naseri-Nosar M, Ziora ZM (2018) Wound dressings from naturally-occurring polymers: a review on homopolysaccharide-based composites. In: Carbohydrate polymers, vol 189. https://doi.org/10.1016/j.carbpol.2018.02.003
55. Hu Y, Zhang Z, Li Y, Ding X, Li D, Shen C, Xu F-J (2018) Dual-crosslinked amorphous polysaccharide hydrogels based on chitosan/alginate for wound healing applications. Macromol Rapid Commun 39(20):1800069. https://doi.org/10.1002/marc.201800069
56. Shabunin A, Yudin V, Dobrovolskaya I, Zinovyev E, Zubov V, Ivan'kova E, Morganti P (2019) Composite wound dressing based on chitin/chitosan nanofibers: processing and biomedical applications. Cosmetics 6(1):16. https://doi.org/10.3390/cosmetics6010016
57. Kaczmarek B, Nadolna K, Owczarek A, Michalska-Sionkowska M, Sionkowska A (2019) The characterization of thin films based on chitosan and tannic acid mixture for potential applications as wound dressings. Polym Test 78(March):106007. https://doi.org/10.1016/j.polymertesting.2019.106007
58. Li N, Yang X, Liu W, Xi G, Wang M, Liang B, Shi C (2018) Tannic acid cross-linked polysaccharide-based multifunctional hemostatic microparticles for the regulation of rapid wound healing. Macromol Biosci 18(11):1–16. https://doi.org/10.1002/mabi.201800209
59. Lu B, Wang T, Li Z, Dai F, Lv L, Tang F, Lan G (2016) Healing of skin wounds with a chitosan–gelatin sponge loaded with tannins and platelet-rich plasma. Int J Biol Macromol 82:884–891. https://doi.org/10.1016/j.ijbiomac.2015.11.009
60. Xu F, Weng B, Gilkerson R, Materon LA, Lozano K (2015) Development of tannic acid/chitosan/pullulan composite nanofibers from aqueous solution for potential applications as wound dressing. Carbohyd Polym 115:16–24. https://doi.org/10.1016/j.carbpol.2014.08.081
61. Abdul Khalil HPS, Chong EWN, Owolabi FAT, Asniza M, Tye YY, Rizal S, Paridah MT (2019) Enhancement of basic properties of polysaccharide-based composites with organic and inorganic fillers: a review. J Appl Polym Sci 136(12):47251. https://doi.org/10.1002/app.47251
62. Li Q, Lu F, Zhou G, Yu K, Lu B, Xiao Y, Lan G (2017) Silver inlaid with gold nanoparticle/chitosan wound dressing enhances antibacterial activity and porosity, and promotes wound healing. Biomacromol 18(11):3766–3775. https://doi.org/10.1021/acs.biomac.7b01180
63. Lu Z, Gao J, He Q, Wu J, Liang D, Yang H, Chen R (2017) Enhanced antibacterial and wound healing activities of microporous chitosan-Ag/ZnO composite dressing. Carbohyd Polym 156:460–469. https://doi.org/10.1016/j.carbpol.2016.09.051
64. Ma B, Huang Y, Zhu C, Chen C, Chen X, Fan M, Sun D (2016) Novel Cu@SiO_2/bacterial cellulose nanofibers: preparation and excellent performance in antibacterial activity. Mater Sci Eng C 62:656–661. https://doi.org/10.1016/j.msec.2016.02.011
65. Singla R, Soni S, Kulurkar PM, Kumari ASM, Patial V, Yadav SK (2017) In situ functionalized nanobiocomposites dressings of bamboo cellulose nanocrystals and silver nanoparticles for accelerated wound healing. Carbohyd Polym 155:152–162. https://doi.org/10.1016/j.carbpol.2016.08.065
66. Tran CD, Prosenc F, Franko M (2018) Facile synthesis, structure, biocompatibility and antimicrobial property of gold nanoparticle composites from cellulose and keratin. J Colloid Interface Sci 510:237–245. https://doi.org/10.1016/j.jcis.2017.09.006
67. Mihai MM, Dima MB, Dima B, Holban AM (2019) Nanomaterials for wound healing and infection control. Materials 12(13):1–16. https://doi.org/10.3390/ma12132176
68. Shao W, Wu J, Wang S, Huang M, Liu X, Zhang R (2017) Construction of silver sulfadiazine loaded chitosan composite sponges as potential wound dressings. Carbohyd Polym 157:1963–1970. https://doi.org/10.1016/j.carbpol.2016.11.087
69. Gao Y, Zhang X, Jin X (2019) Preparation and properties of minocycline-loaded carboxymethyl chitosan gel/alginate nonwovens composite wound dressings. Marine Drugs 17(10):575. https://doi.org/10.3390/md17100575
70. Carvalho T, Guedes G, Sousa FL, Freire CSR, Santos HA (2019) Latest advances on bacterial cellulose-based materials for wound healing, delivery systems, and tissue engineering. Biotechnol J 14(12):1900059. https://doi.org/10.1002/biot.201900059
71. Portela R, Leal CR, Almeida PL, Sobral RG (2019) Bacterial cellulose: a versatile biopolymer for wound dressing applications. Microb Biotechnol 12(4):586–610. https://doi.org/10.1111/1751-7915.13392

72. Wichai S, Chuysinuan P, Chaiarwut S, Ekabutr P, Supaphol P (2019) Development of bacterial cellulose/alginate/chitosan composites incorporating copper (II) sulfate as an antibacterial wound dressing. J Drug Deliv Sci Technol 51:662–671. https://doi.org/10.1016/j.jddst.2019.03.043
73. Ye S, Jiang L, Su C, Zhu Z, Wen Y, Shao W (2019) Development of gelatin/bacterial cellulose composite sponges as potential natural wound dressings. Int J Biol Macromol 133:148–155. https://doi.org/10.1016/j.ijbiomac.2019.04.095
74. Volova TG, Shumilova AA, Nikolaeva ED, Kirichenko AK, Shishatskaya EI (2019) Biotechnological wound dressings based on bacterial cellulose and degradable copolymer P(3HB/4HB). Int J Biol Macromol 131:230–240. https://doi.org/10.1016/j.ijbiomac.2019.03.068
75. Rezvanian M, Mohd Amin MCI, Ng SF (2016) Development and physicochemical characterization of alginate composite film loaded with simvastatin as a potential wound dressing. Carbohyd Polym 137:295–304. https://doi.org/10.1016/j.carbpol.2015.10.091
76. Ma R, Wang Y, Qi H, Shi C, Wei G, Xiao L, Guo Z (2019) Nanocomposite sponges of sodium alginate/graphene oxide/polyvinyl alcohol as potential wound dressing: in vitro and in vivo evaluation. Compos Part B: Eng 167(December 2018):396–405. https://doi.org/10.1016/j.compositesb.2019.03.006
77. Varaprasad K, Jayaramudu T, Kanikireddy V, Toro C, Sadiku ER (2020) Alginate-based composite materials for wound dressing application: a mini review. Carbohydr Polym 116025. https://doi.org/10.1016/j.carbpol.2020.116025
78. Chen K, Wang F, Liu S, Wu X, Xu L, Zhang D (2020) In situ reduction of silver nanoparticles by sodium alginate to obtain silver-loaded composite wound dressing with enhanced mechanical and antimicrobial property. Int J Biol Macromol 148:501–509. https://doi.org/10.1016/j.ijbiomac.2020.01.156
79. Burdick JA, Prestwich GD (2011) Hyaluronic acid hydrogels for biomedical applications. Adv Mater 23(12):H41–H56. https://doi.org/10.1002/adma.201003963
80. Longinotti C (2014) The use of hyaluronic acid based dressings to treat burns: a review. Burns Trauma 2(4):162. https://doi.org/10.4103/2321-3868.142398
81. Eskandarinia A, Kefayat A, Rafienia M, Agheb M, Navid S, Ebrahimpour K (2019) Cornstarch-based wound dressing incorporated with hyaluronic acid and propolis: in vitro and in vivo studies. Carbohyd Polym 216:25–35. https://doi.org/10.1016/j.carbpol.2019.03.091
82. Makvandi P, Ali GW, Della Sala F, Abdel-Fattah WI, Borzacchiello A (2019) Biosynthesis and characterization of antibacterial thermosensitive hydrogels based on corn silk extract, hyaluronic acid and nanosilver for potential wound healing. Carbohyd Polym 223:115023. https://doi.org/10.1016/j.carbpol.2019.115023
83. Yin F, Lin L, Zhan S (2019) Preparation and properties of cellulose nanocrystals, gelatin, hyaluronic acid composite hydrogel as wound dressing. J Biomater Sci Polym Ed 30(3):190–201. https://doi.org/10.1080/09205063.2018.1558933
84. Lin Z, Wu T, Wang W, Li B, Wang M, Chen L, Zhang T (2019) Biofunctions of antimicrobial peptide-conjugated alginate/hyaluronic acid/collagen wound dressings promote wound healing of a mixed-bacteria-infected wound. Int J Biol Macromol 140:330–342. https://doi.org/10.1016/j.ijbiomac.2019.08.087
85. Simões D, Miguel SP, Ribeiro MP, Coutinho P, Mendonça AG, Correia IJ (2018) Recent advances on antimicrobial wound dressing: a review. Eur J Pharm Biopharm 127:130–141. https://doi.org/10.1016/j.ejpb.2018.02.022

Biomedical Nanocomposites

Amit Kumar Nayak, Saad Alkahtani, and Md Saquib Hasnain

1 Introduction

A huge volume of research endeavors has been covered for the advancements in material science engineering and technologies to develop several kinds of biomaterials for biomedical applications [1–11]. These advanced biomaterials for biomedical applications cover the designing of gels, hydrogels, membranes, films, patches, polymer-blends, composites and nanocomposites, scaffolds, etc. [12–23]. The latest studies in the domain of nanotechnology have been recognized as the popular most and flourishing areas for the preparation of a variety of nanomaterials for biomedical applications [24–29]. Some of these key biomedical nanomaterials are polymeric nanoparticles and nanocapsules [30–32], lipid nanoparticles [33], metallic nanoparticles [27, 34], ceramic nanoparticles [35], nanogels [14, 36, 37], nanovesicles [38–40], nanotubes [41–43], nanocomposites [44, 45], etc.

Amongst these nanomaterials, nanocomposites are consisting of a number of nanoscale substances or nanoscale substances incorporated into the bulk substances [46, 47]. Since past few decades, a number of nanocomposites (i.e., organic-organic, inorganic-inorganic, and organic–inorganic nanocomposites) are being prepared for biomedical applications (for example drug delivery, wound healing, gene delivery, tissue regeneration, dentistry, antimicrobial, bioimaging, biosensors, etc. [45–54].

A. K. Nayak (✉)
Department of Pharmaceutics, Seemanta Institute of Pharmaceutical Sciences, Mayurbhanj, Odisha, India
e-mail: amitkrnayak@yahoo.co.in

S. Alkahtani
Department of Zoology, College of Science, King Saud University, P.O. Box 2455, Riyadh 11451, Saudi Arabia

M. S. Hasnain
Department of Pharmacy, Palamau Institute of Pharmacy, Chianki, Daltonganj, Jharkhand, India

A. K. Nayak et al. (eds.), *Biomedical Composites*, Materials Horizons: From Nature to Nanomaterials, https://doi.org/10.1007/978-981-33-4753-3_3

This chapter presents an inclusive review on the nanocomposites for biomedical applications.

2 Nanocomposites

In general, nanocomposites belong to a category of nanoengineered materials possessing the structures modulated in the nanosizing scales [46, 47]. In general, nanocomposites differ from the conventional composites owing to the higher ratio of surface to volume for the reinforcing-phase, augmented ductility without the decreasing of material strength profile as well as scratching resistances [55]. During the past few decades, numerous kinds of nanocomposites are being investigated and developed. Most of these nanocomposites are made of natural, semi-synthetic and synthetic polymers, metals and their oxides, ceramics, structured carbons, and other inorganic components [47–54]. Various materials are usually reinforced to improve physical, physico-chemical, mechanical, and biomedical properties of the composite/nanocomposite matrices. The materials reinforced for the syntheses of nanocomposites are polymeric nanoparticles, inorganic nanomaterials (e.g., nanostructured carbons, metallic nanoparticles, metal oxide nanoparticles, nanoceramics, etc.), nanofibres (e.g., electrospun fibers, etc.), nanotubes (carbon nanotubes, etc.), nanosheets (e.g., exfoliated clay stacks), etc. [56–59]. Various nanocomposites can be classified on the basis of composition and matrix-reinforcement: (i) organic-organic nanocomposites (e.g., polymeric nanocomposites, etc.), (ii) organic–inorganic nanocomposites (e.g., polymeric-ceramic nanocomposites, polymeric-metallic nanocomposites, etc.), (iii) and inorganic-inorganic nanocomposites (e.g., metallic nanocomposites, ceramic nanocomposites, metallic-ceramic nanocomposites, carbon nanotube-ceramic nanocomposites, carbon nanotube-metallic nanocomposites, etc.).

A number of synthesis methodologies are being used to synthesize many kinds of nanocomposites. The extensively used synthesis methodologies of nanocomposites are: physically mixing, film-casting, dip-coating, in situ preparation, co-precipitation, ionic-gelation, covalent coupling, electro-spinning, colloidal assembly, layer-by-layer assembly, etc. [48, 58]. These synthesis methodologies facilitate the advantages of hydrogen bonding, electrostatic interactions, coulombic interactions, ionic interactions, covalent bonding, hydrophobic effects, assembly formation, etc. [48]. The challenges in synthesis of different nanocomposites involve the controlling of synthesis methodologies, assurance of compatibility with the component materials, achieving the product of unique as well as desirable material properties [56, 58]. Recent years, the applications of nanocomposites direct a variety of newer technologies along with the functional and classy viewpoints for numerous fields including biomedical area [45–58].

3 Nanocomposites for Biomedical Applications

Nanocomposites designed for the use in various biomedical applications are usually well-known as 'biomedical nanocomposites'. The most important applications of biomedical nanocomposites include drug delivery, gene delivery, antimicrobial, tissue regeneration, wound healing, dentistry, bioimaging, biosensors, etc. [45–58]. Various important applications of biomedical nanocomposites are illustrated in Fig. 1.

3.1 *Drug Delivery*

The area of drug delivery applications demands the designing and development of efficient dosage forms capable of delivering different drugs to the body for management as well as treatment of various diseases and ailments [60]. Since long, drug delivery has been recognized as a most important healthcare area where the delivering of different kinds of drugs via the administration dosage forms is requisite to accomplish the optimum therapeutics to treat various diseases and ailments [61, 62]. Many biomaterials like natural, synthetic and semi-synthetic biopolymers, bioceramics, metal and metal oxides, etc., are being employed to formulate drug delivery systems [61, 63–84]. Using these biomaterials, various dosage forms like tablets [85–88], capsules [89, 90], topical gels [91], hydrogels [92–96], in situ gels [14], beads [97–109], microparticles [110–114], nanoparticles or nanocapsules [25, 31,

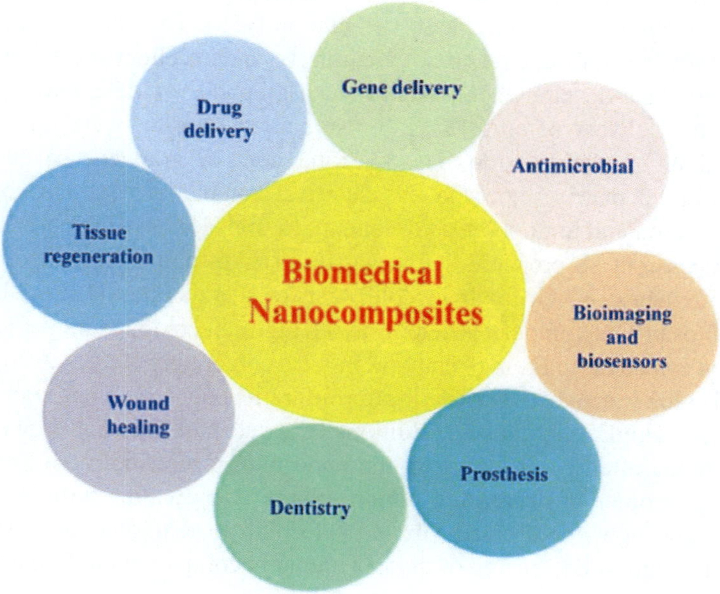

Fig. 1 Important applications of biomedical nanocomposites

32, 115], buccal patches [21, 116], dental pastes [117, 118], emulsions [118–120], implants [121–125], etc., are being formulated for a variety of drugs and used for treatment or management of many diseases. In recent times, nanocomposite has become a promising carrier system for drug delivery because of their improved improve physical, physico-chemical, mechanical, and biomedical properties [47, 48, 58, 59, 126]. In addition, high drug loading ability and nanoscale reinforcement have further unlocked new realms in the field of drug delivery.

In a research, alginate-montmorillonite (MMT) nanocomposite bead system loaded with irinotecan was developed by Illiescu et al. [127]. These alginate-MMT nanocomposite beads of irinotecan were synthesized via ionic gelation method. Both air-drying and freeze-drying were used for drying these nanocomposite beads of irinotecan. Irinotecan-loaded nanocomposite beads prepared by air-dried process were found as smaller sized as compared to that of nanocomposite beads prepared by freeze-dried process. The reinforcement of MMT within the alginate matrix exhibited the sustained release of loaded irinotecan over a longer period. In another research, the same research group synthesized MMT-alginate nanocomposite beads loaded with carboplatin via ionic gelation method [128]. These carboplatin-loaded MMT-alginate nanocomposite beads exhibited sustained drug release profile over a longer period.

5-fluorouracil releasing alginate-chitosan-MMT nanocomposite system was synthesized by Azhar and Olad [129]. The synthesized nanocomposite systems containing 30% wt MMT demonstrated a sustained releasing profile of loaded 5-fluorouracil, in vitro. The in vitro 5-fluorouracil releasing from the alginate-chitosan-MMT nanocomposite system was found to follow Korsmeyer–Peppas kinetic model. MMT-chitosan nanocomposite hydrogel was prepared for controlled release application of vitamin B_{12} under electro-stimulation [130]. The results of in vitro study demonstrated the pulsatile release of vitamin B_{12} and excellent anti-fatigue behavior. Justin and Chen developed chitosan-reduced graphene oxide nanocomposite form transdermal delivery of drug [131]. These nanocomposites were prepared via the reduction of graphene oxide. The reinforcement of graphene oxide within the chitosan-based matrix improved the electrical conductivity, which could help in electroporation and iontophoresis for enhanced transdermal drug permeation.

Venkatesan et al. prepared hydroxyapatite (HAp)-chitosan nanocomposites for colonic delivery of celecoxib in colon cancer therapeutics [132]. The prepared HAp-chitosan nanocomposites demonstrated high celecoxib loading efficiencies and sustained celecoxib releasing behavior. In vitro cell culture studies using HT 29 and HCT 15 colon cancer cells revealed antiproliferation, apoptosis, and significant time-dependent cytoplasmic uptake. Additionally, a substantial tumor growth inhibition on the xenografted colon tumor in the nude mouse was indicated by the in vivo studies. The order of decreasing tumor volume after treating with free celecoxib, nanocomposites without celecoxib and celecoxib loaded nanocomposites demonstrated the highest decreasing of tumor volume in comparison to free celecoxib and nanocomposites without celecoxib.

Nanda et al. developed chitosan-polylactide nanocomposites using cloisite 30B and investigated the releasing potential of paclitaxel (an anticancer drug) from these developed nanocomposites [133]. The results of in vitro drug release study demonstrated that the paclitaxel releasing from these nanocomposites was found dependent on the polymer matrix and pH of drug release media. In addition, in vitro drug release study suggested that the excellent release could be achieved in the basic pH milieu. For the development of controlled doxorubicin (an anticancer drug) releasing in cancer therapeutics, Wu et al. fabricated chitosan-based mesoporous magnetic nanocomposite using silica (SiO_2) and magnetite [134]. In this chitosan-based mesoporous magnetic nanocomposite, chitosan was employed as blocking excipient for the prevention of early releasing of doxorubicin. The doxorubicin releasing from these magnetic nanocomposite system was found pH dependent.

Like other organs, bones are also susceptible to many orthopaedic diseases like osteoarthritis, osteoporosis, osteomyelitis, bone metastasis, osteosarcoma, etc. [135]. Since past few decades, huge volume of research and development in the orthopaedic drug delivery has been noted to treat the said orthopaedic diseases [121–125]. As a result of this, a variety of nanocomposites have been reported as orthopaedic drug delivery carrier matrices. In a research, a nanocomposite of calcium sulfate and nanocrystalline apatite was synthesized by Hesaraki et al. for orthopaedic release of indomethacin [136]. The synthesized nanocomposite exhibited a slower sustained releasing profile of indomethacin. The setting time and of injectability of this nanocomposite system was found relatively superior than those of the pure calcium sulfate. In addition, the indomethacin loading within the calcium sulfate-nanocrystalline apatite nanocomposite system slightly enhanced the setting time and of injectability, which did not produce any influence on the compressive strength. The results of in vitro biocompatibility evaluation of indomethacin-loaded calcium sulfate-nanocrystalline apatite nanocomposites demonstrated good cytocompatibility, when tested on mouse fibroblast L929 cell line.

Mesoporous silicate MCM-48/HAp nanocomposite loaded with ibuprofen was synthesized by Aghaei et al. via the in situ synthesis method [137]. The in vitro ibuprofen release from this nanocomposite was found slightly rapid when ibuprofen loading was done using 25 mg/ml ibuprofen concentration. In vitro cytocompatibility of the mesoporous silicate MCM-48/HAp nanocomposite was tested by MTT [3-(4, 5-dimethylthiazolyl-2)-2, 5-diphenyltetrazolium bromide] assay using osteosarcoma cell line. The nanocomposite loaded with ibuprofen showed above 40% toxicity indicating the cytocompatibility of the ibuprofen-loaded nanocomposites for orthopedic drug delivery application.

In another research, ciprofloxacin-loaded HAp-polycaprolactone nanocomposite-based film for orthopedic implantable drug delivery in treatment of osteomyelitis was developed by Nithya et al. [138]. The nanocrystalline HAp was synthesized via the precipitation process using eggshell and used to develop HAp-polycaprolactone nanocomposite-based film. The ciprofloxacin-loaded HAp-polycaprolactone nanocomposite implantable film was prepared via solvent evaporation procedure. In vitro drug releasing profile indicated its capacity for faster ciprofloxacin releasing. In addition, the results of in vitro cytotoxicity evaluation using fibroblast NIH-3T3 cell line and osteoblast MG 63 cell line indicated that this ciprofloxacin-loaded HAp-polycaprolactone nanocomposite film possessed excellent biocompatibility. Some recently reported nanocomposite systems used in drug delivery applications are presented in Table 1.

3.2 *Gene Delivery*

At present, gene delivery has expansively been introduced and employed for the treatment/management of various kinds of diseases, such as genetic diseases, cystic fibrosis, cancers, autoimmune diseases, etc. [161]. In this context, the delivery option of entrapped gene within the nanoscale carriers like nanoparticles and nanocomposites can improve cellular uptake via the endocytosis and the prevention of the premature releasing of gene into the non-targeted organs [162]. In recent years, an extensive volume of researches on the applications of nanocomposites has been performed for the delivery of gene [163].

In a research, Kashkouli et al. developed and investigated aminotetrazole-functionalized magnetic chitosan nanocomposite for targeted gene delivery [164]. This magnetic nanocomposite was fabricated using Fe_3O_4/chitosan grafted with organosilane modified 5-amino-1H-tetrazole via the chemical modification process. The magnetic nanocomposite demonstrated the capacities of high loading and targeted delivery of plasmid as well. An improved releasing of plasmid from the nanocomposite at pH 7.4 was found with the enhancement of gene expression, when studied in HECK-293 T cell line.

In another report, Xie et al. developed anionic-charged hybrid nanocomposite for delivery of siRNA [165]. This nanocomposite was synthesized using polyethylene grafted carboxymethyl and calcium phosphate via the self-assembly procedure in an aqueous environment. Zhu et al. reported doxorubicin loaded cholesterol siRNA/low-density lipoprotein coupled N-succinyl chitosan nanocomposite for delivery of siRNA [166]. Yan et al. developed Tat tagged and folate modified N-succinyl chitosan nanocomposite for tumor targeted gene therapy [167]. This Tat tagged and folate modified nanocomposite was prepared via the self-assembly procedure. Some recently reported nanocomposites for gene delivery are presented in Table 2.

3.3 *Antimicrobials*

In recent times, many antimicrobial biomaterials are being developed to offer promising microbial inactivation [175]. In general, antimicrobial agents are used to develop these kinds of biomaterials. In most of the cases, the loading or impregnating or reinforcing of antimicrobial agents with other materials has been used to develop many effective antimicrobial biomaterials [176]. In recent years, numerous antimicrobial nanocomposites have been developed by many researchers and scientists.

Table 1 Some recently reported nanocomposite systems used in drug delivery applications

Nanocomposite systems	Drug released	References
Carboxymethyl cellulose/starch/zinc oxide nanocomposite hydrogel beads	Doxorubicin	Gholamali and Yadollahi [139]
pH-sensitive nanocomposite beads of folic acid intercalated layered double hydroxide and chitosan	Folic acid	Mallakpour and Hatami [140]
Pectin-zinc oxide hybrid nanocomposite	Donepezil	Kodoth et al. [141]
Oral colon-specific drug delivery system based on the pectin/modified nano-carbon sphere nanocomposite gel films	5-Fluorouracil	Wang et al. [142]
Magnetite/silica nanocomposites	Doxorubicin	Taufiq et al. [143]
Chitosan-based magnetic/fluorescent nanocomposites	Doxorubicin	Ding et al. [144]
Cassava starch acetate–PEG/gelatin nanocomposites	Cisplatin	Raj and Prabha [145]
Alginate-polyvinyl pyrrolidone-nanoHAp composite matrices	Diclofenac sodium	Hasnain et al. [146]
Sodium caseinate-magnesium aluminum silicate nanocomposite films for modified-release tablets	Acetaminophen	Kajthunyakarn et al. [147]
Magnesium aluminium silicate-polyethylene oxide nanocomposite matrices	Diltiazem HCl	Asare-Addo et al. [148]
HAp-alginate nanocomposite beads	Ofloxacin	Roul et al. [149]
Polycaprolactone-forsterite nanocomposite fibrous membranes	Dexamethasone	Kharaziha et al. [150]
Chitosan-starch nanocomposite particles	Bis-desmethoxy curcumin analog	Bala Subramanian et al. [151]
Dextran sulfate-modified pH-sensitive layered double hydroxide nanocomposites	Methotrexate	Wang et al. [152]
Agarose encapsulated mesoporous carbonated HAp nanocomposites powder	5-fluorouracil and amoxcillin	Kolanthai et al. [153]
MMT-alginate nanocomposites	Vit. B1 and B6	Kevadiya et al. [154]
Chitosan-poly(aminopropyl/phenylsilsesquioxane) hybrid nanocomposite membranes	5-fluorouracil	Kummari et al. [155]
PEG-chitosan-iron oxide nanocomposites	Methotraxate	Lin et al. [156]
Chitosan/ZnO bio-nanocomposite hydrogel beads	Ibuprofen	Yadollahi et al. [157]
Chitosan-fibrin nanocomposites	Methotraxate	Vedakumari et al. [158]
Hollow chitosan nanocomposites	Ramipril	Basu et al. [159]
Sodium alginate/chitosan/HAp nanocomposite hydrogels	Doxorubicin	Taleb et al. [160]

Table 2 Some recently reported nanocomposites for gene delivery

Nanocomposite systems	Gene delivery	References
Bioactive nanocomposite coatings based on collagen/gold nanoparticles under visible light illumination	Surface-mediated gene delivery	Yao et al. [168]
Collagen-silica nanocomposites	Modulating inflammation in a cutaneous chronic wound model by IL-10 released	Wang et al. [169]
Reduced graphene oxide-polyethylenimine nanocomposite	Photothermally controlled gene delivery	Kim and Kim [170]
DNA-amorphous calcium phosphate nanocomposite spheres	Surface-mediated gene delivery	Oyane et al. [171]
Mannose-conjugated layered double hydroxide nanocomposite	Targeted siRNA delivery	Li et al. [172]
Stabilized calcium phosphate hybrid nanocomposite using a benzoxaborole-containing polymer	pH-responsive siRNA delivery	Zhou et al. [173]
Graphene oxide-HAp nanocomposites	HSV-TK suicide gene delivery to inhibit human breast cancer growth	Cheang et al. [174]

Butchosa et al. developed bacterial cellulose-based nanocomposites possessing strong antibacterial activity [177]. These bacterial cellulose-based antibacterial nanocomposites were synthesized by post-synthetic modification via mixing aqueous suspension of partially deacetylated chitosan nanocrystal with bacterial cellulose. The antibacterial action of these bacterial cellulose-partially deacetylated chitosan nanocrystal nanocomposites was found to be increased with the increment of partially deacetylated chitosan nanocrystal concentrations.

Rinehart et al. developed antimicrobial nanocomposite loaded chitosan/polyvinyl alcohol (PVA) hydrogel, which showed remarkable bactericidal action against both Gram-positive bacteria (*Staphylococcus aureus*) and Gram-negative bacteria (*Escherichia coli*) [178]. In addition to improve the mechanical stability, this antimicrobial nanocomposite was found capable to prevent the formation of biofilm and also, to enhance the growth of fibroblasts. In a research, Elbarbary and El-Sawy developed antimicrobial nanocomposite membranes of chitosan, polyvinyl alcohol and silver [179]. These antimicrobial nanocomposite membranes were synthesized by γ-irradiation. These membranes were found to be antibacterial, nonthrombogenic, and haemolytic in nature. Youssef et al. developed chitosan-gold and chitosan-silver nanocomposite films for antimicrobial applications [180]. These nanocomposite films exhibited considerable antimicrobial actions against Gram-positive bacteria (*Staphylococcus aureus*) and Gram-negative bacteria (*Pseudomonas aerugenosa*). Likewise, these chitosan-gold and chitosan-silver nanocomposite films also showed their effectiveness against fungi (*Aspergillus niger*) and yeast (*Candida albicans*). Mohamed and Sabaa reported antimicrobial nanocomposite of silver nanoparticles

Table 3 Some recently reported antimicrobial nanocomposites

Antimicrobial nanocomposites	References
Graphene oxide-chitosan and graphene oxide-ethylene diamine tetraacetic acid nanocomposites	Khalil et al. [183]
Silver nanoparticles-loaded chitosan nanocomposite	Chen et al. [184]
Zinc-mineralized alginate nanocomposites	Malagurski et al. [185]
Chitosan based silver nano-biocomposites	Davoodbasha et al. [186]
Bacterial cellulose-lignin-cellulose nanocrystal nanocomposite films	Sá et al. [187]
N,N,N-trimethyl chitosan chloride/poly (acrylic acid)/silver nanocomposites	Mahmoud et al. [188]
Polyvinyl alcohol/silver nanocomposite films	Mathew et al. [189]
Polystyrene-silver nanocomposite	Jabbar et al. [190]
Silver-zinc oxide nanocomposites	Noohpisheh et al. [191]
Multivalent and synergistic chitosan oligosaccharide-silver nanocomposites	Mei et al. [192]
Chitosan-titanium dioxide nanocomposite film	Siripatrawan et al. [193]
Antibiotic-functionalized silver nanocomposites	Guom et al. [194]
Grafted sugarcane bagasse/silver nanocomposites	Abdelwahab and Shukry [195]
Streptomycin-loaded chitosan-coated magnetic nanocomposites	El Zowalaty et al. [196]
Oleo-polyurethane-carbon nanocomposites	Ahmadi et al. [197]
Nanocellulose/carboxymethyl cellulose and nanochitosan/carboxymethyl cellulose composite films	Jannatyha et al. [198]

with chitosan [181]. This kind of nanocomposite exhibited its effectiveness against several bacteria, namely *Staphylococcus aureus*, *Streptococcus faecalis*, *Pseudomonas aerugenosa, Escherichia coli*, *Bacillus subtilis*, and *Neisseria gonorrhoeae* and fungal species like *Candida albicans*. Farhoudian et al. developed antimicrobial nanocomposite composed of CuO nanoparticles within chitosan hydrogel beads [182]. These chitosan hydrogel beads were prepared by ionic-gelation reaction using sodium tripolyphosphate as ionic cross-linker. The nanocomposite showed significant antibacterial action against Gram-positive bacteria (*Staphylococcus aureus*) and Gram-negative bacteria (*Escherichia coli*). Some recently reported antimicrobial nanocomposites are presented in Table 3.

3.4 Tissue Regeneration

In the regenerative medicine, tissue regeneration is well recognized for aiming to repair the diseased/damaged tissues of the body [199]. The tissue regeneration strategies deal with the development of scaffold materials via the exploitation of different biomaterials (such as natural, synthetic and semi-synthetic biopolymers, bioceramics, metals, etc.) and these scaffold systems are being implanted in the diseased/damaged sites for tissue repair through regeneration [9, 47, 48, 199]. Generally, scaffold materials are the fundamental constituents of the tissue regeneration approaches as these scaffold materials present an architectural milieu, wherein extracellular matrix, cell–cell interactions, and cell-growth factor interactions combine to ease the regenerative niche at the implanted sites [9, 200]. Therefore, the tissue regeneration approaches mimic the extracellular matrix and also facilitates cell attachment signaling, proliferations and differentiations of cells to meet up the requisite for the tissue regeneration.

Meskinfam et al. reported the synthesis of nanohydroxyapatite (nHAp)-starch composites for tissue regeneration [201]. In this work, the nanocomposites were synthesized by a biomimetic process. The in vitro biocompatibility was tested via MTT assay demonstrated that the nHAp reinforcement with starch influenced the cell proliferation. These nanocomposites did not display any adverse outcome on the cultured cell structure, cell viability, and cell proliferation, in vitro. Similar kinds of starch-based nanocomposite scaffolds for bone tissue regeneration were reported to be developed by Sadjadi et al. [202]. These starch-based nanocomposite scaffolds were also synthesized by in situ biomimetic methodology [202]. In another research, Huang et al. developed an injectable nanocomposite hydrogel composed of hyaluronic acid, glycol chitosan, and nHAp [203]. The injectable nanocomposite hydrogel exhibited a good promise for its probable use in bone tissue regeneration. Chae et al. developed and tested alginate/HAp nanocomposite-based scaffolds for bone tissue regeneration [204]. These nanocomposite-based scaffolds were prepared by electro-spinning and in situ synthesis of HAp mimicking bone mineralized collagen fibrils. The prepared alginate/HAp nanocomposite-based scaffolds were fibrous in nature.

Liu et al. synthesized alginate/halloysite nanotubes composite scaffolds for bone tissue regeneration [205]. These nanocomposite scaffolds were prepared by employing solution-mixing and freeze-drying methodologies. The in vitro biocompatibility was tested via MTT assay using mouse fibroblast cells and the results of these testing demonstrated active mitochondrial activities of the living cells, when treated by these nanocomposite scaffolds. The overall results of in vitro biocompatibility testing suggested its use in bone tissue regeneration. In another research, Kawaguchi et al. synthesized alginate-carbon nanotube nanocomposite scaffolds for tissue regeneration [206].

Correia et al. fabricated chitosan-HAp nanocomposite for the use in cartilage tissue repair [207]. This kind of chitosan-HAp nanocomposite possessed the network configuration and was found to improve the production of extracellular matrix. Some recently reported nanocomposite systems for tissue regeneration applications are presented in Table 4.

Table 4 Some recently reported nanocomposite systems for tissue regeneration applications

Nanocomposite systems	Tissue regeneration types	References
Bioactive electrospun nanocomposite scaffolds of poly(lactic acid)/cellulose nanocrystals	Bone tissue	Patel et al. [208]
Fibrin hydrogel incorporated with graphene oxide functionalized nanocomposite scaffolds	Bone tissue	Pathmanapan et al. [209]
Biocompatible nanocomposite scaffolds based on copolymer-grafted chitosan	Bone tissue	Saber-Samandari and Saber-Samandari [210]
Graphene and HAp self-assemble free standing nanocomposite hydrogels	Bone tissue	Xie et al. [211]
Zirconium oxide nanoceramic modified chitosan-based porous nanocomposites	Bone tissue	Bhowmick et al. [212]
Organically modified clay supported chitosan/hydroxyapatite-zinc oxide nanocomposites	Bone tissue	Bhowmick et al. [213]
3D fiber-deposited poly(ε-caprolactone)/iron-doped hydroxyapatite nanocomposite magnetic scaffolds	Bone tissue	De Santis et al. [214]
Collagen-inspired mineral-hydrogel nanocomposites	Bone tissue	Patel et al. [215]
Chitosan–gelatin-alginate-HAp nanocomposite scaffold	Bone tissue	Sharma et al. [216]
Bioinspired collagen-apatite nanocomposites	Bone tissue	Liu et al. [217]
Nano-hydroxyapatite/β-cyclodextrin/chitosan nanocomposite	Bone tissue	Shakir et al. [218]
Osteoblast-conditioned nHAp/gelatin composite scaffold	Bone tissue	Samadikuchaksaraei et al. [219]
nHAp–pullulan/dextran composite macroporous material	Bone tissue	Fricain et al. [220]
Intercalated chitosan/HAp nanocomposites	Bone tissue	Nazeer et al. [221]
Bioactive gum Arabic/κ-carrageenan-incorporated nHAp Nanocomposites	Bone tissue	Mirza et al. [222]
Argon plasma modified nanocomposite polyurethane scaffolds	Cartilage tissue	Griffin et al. [223]
Photopolymerized maleilated chitosan/methacrylated silk fibroin micro/nanocomposite hydrogels	Cartilage tissue	Zhou et al. [224]
Poly 3-hydroxybutyrate-chitosan-multiwalled carbon nanotube/silk nano-micro composite scaffold	Cartilage tissue	Mirmusavi et al. [225]

(continued)

Table 4 (continued)

Nanocomposite systems	Tissue regeneration types	References
Porous poly(d,l-lactic acid)/vertically aligned carbon nanotubes/nHAp Scaffolds	Osteochondral tissue	Stocco et al. [226]
Nanocomposites complexed with gold nanoparticles on polyaniline	Nerve	Kim et al. [227]
Electroactive alginate hydrogel nanocomposite reinforced by functionalized graphite nanofilaments	Nerve	Homaeigohar et al. [228]
Chitosan-selenium biodegradable nanocomposite	Nerve	Dolkhani et al. [229]
Carbon nanotube-polyurethane nanocomposite	Cardiac tissue	Shokraei et al. [230]

3.5 Prosthesis

In prosthetic applications, a variety of nanomaterials are being used since past few decades. Amongst these prosthetic nanomaterials, nanocomposites are recently being used in cardiac prosthetics, orthopaedic prosthetics, dental and maxillofacial prosthetics, etc. [231, 232]. In recent years, many polymeric nanocomposites are being developed via the reinforcement of fibers [50]. These fiber-reinforced polymeric nanocomposites are being used in a variety of orthopaedic prosthetic devices because of the capability of their higher mechanical strength [231]. In a work, Ghanbari et al. investigated novel kinds of cardiac prosthesis made of modified polyhedral oligomeric silsesquioxane-nanocomposite [232]. This nanocomposite material showed self-endothelialization potential for the use in heart valve prosthesis. Some recently reported biomedical nanocomposites used in prosthetics are presented in Table 5.

3.6 Dentistry

During past few years, numerous kinds of nanomaterials are being developed and used for dentistry applications like dental drug delivery, dental tissue regeneration, dental pulp capping, pulpotomy, etc. [51–53]. Amongst these nanomaterials, dental nanocomposites are the widely used nanocandidate category, which are designed and investigated for periodontal drug delivery, dental restoration, dentin-pulp regeneration, enamel substitution, etc. [51–54].

In a research, nanocomposite microspheres for periodontal drug delivery were synthesized by Pataquiva Mateus et al. using nHAp using sodium alginate [245]. Within these nHAp-alginate nanocomposite microspheres, two different kinds of antibiotics namely, erythromycin and amoxicillin were loaded. These antibiotics-loaded nanocomposite microspheres showed a promise for their sustained drug

Table 5 Some recently reported nanocomposites used in prosthetic devices

Biomedical uses in prosthetics	Nanocomposites	References
Orthopedic prosthetics	Nano-alumina/multi-walled carbon nanotubes/high-density polyethylene hybrid nanocomposites for hip joint replacement	Dabees et al. [233]
	Bioresorbable β-TCP-FeAg nanocomposites for load bearing bone implants	Swain et al. [234]
	Pullulan/dextran/nHAp macroporous composite beads for repairing of a femoral condyle defect in rats	Schlaubitz et al. [235]
	Biomimetic nanocomposites of carboxymethyl cellulose-HAp as three dimensional load bearing bone grafts	Garai and Sinha [236]
	Osteoblast-seeded bioglass/gelatin nanocomposite for critical-size calvarial defect repair in rat	Johari et al. [237]
	Carbon nanotube reinforced natural rubber nanocomposite for anthropomorphic prosthetic foot purpose	Medupin et al. [238]
Dental prosthetics	Nanocomposite fibrous scaffold of silica coated nanoHAp-gelatin reinforced with electrospun poly(L-lactic acid) to promote new bone formation and osseointegration in mandibular defect	Manju et al. [239]
	Chemically cross-linked poly(ε-caprolactone)-HAp nanocomposite scaffold for repairing of a mandibular bone defect	Liu et al. [240]
	Ceria-stabilized zirconia/alumina nanocomposite for fabricating the frameworks of removable dental prostheses	Hagiwara and Nakajima [241]
Cardiac prosthetics	Nanocrystalline cellulose-fibrin nanocomposites for artificial vascular graft applications	Brown et al. [242]
	Poly(vinyl alcohol)-bacterial cellulose nanocomposite for mechanical aortic heart valve prosthesis	Mohammadi et al. [243]

(continued)

Table 5 (continued)

Biomedical uses in prosthetics	Nanocomposites	References
	Barium sulfate/poly-L-lactide-based nanocomposites for coronary stents	Ang et al. [244]

releasing prospects in the treatment of periodontitis. In another work, similar kinds of erythromycin and amoxicillin-loaded nanocomposite microspheres were developed Ferraz et al. [246]. The in vitro drug releasing from these nanocomposite microspheres, antimicrobial activities, and in vitro cytocompatibility with osteoblasts were assessed. The outcome of this research suggested the use of erythromycin and amoxicillin-loaded nanocomposite microspheres for periodontal drug delivery carriers. Madhumathil et al. developed antibiotic-loaded gelatin-alginate/apatite nanocomposite films [247]. These antibiotic-loaded nanocomposite films were found bioactive in nature and can be used in the treatment of periodontal infrabony defects.

In recent years, different dental nanocomposites are being developed and investigated for the fortification of multifaceted vital dentin pulp [52]. Dental nanocomposites composed of biopolymers and nanoHAp and/or inorganic metallic components have been developed and investigated to fill the tooth perforations [248–250]. Dental pulp capping nanocomposites made of polyglutamic acid (PGA) and ∞-melanocyte stimulating hormone (α-MSH, an anti inflammatory hormone), which demonstrated the acceleration of the regeneration of pulp connective tissue resulting in adhesion as well as growth of pulp fibroblasts [251]. Lipopolysaccharide was used to motivate the growth of pulp fibroblasts via incubating the PGA-α-MSH nanocomposites. These PGA-α-MSH nanocomposites have been proved to stimulate the adhesion of pulp fibroblast along with improved proliferation for fibroblast cell. The outcome of this research indicated that these nanocomposites are capable to lessen the inflammatory condition of lipopolysaccharide restorative dental pulp fibroblasts, which commonly occurred during infections by Gram-negative bacteria. These PGA-α-MSH nanocomposites can be used for the treatment of endodontic injury as well as lesions.

Panahi et al. prepared tricalcium silicate-based nanocomposites for root-end dental application [252]. Besides tricalcium silicate, chitosan and dicalcium phosphate were used for the fabrication of these dental nanocomposites. The researchers noticed the occurrence of synergic influence of chitosan and dicalcium phosphate on these tricalcium silicate-based dental nanocomposites. Some recently reported dental nanocomposites are presented in Table 6.

Table 6 Some recently reported dental nanocomposites

Dental nanocomposite systems	Purpose	References
Bioactive upconversion nanocomposites containing chlorin e6	Treatment of periodontal diseases	Zhang et al. [253]
Amoxicillin-loaded electrospun nanocomposite membranes based on poly(ε-caprolactone) and nHAp	Fibrous-based antibiotic carrier system for dental and tissue engineering applications	Furtos et al. [254]
Porous tri-layered nanocomposite hydrogel scaffold composed of chitin-poly(lactic-co-glycolic acid)/nanobioactive glass ceramic /cementum protein 1	Concurrent regeneration of cementum, periodontal ligament, and alveolar bone	Sowmya et al. [255]
Piroxicam loaded biodegradable chitosan/poly(vinyl alcohol)/HAp electrospun nanocomposite scaffolds	Periodontal regeneration	Farooq et al. [256]
2-methacryloyloxyethyl phosphorylcholine-dimethylaminohexadecyl methacrylate Class V nanocomposite	For inhibition of periodontal pathogens, combat periodontitis and protect the periodontium	Wang et al. [257]
Alginate/HAp-based nanocomposite	Dental pulp biomineralization and differentiation	Sancilio et al. [258]

3.7 Wound Healing

During past few years, a considerable amount of interest has been paid in the research and development for the advancements of wound healing therapeutics [8]. Many groups of researchers and scientists are trying to develop improved and efficient wound healing biomaterials, which should possess biocompatibility, ability to stop bleeding, nonallergic and noninfective properties, permeability for gases, ability to retain moisture and to absorb exudates, capability of promoting skin regeneration, and fastening wound healing process [8, 259, 260]. In recent years, many nanocomposite biomaterials are being developed and employed for the use in wound healing.

Shah et al. developed moxifloxacin (a fluoroquinolone derivative antibiotic) loaded chitosan-silver-sericin nanocomposite films for wound healing applications [261]. These moxifloxacin loaded nanocomposite films were prepared by solvent casting technique. These moxifloxacin loaded nanocomposite films demonstrated a potential antibacterial action against both Gram-positive bacteria (*Staphylococcus aureus* and *Staphylococcus epidermidis*) and Gram-negative bacteria (*Acinetobacter baumannii* and *Pseudomonas aeruginosa*). In addition, moxifloxacin loaded films showed their antibacterial efficacy against methicillin-resistant *Staphylococcus aureus* (MRSA) mediated hospital-acquired skin infections. In the in vivo testing on the burn-injured wound model developed Sprague Dawley rates, these moxifloxacin loaded chitosan-silver-sericin nanocomposite films displayed significantly faster wound healing results. In another research, Kumar et al. developed chitosan hydrogel-zinc oxide nanocomposite bandages for wound healing applications [262]. These nanocomposite bandages showed significant antibacterial efficacy against both Gram-positive bacteria (*Staphylococcus aureus*) and Gram-negative bacteria (*Escherichia coli*). Sandri et al. fabricated nanocomposite composed of chitosan and halloysite [263]. This kind of chitosan-based nanocomposite showed remarkable wound healing results in cases of burns and skin lesions types of wounds. Anisha et al. prepared nanocomposite sponge composed of chitosan-hyaluronan/chondroitin sulfate nanoparticles for wound healing uses [264]. In addition, the nanocomposite sponge demonstrated enhanced blood clotting and platelet activation abilities with a controlled swelling and biodegradation. The results of in vitro cytotoxicity testing on human dermal fibroblast cells indicated the biocompatible (nontoxic) nature of this nanocomposite sponge.

Figueiredo et al. developed bacterial cellulose-based nanocomposite films for wound dressings in combination with poly(2-hydroxyethyl methacrylate) [265]. These nanocomposite films were synthesized via an in situ radical polymerization of 2-hydroxyethyl methacrylate and in this synthesis, poly(ethylene glycol) diacrylate was employed as cross-linking agent. The bacterial cellulose-poly(2-hydroxyethyl methacrylate) nanocomposite films exhibited desirable biocompatibility with improved cell adhesion and cell proliferation for human adipose-derived mesenchymal stem cells. Some recently reported nanocomposites for wound healing applications are presented in Table 7.

3.8 *Bioimaging and Biosensor*

In recent years, nanocomposites are broadly used in many bioimaging and biosensor applications [284]. Lin et al. prepared polyethylene glycol (PEG)-chitosan-iron oxide nanocomposite containing fluorescent cyanin dye for bioimaging applications. The prepared PEG-chitosan-iron oxide nanocomposite was found potential for fluorescence imaging and magnetic imaging (MRI) [285].

Fu et al. developed gold-embedded chitosan nanocomposite for the use as surface-enhanced Raman scattering sensor (SERS) [286]. The 3D architecture of gold-embedded chitosan nanocomposite was noticed to serve as an outstanding SERS substrate for detecting 4-mercaptobenzoic acid. In addition, because of the pH-sensitive characteristics, the chitosan nanocomposite-based SERS was found capable of detecting the charge of dye molecules. In another research, Liu et al. reported the fabrication of chitosan-carbon nanotubes composite for biosensing applications [287]. This nanocomposite was found efficient as glucose electrochemical sensor by immobilizing the glucose oxidase. Some recently reported nanocomposites used in bioimaging and biosensors are presented in Table 8.

Table 7 Some recently reported nanocomposites for wound healing applications

Nanocomposites for wound healing applications	References
Chitosan-laponite nanocomposite scaffolds	Gonzaga et al. [266]
Basil seed (*Ocimum basilicum* L.) mucilage-zinc oxide nanocomposite	Tantiwatcharothai and Prachayawarakorn [267]
Curcumin-loaded electrospun polycaprolactone/MMT nanocomposite	Sadeghianmaryan et al. [268]
Bacterial cellulose-zinc oxide nanocomposites	Khalid et al. [269]
Chitosan/banana peel powder nanocomposites	Kamel et al. [270]
MMT-norfloxacin nanocomposite	García-Villén et al. [271]
A novel bilayer zein/MMT nanocomposite	Gunes et al. [272]
Titanium dioxide nanotubes incorporated gellan gum bio-nanocomposite film	Razali et al. [273]
Alginate/acacia based nanocomposites of zinc oxide nanoparticles	Manuja et al. [274]
Citric acid cross-linked carboxymethyl guar gum nanocomposite films loaded with ciprofloxacin	Prabhakar and Matta [275]
Bacterial Cellulose/methylglyoxal nanocomposite	Yang et al. [276]
Edaravone-loaded alginate-based nanocomposite hydrogel	Fan et al. [277]
Alginate/silver/nicotinamide nanocomposites for treating diabetic wounds	Montaser et al. [278]
Cellulose based nanocomposite hydrogel films consisting of sodium carboxymethylcellulose-grapefruit seed extract nanoparticles	Koneru et al. [279]
Incorporated plant extract fabricated silver/poly-D,l-lactide-co-glycolide nanocomposites	Renu et al. [280]
Keratin-chitosan/zinc oxide nanocomposite hydrogel	Zhai et al. [281]
Bioactive antibacterial silica-based nanocomposites hydrogel scaffolds	Li et al. [282]
Polyurethane nanocomposite impregnated with chitosan-modified graphene oxide	Najafabadi et al. [283]

4 Conclusion

In recent years, there has been a growing interest for the development of different nanomaterials for various biomedical applications. Amongst these nanomaterials, different nanocomposites, such as organic-organic, inorganic-inorganic, and

Table 8 Some recently reported nanocomposites used in bioimaging and biosensors

Nanocomposites	Applications (bioimaging and/or biosensors)	References
Bi/phthalocyanine manganese nanocomposite	Trimodal imaging directed photodynamic and photothermal therapy	Wang et al. [288]
Fluorescently-labeled magnetic nanocomposites	Optical and magnetic resonance imaging	Zhu et al. [289]
Fluorescent gold nanocrystals-silica hybrid nanocomposite	Bioimaging	Kim et al. [290]
MnO_2-DNAzyme-photosensitizer nanocomposite	Cell imaging	Wang et al. [291]
Magneto-fluorescent perovskite nanocomposites	Directed cell motion and imaging	Tan et al. [292]
Graphene/polyvinylpyrrolidone/polyaniline nanocomposite	Novel paper-based cholesterol biosensor	Ruecha et al. [293]
Graphene–polyaniline nanocomposite	Biosensor for detection of antimalarial drug artesunate in pharmaceutical formulation and biological fluids	Radhapyari et al. [294]
Multifunctional zirconia-reduced graphene oxide-thionine nanocomposite	Ultrasensitive electrochemical DNA biosensor	Chen et al. [295]
Polyaniline–bismuth oxide nanocomposite	Biosensor for quantification of anti-parkinson drug pramipexole in solubilized system	Jain et al. [296]
Mushroom-like polyaniline and gold nanoparticle nanocomposite	Detection of silver ions and DNA biosensor	Yang et al. [297]
Polyaniline capped Bi_2S_3 nanocomposite	Impedimetric DNA biosensor	Zhu et al. [298]
Polyaniline@nickel metal–organic framework nanocomposite	Electrochemical biosensor	Sheta et al. [299]
Cysteine-silver nanoparticles/graphene oxide nanocomposite	Biosensor	Wang et al. [300]

organic–inorganic nanocomposites are being developed for biomedical applications. From this chapter, we can conclude that these biomedical nanocomposites possess several unique properties, such as physical, physico-chemical, mechanical and biomedical properties for the inclusive range of applications, such as drug delivery, gene delivery, antimicrobial, tissue regeneration, wound healing, dentistry, bioimaging, biosensors, etc.

References

1. Sah MK, Pramanik K (2012) Surface modification and characterisation of natural polymers for orthopaedic tissue engineering: a review. Int J Biomed Eng Technol 9:101–121
2. Hasnain MS, Nayak AK (2019) Background: carbon nanotubes for targeted drug delivery. In: Hasnain MS, Nayak AK (eds) Carbon nanotubes for targeted drug delivery. Springer, Singapore, pp 1–9
3. Nayak AK, Hasnain MS (2019) Some other plant polysaccharide based multiple units for oral drug delivery. In: Nayak AK, Hasnain MS (eds) Plant polysaccharides-based multiple-unit systems for oral drug delivery. Springer, Singapore, pp 123–1128
4. Hasnain MS, Nayak AK (2019) Carbon nanotubes as quantum dots for therapeutic purpose. In: Hasnain MS, Nayak AK (eds) Carbon nanotubes for targeted drug delivery. Springer, Singapore, pp 59–64
5. Hasnain MS, Nayak AK (2019) Carbon nanotubes in vaccine delivery. In: Hasnain MS, Nayak AK (eds) Carbon nanotubes for targeted drug delivery. Springer, Singapore, pp 69–73
6. Hasnain MS, Nayak AK (2019) Carbon nanotubes in gene delivery. In: Hasnain MS, Nayak AK (eds) Carbon nanotubes for targeted drug delivery. Springer, Singapore, pp 75–87
7. Mohanta BC, Javed MN, Hasnain MS, Nayak AK (2020) Polyelectrolyte complexes of alginate for controlling drug release. In: Nayak AK, Hasnain MS (eds) Alginates in drug delivery. Academic Press, Elsevier Inc., USA, pp 297–321
8. Kurakula M, Rao GSNK, Kiran V, Hasnain MS, Nayak AK (2020) Alginate-based hydrogel systems for drug releasing in wound healing. In: Nayak AK, Hasnain MS (eds) Alginates in drug delivery. Academic Press, Elsevier Inc., USA, pp 323–358
9. Nayak AK, Mohanta BC, Hasnain MS, Hoda MN, Tripathi G (2020) Alginate-based scaffolds for drug delivery in tissue engineering. In: Nayak AK, Hasnain MS (eds) Alginates in drug delivery. Academic Press, Elsevier Inc., USA, pp 359–386
10. Hasnain MS, Kiran V, Kurakula M, Rao GSNK, Tabish M, Nayak AK (2020) Use of alginates for drug delivery in dentistry. In: Nayak AK, Hasnain MS (eds) Alginates in drug delivery. Academic Press, Elsevier Inc., USA, pp 387–404
11. Pal D, Nayak AK, Saha S (2019) Cellulose-based hydrogels: present and future. In: Akhtar MS, Swamy MK, Sinniah UR (eds) Natural bio-active compounds. Singapore, Springer, pp 285–332
12. Chowdhury S, Chakraborty S, Maity M, Hasnain MS, Nayak AK (2020) Biocomposites of alginates in drug delivery. In: Nayak AK, Hasnain MS (eds) Alginates in drug delivery. Academic Press, Elsevier Inc., USA, pp 153–185
13. Kenawy ER, Hasnain MS (2020) Nanofibers for filtration applications. In: Nayak AK, Hasnain MS (eds) Advanced biopolymeric systems for drug delivery. Springer, Cham, pp 361–371
14. Nayak AK, Bera H (2019) *In situ* polysaccharide-based gels for topical drug delivery applications. In: Maiti S, Jana S (eds) Polysaccharide carriers for drug delivery. Woodhead Publishing, Elsevier Inc., pp 615–638
15. Hasnain MS, Rishishwar P, Ali S, Alkahtani S, Tabish M, Milivojevic M, Ansari MT, Nayak AK (2020) Formulation and *ex vivo* skin permeation of lidocaine HCl topical gels using dillenia (*Dillenia indica* L.) fruit gum. Revista Mexicana de Ingeniería Química 19:1465–1476
16. Hasnain MS, Rishishwar P, Ali S (2017) Use of cashew bark exudate gum in the preparation of 4 % lidocaine HCL topical gels. Int J Phar Pharm Sci 9(8):146–150
17. Hasnain MS, Rishishwar P, Rishishwar S, Ali S, Nayak AK (2018) Isolation and characterization of *Linum usitatisimum* polysaccharide to prepare mucoadhesive beads of diclofenac sodium. Int J Biol Macromol 116:162–172
18. Nayak AK, Pal D, Santra K (2016) Swelling and drug release behavior of metformin HCl-loaded tamarind seed polysaccharide-alginate beads. Int J Biol Macromol 82:1023–1027
19. Jana S, Ali SA, Nayak AK, Sen KK, Basu SK (2014) Development and optimization of topical gel containing aceclofenac-crospovidone solid dispersion by "Quality by Design" approach. Chem Eng Res Des 92:2095–2105

20. Nayak AK, Pal D (2013) Ionotropically-gelled mucoadhesive beads for oral metformin HCl delivery: formulation, optimization and antidiabetic evaluation. J Sci Ind Res 72:15–22
21. Hasnain MS, Guru PR, Rishishwar P, Ali S, Ansari MT, Nayak AK (2020) Atenolol-releasing buccal patches made of *Dillenia indica* L. fruit gum: preparation and *ex vivo* evaluations. SN Appl Sci 2(1):57
22. Pal D, Nayak AK (2015) Interpenetrating polymer networks (IPNs): natural polymeric blends for drug delivery. In: Mishra M (ed) Encyclopedia of biomedical polymers and polymeric biomaterials, vol VI. Taylor & Francis Group, USA, pp 4120–4130
23. Nanda SS, Yi DK, Hasnain MS, Nayak AK (2019) Hydroxyapatite-alginate composites in drug delivery. In: Hasnain MS, Nayak AK (eds) Alginate: versatile polymer in biomedical applications and therapeutics. Apple Academic Press, USA, pp 483–503
24. Alam MS, Javed MN, Pottoo FH, Waziri A, Almalki FA, Hasnain MS, Garg A, Saifullah MK (2019) QbD approached comparison of reaction mechanism in microwave synthesized gold nanoparticles and their superior catalytic role against hazardous nirto-dye. Appl Organomet Chem 33(9):e5071
25. Hasnain MS, Nayak AK, Kurakula M, Hoda MN (2020) Alginate nanoparticles in drug delivery. In: Nayak AK, Hasnain MS (eds) Alginates in drug delivery. Academic Press, Elsevier Inc., USA, pp 129–152
26. Nayak AK, Bera H, Hasnain MS (2020) Particulate matrices of ionotropically gelled alginate- and plant-derived starches for sustained drug release. In: Nayak AK, Hasnain MS (eds) Alginates in drug delivery. Academic Press, Elsevier Inc., USA, pp 257–295
27. Majeed S, Shamsuddin AF, Hasnain MS (2018) Novel green approach for synthesis of metallic nanoparticles and its biomedical application. Current Nanomed 8(3):177–183
28. Hasnain MS, Ahmad SA, Hoda MN, Rishishwar S, Rishishwar P, Nayak AK (2019) Stimuli-responsive carbon nanotubes for targeted drug delivery. In: Makhlouf ASH, Abu-Thabit NY (eds) Stimuli responsive polymeric nanocarriers for drug delivery applications: vol 2: advanced nanocarriers for therapeutics. Woodhead Publishing Series in Biomaterials, Elsevier Ltd., pp 321–344
29. Panigrahi BK, Nayak AK (2020) Carbon nanotubes: an emerging drug delivery carrier in cancer therapeutics. Curr Drug Deliv 17:000–000
30. Ray S, Sinha P, Laha B, Maiti S, Bhattacharyya UK, Nayak AK (2018) Polysorbate 80 coated crosslinked chitosan nanoparticles of ropinirole hydrochloride for brain targeting. J Drug Deliv Sci Technol 48:21–29
31. Jana S, Gangopadhaya A, Bhowmik BB, Nayak AK, Mukhrjee A (2015) Pharmacokinetic evaluation of testosterone-loaded nanocapsules in rats. Int J Biol Macromol 72:28–30
32. Jana S, Maji N, Nayak AK, Sen KK, Basu SK (2013) Development of chitosan-based nanoparticles through inter-polymeric complexation for oral drug delivery. Carbohyd Polym 98:870–876
33. Waghule T, Rapalli VK, Singhvi G, Manchanda P, Hans N, Dubey SK, Hasnain MS, Nayak AK (2019) Voriconazole loaded nanostructured lipid carriers based topical delivery system: QbD based designing, characterization, *in-vitro* and *ex-vivo* evaluation. J Drug Deliv Sci Technol 52:303–315
34. Hasnain MS, Javed MN, Alam MS, Rishishwar P, Rishishwar S, Ali S, Nayak AK, Beg S (2019) Purple heart plant leaves extract-mediated silver nanoparticle synthesis: optimization by Box-Behnken design. Mater Sci Eng, C 99:1105–1114
35. Nayak AK (2010) Hydroxyapatite synthesis methodologies: an overview. Int J Chem Tech Res 2(2):903–907
36. Hasnain MS, Nayak AK (2019) Recent progress in responsive polymer-based drug delivery systems. In: Makhlouf ASH, Abu-Thabit NY (eds) Stimuli responsive polymeric nanocarriers for drug delivery applications: vol 2: advanced nanocarriers for therapeutics. Woodhead Publishing Series in Biomaterials, Elsevier Ltd., pp 569–595
37. Hasnain MS, Nayak AK (2018) Chitosan as responsive polymer for drug delivery applications. In: Makhlouf ASH, Abu-Thabit NY (eds) Stimuli responsive polymeric nanocarriers for drug delivery applications, vol 1. Types and triggers. Woodhead Publishing Series in Biomaterials, Elsevier Ltd., pp 581–605

38. Bhattacharjee A, Das PJ, Dey S, Nayak AK, Roy PK, Chakrabarti S, Marbaniang D, Das SK, Ray S, Chattopadhyay P, Mazumder B (2020) Development and optimization of besifloxacin hydrochloride loaded liposomal gel prepared by thin film hydration method using 3^2 full factorial design. Colloids Surf, A 585:124071
39. Das B, Sen SO, Maji R, Nayak AK, Sen KK (2017) Transferosomal gel for transdermal delivery of risperidone: formulation optimization and *ex vivo* permeation. J Drug Deliv Sci Technol 38:59–71
40. Malakar J, Sen SO, Nayak AK, Sen KK (2012) Preparation, optimization and evaluation of transferosomal gel for transdermal insulin delivery. Saudi Pharma J. 20:355–363
41. Hasnain MS, Nayak AK (2019) Absorption and transportation of carbon nanotubes. In: Hasnain MS, Nayak AK (eds) Carbon nanotubes for targeted drug delivery. Springer, Singapore, pp 65–68
42. Hasnain MS, Nayak AK (2019) Applications of carbon nanotubes. In: Hasnain MS, Nayak AK (eds) Carbon nanotubes for targeted drug delivery. Springer, Singapore, pp 33–36
43. Hasnain MS, Nayak AK (2019) Targeted delivery with carbon nanotubes. In: Hasnain MS, Nayak AK (eds) Carbon nanotubes for targeted drug delivery. Springer, Singapore, pp 37–50
44. Hasnain MS, Ahmed SA, Behera A, Alkahtani S, Nayak AK (2020) Inorganic materials–alginate composites in drug delivery. In: Nayak AK, Hasnain MS (eds) Alginates in drug delivery. Academic Press, Elsevier Inc., USA, pp 231–256
45. Das S, Pattanayak D, Nayak AK, Yi DK, Nanda SS, Ansari MT, Hasnain MS (2020) Alginate–montmorillonite composite systems as sustained drug delivery carriers. In: Nayak AK, Hasnain MS (eds) Alginates in drug delivery. Academic Press, Elsevier Inc., USA, pp 187–201
46. Nayak AK, Pal D (2012) Natural polysaccharides for drug delivery in tissue engineering. Everyman's Sci XLVI:347–352
47. Hasnain MS, Ahmad SA, Chaudhary N, Hoda MN, Nayak AK (2019) Biodegradable polymer matrix nanocomposites for bone tissue engineering. In: Inamuddin, Asiri AM, Mohammad A (eds) Applications of nanocomposite materials in orthopaedics. Woodhead Publishing Series in Biomaterials, Elsevier Inc., pp 1–37
48. Hasnain MS, Nayak AK (2019) Nanocomposites for improved orthopedic and bone tissue engineering applications. In: Inamuddin, Asiri AM, Mohammad A (eds) Applications of nanocomposite materials in orthopaedics. Woodhead Publishing Series in Biomaterials, Elsevier Inc., pp 145–177
49. Ullah H, Wahid F, Santos HA, Khan T (2016) Advances in biomedical and pharmaceutical applications of functional bacterial cellulose-based nanocomposites. Carbohydr Polym 150:330–352
50. Hasnain MS, Ahmad SA, Minhaj MA, Ara TJ, Nayak AK (2019) Nanocomposite materials for prosthetic devices. In: Inamuddin, Asiri AM, Mohammad A (eds) Applications of nanocomposite materials in orthopaedics. Elsevier Inc., pp 127–144
51. Nayak AK, Mazumder S, Ara TJ, Ansari MT, Hasnain MS (2019) Calcium fluoride-based dental nanocomposites. In: Inamuddin, Asiri AM, Mohammad A (eds) Applications of nanocomposite materials in dentistry. Elsevier Inc., pp 27–45
52. Rani P, Pal D, Hoda MN, Ara TJ, Beg S, Hasnain MS, Nayak AK (2019) Dental pulp capping nanocomposites. In: Inamuddin, Asiri AM, Mohammad A (eds) Applications of nanocomposite materials in dentistry. Elsevier Inc., pp 65—91
53. Hasnain MS, Ahmad SA, Chaudhary N, Minhaj MA, Nayak AK (2019) Degradation and failure of dental composite materials. In: Inamuddin, Asiri AM, Mohammad A (eds) Applications of nanocomposite materials in dentistry. Elsevier Inc., pp 108–121
54. Mazumder S, Nayak AK, Ara TJ, Hasnain MS (2019) Hydroxyapatite composites for dentistry. In: Inamuddin, Asiri AM, Mohammad A (eds) Applications of nanocomposite materials in dentistry. Woodhead Publishing Series in Biomaterials, Elsevier Inc., pp 108–121
55. Kumar SK, Krishnamoorti R (2010) Nanocomposites: structure, phase behavior, and properties. Ann Rev Chem Biomol Eng 1:37–58
56. Camargo PHC, Satyanarayana KG, Wypych F (2009) Nanocomposites: synthesis, structure, properties and new application opportunities. Mater Res 12:1–39

57. Ashori A, Sheykhnazari S, Tabarsa T, Shakeri A, Golalipour M (2012) Bacterial cellulose/silica nanocomposites: preparation and characterization. Carbohydr Polym 90:413–418
58. Zheng Y, Monty J, Linhardt RJ (2015) Polysaccharide-based nanocomposites and their applications. Carbohydr Res 405:23–32
59. Hu W, Chen S, Yang J, Li Z, Wang H (2014) Functionalized bacterial cellulose derivatives and nanocomposites. Carbohydr Polym 101:1043–1060
60. Nayak AK, Ahmad SA, Beg S, Ara TJ, Hasnain MS (2018) Drug delivery: present, past and future of medicine. In: Inamuddin, Asiri AM, Mohammad A (eds) Applications of nanocomposite materials in drug delivery. Woodhead Publishing Series in Biomaterials, Elsevier Inc., pp 255–282
61. Hasnain MS, Ahmed SA, Alkahtani S, Milivojevic M, Kandar CC, Dhara AK, Nayak AK (2020) Biopolymers for drug delivery. In: Nayak AK, Hasnain MS (eds) Advanced biopolymeric systems for drug delivery. Springer, Cham, pp 1–29
62. Nayak AK, Hasnain MS (2019) Background: multiple units in oral drug delivery. In: Nayak AK, Hasnain MS (eds) Plant polysaccharides-based multiple-unit systems for oral drug delivery. Springer, Singapore, pp 1–17
63. Nayak AK, Ansari MT, Sami F, Balvir Singh HK, Hasnain MS (2020) Alginates as drug delivery excipients. In: Nayak AK, Hasnain MS (eds) Alginates in drug delivery. Academic Press, Elsevier Inc., USA, pp 19–39
64. Bera H, Maiti S, Saha S, Nayak AK (2019) Biopolymers-based gastroretentive buoyant systems for therapeutic management of *Helicobacter pylori* infection. In: Maiti S, Jana S (eds) Polysaccharide carriers for drug delivery. Woodhead Publishing, Elsevier Inc., pp 713–736
65. Nayak AK, Hasnain MS (2019) Plant polysaccharides in drug delivery applications. In: Nayak AK, Hasnain MS (eds) Plant polysaccharides-based multiple-unit systems for oral drug delivery. Springer, Singapore, pp 19–23
66. Nayak AK, Hasnain MS (2019) Gum Arabic based multiple units for oral drug delivery. In: Nayak AK, Hasnain MS (eds) Plant polysaccharides-based multiple-unit systems for oral drug delivery. Springer, Singapore, pp 25–30
67. Nayak AK, Hasnain MS (2019) Tamarind polysaccharide based multiple units for oral drug delivery. In: Nayak AK, Hasnain MS (eds) Plant polysaccharides-based multiple-unit systems for oral drug delivery. Springer, Singapore, pp 31–59
68. Nayak AK, Hasnain MS (2019) Locust bean gum based multiple units for oral drug delivery. In: Nayak AK, Hasnain MS (eds) Plant polysaccharides-based multiple-unit systems for oral drug delivery. Springer, Singapore, pp 61–66
69. Nayak AK, Hasnain MS (2019) Sterculia gum based multiple units for oral drug delivery. In: Nayak AK, Hasnain MS (eds) Plant polysaccharides-based multiple-unit systems for oral drug delivery. Springer, Singapore, pp 67–82
70. Nayak AK, Hasnain MS (2019) Okra gum based multiple units for oral drug delivery. In: Nayak AK, Hasnain MS (eds) Plant polysaccharides-based multiple-unit systems for oral drug delivery. Springer, Singapore, pp 83–92
71. Nayak AK, Hasnain MS (2019) Fenugreek seed mucilage based multiple units for oral drug delivery. In: Nayak AK, Hasnain MS (eds) Plant polysaccharides-based multiple-unit systems for oral drug delivery. Springer, Singapore, pp 93–112
72. Nayak AK, Hasnain MS (2019) Potato starch based multiple units for oral drug delivery. In: Nayak AK, Hasnain MS (eds) Plant polysaccharides-based multiple-unit systems for oral drug delivery. Springer, Singapore, pp 113–116
73. Nayak AK, Hasnain MS (2019) Linseed polysaccharide based multiple units for oral drug delivery. In: Nayak AK, Hasnain MS (eds) Plant polysaccharides-based multiple-unit systems for oral drug delivery. Springer, Singapore, pp 117–121
74. Milivojevic M, Pajic-Lijakovic I, Bugarski B, Nayak AK, Hasnain MS (2019) Gellan gum in drug delivery applications. In: Hasnain MS, Nayak AK (eds) Natural polysaccharides in drug delivery and biomedical applications. Academic Press, Elsevier Inc., pp 145–186
75. Bera H, Abbasi YF, Hasnain MS, Nayak AK (2019) Sterculia gum in drug delivery applications. In: Hasnain MS, Nayak AK (eds) Natural polysaccharides in drug delivery and biomedical applications. Academic Press, Elsevier Inc., pp 223–247

76. Nayak AK, Ansari MT, Sami F, Bera H, Hasnain MS (2019) Cashew gum in drug delivery applications. In: Hasnain MS, Nayak AK (eds) Natural polysaccharides in drug delivery and biomedical applications. Academic Press, Elsevier Inc., pp 263–283
77. Dey S, Nandy BC, De JN, Hasnain MS, Nayak AK (2019) Tamarind gum in drug delivery applications. In: Hasnain MS, Nayak AK (eds) Natural polysaccharides in drug delivery and biomedical applications. Academic Press, Elsevier Inc., pp 285–306
78. Samanta A, De A, Hasnain MS, Bera H, Nayak AK (2019). Gum odina as pharmaceutical excipient. In: Hasnain MS, Nayak AK (eds) Natural polysaccharides in drug delivery and biomedical applications. Academic Press, Elsevier Inc., pp 327–337
79. Jana S, Maiti S, Jana S, Sen KK, Nayak AK (2019) Guar gum in drug delivery applications. In: Hasnain MS, Nayak AK (eds) Natural polysaccharides in drug delivery and biomedical applications. Academic Press, Elsevier Inc., pp 187–201
80. Pal D, Nayak AK, Saha S (2019) Interpenetrating polymer network hydrogels of chitosan: applications in controlling drug release. In: Mondal MIH (ed) Cellulose-based superabsorbent hydrogels. Springer, Cham, pp 1727–1768
81. Nayak AK, Pal D (2018) Functionalization of tamarind gum for drug delivery. In: Thakur VK, Thakur MK (eds) Functional biopolymers. Springer International Publishing, Switzerland, pp 35–56
82. Hasnain MS, Nayak AK (2018) Alginate-inorganic composite particles as sustained drug delivery matrices. In: Inamuddin, Asiri AM, Mohammad A (eds) Applications of nanocomposite materials in drug delivery. Elsevier Inc., pp 39–74
83. Nayak AK, Ara TJ, Hasnain MS, Hoda N (2018) Okra gum-alginate composites for controlled releasing drug delivery. In: Inamuddin, Asiri AM, Mohammad A (eds) Applications of nanocomposite materials in drug delivery. Woodhead Publishing Series in Biomaterials, Elsevier Inc., pp 761–785
84. Dey S, Roy S, MS, Nayak AK (2020) Grafted alginates in drug delivery. Nayak AK, Hasnain MS (eds) Alginates in drug delivery. Academic Press, Elsevier Inc., USA, pp 71–100
85. Gupta S, Hasnain MS, Agarwal SS (2012) Formulation and evaluation of oral disintegrating tablets of itopride hydrochloride using ion exchange resins as drug carrier. Asian J Pharm Sci 7(3):207–218
86. Hasnain MS, Rishishwar P, Ali S (2017) Floating-bioadhesive matrix tablets of hydralazine HCL made of cashew gum and HPMC K4M. Int J Phar Pharm Sci 9(7):124–129
87. Malakar J, Das K, Nayak AK (2014) In situ cross-linked matrix tablets for sustained salbutamol sulfate release—formulation development by statistical optimization. Polymers Med 44:221–230
88. Malakar J, Nayak AK (2013) Floating bioadhesive matrix tablets of ondansetron HCl: optimization of hydrophilic polymer-blends. Asian J Pharma 7:174–183
89. Nayak AK, Das B, Maji R (2013) Gastroretentive hydrodynamically balanced system of ofloxacin: formulation and *in vitro* evaluation. Saudi Pharma J 21:113–117
90. Verma A, Dubey J, Verma N, Nayak AK (2017) Chitosan-hydroxypropyl methylcellulose matrices as carriers for hydrodynamically balanced capsules of moxifloxacin HCl. Curr Drug Deliv 14:83–90
91. Das B, Nayak AK, Nanda U (2013) Topical gels of lidocaine HCl using cashew gum and Carbopol 940: preparation and *in vitro* skin permeation. Int J Biol Macromol 62:514–517
92. Ray P, Maity M, Barik H, Sahoo GS, Hasnain MS, Hoda MN, Nayak AK (2020) Alginate-based hydrogels for drug delivery applications. In: Nayak AK, Hasnain MS (eds) Alginates in drug delivery. Academic Press, Elsevier Inc., USA, pp 41–70
93. Nayak AK, Hasnain MS (2020) Ionotropically gelled alginate particles in sustained drug release. In: Nayak AK, Hasnain MS (eds) Alginates in drug delivery. Academic Press, Elsevier Inc., USA, pp 203–230
94. Nayak AK, Hasnain MS, Pal K, Banerjee I, Pal D (2020) Gum-based hydrogels in drug delivery. In: Pal K, Banerjee I, Sarkar P, Kim D, Deng W-P, Dubey NK, Majumder K (eds) Biopolymer-based formulations, biomedical and food applications. Elsevier Inc., pp 605–645

95. Nayak AK, Das B (2018) Introduction to polymeric gels. In: Pal K, Bannerjee I (eds) Polymeric gels, characterization, properties and biomedical applications. Woodhead Publishing Series in Biomaterials, Elsevier Ltd., pp 3–27
96. Nayak AK, Pal D (2016) Sterculia gum-based hydrogels for drug delivery applications. In: Kalia S (ed) Polymeric hydrogels as smart biomaterials. Springer Series on polymer and composite materials. Springer International Publishing, Switzerland, pp 105–151
97. Guru PR, Bera H, Das M, Hasnain MS, Nayak AK (2018) Aceclofenac-loaded *Plantago ovata* F. husk mucilage-Zn^{+2}-pectinate controlled-release matrices. Starch—Stärke 70:1700136
98. Nayak AK, Pal D, Santra K (2014) Ispaghula mucilage-gellan mucoadhesive beads of metformin HCl: development by response surface methodology. Carbohyd Polym 107:41–50
99. Nayak AK, Pal D (2014) *Trigonella foenum-graecum* L. seed mucilage-gellan mucoadhesive beads for controlled release of metformin HCl. Carbohyd Polym 107:31–40
100. Nayak AK, Pal D, Santra K (2014) *Artocarpus heterophyllus* L. seed starch-blended gellan gum mucoadhesive beads of metformin HCl. Int J Biol Macromole 65:329–339
101. Nayak AK, Kalia S, Hasnain MS (2013) Optimization of aceclofenac-loaded pectinate-poly (vinyl pyrrolidone) beads by response surface methodology. Int J Biol Macromol 62:194–202
102. Nayak AK, Pal D, Pradhan J, Hasnain MS (2013) Fenugreek seed mucilage-alginate mucoadhesive beads of metformin HCl: design, optimization and evaluation. Int J Biol Macromol 54:144–154
103. Nayak AK, Pal D (2013) Formulation optimization of jackfruit seed starch-alginate mucoadhesive beads of metformin HCl. Int J Biol Macromol 59:264–272
104. Nayak AK, Pal D, Hasnain MS (2013) Development, optimization and *in vitro-in vivo* evaluation of pioglitazone-loaded jackfruit seed starch-alginate beads. Curr Drug Deliv 10:608–619
105. Nayak AK, Pal D (2013) Blends of jackfruit seed starch-pectin in the development of mucoadhesive beads containing metformin HCl. Int J Biol Macromol 62:137–145
106. Nayak AK, Pal D, Das S (2013) Calcium pectinate-fenugreek seed mucilage mucoadhesive beads for controlled delivery of metformin HCl. Carbohyd Polym 96:349–357
107. Nayak AK, Pal D, Malakar J (2013) Development, optimization and evaluation of emulsion-gelled floating beads using natural polysaccharide-blend for controlled drug release. Polym Eng Sci 53:338–350
108. Nayak AK, Pal D, Santra K (2014) Development of pectinate-ispagula mucilage mucoadhesive beads of metformin HCl by central composite design. Int J Biol Macromol 66:203–221
109. Nayak AK, Pal D, Santra K (2014) Tamarind seed polysaccharide-gellan mucoadhesive beads for controlled release of metformin HCl. Carbohyd Polym 103:154–163
110. Pal D, Nayak AK (2015) Alginates, blends and microspheres: controlled drug delivery. In: Mishra M (ed) Encyclopedia of biomedical polymers and polymeric biomaterials, vol I. Taylor & Francis Group, USA, pp 89–98
111. Ali SA, Nayak AK, Sen KK, Prabhakar T (2019) Preparation and characterization of vetiver oil encapsulated polymeric microcapsules for sedative and hypnotic activity. Int J Res Pharma Sci 10(4):3616–3625
112. Nayak AK, Beg S, Hasnain MS, Malakar J, Pal D (2018) Soluble starch-blended Ca^{2+}-Zn^{2+}-alginate composites-based microparticles of aceclofenac: formulation development and *in vitro* characterization. Future J Pharma Sci 4:63–70
113. Das B, Dutta S, Nayak AK, Nanda U (2014) Zinc alginate-carboxymethyl cashew gum microbeads for prolonged drug release: development and optimization. Int J Biol Macromol 70:505–515
114. Jana S, Saha A, Nayak AK, Sen KK, Basu SK (2013) Aceclofenac-loaded chitosan-tamarind seed polysaccharide interpenetrating polymeric network microparticles. Colloids Surf B Biointerf 105:303–309
115. Rapalli VK, Singhvi G, Gorantla S, Waghule T, Dubey SK, Saha RN, Hasnain MS, Nayak AK (2019) Stability indicating liquid chromatographic method for simultaneous quantification of betamethasone valerate and tazarotene in *in-vitro* and *ex-vivo* studies of complex nanoformulation. J Sep Sci 42(22):3413–3420

116. Rath Adhikari SN, Nayak BS, Nayak AK, Mohanty B (2010) Formulation and evaluation of buccal patches for delivery of atenolol. AAPS PharmSciTech 11(3):1034–1044
117. Hasnain MS, Rishishwar P, Rishishwar S, Ali S, Nayak AK (2018) Extraction and characterization of cashew tree (*Anacardium occidentale*) gum; use in aceclofenac dental pastes. Int J Biol Macromol 116:1074–1081
118. Hasnain MS, Rishishwar P, Ali S, Nayak AK (2020) Preparation and evaluation of aceclofenac dental pastes using dillenia fruit gum for periodontitis treatment. SN Appl Sci 2(3):1–8
119. Jena AK, Nayak AK, De A, Mitra D, Samanta A (2018) Development of lamivudine containing multiple emulsions stabilized by gum odina. Future J Pharma Sci 4:71–79
120. Malakar J, Basu A, Nayak AK (2014) Candesartan cilexetil microemulsions for transdermal delivery: formulation, *in-vitro* skin permeation and stability assessment. Curr Drug Deliv 11:313–321
121. Ray P, Hasnain MS, Koley A, Nayak AK (2019) Bone-implantable devices for drug delivery applications. In: Pal K, Kraatz H-H, Li C, Khasnobish A, Bag S, Banerjee I, Kuruganti U (eds) Bioelectronics and medical devices, from materials to devices—fabrication, applications and reliability. Woodhead Publishing Series in Electronic and Optical Materials, Elsevier Inc., pp 333–392
122. Nayak AK, Hasnain MS, Malakar J (2013) Development and optimization of hydroxyapatite-ofloxacin implants for possible bone-implantable delivery in osteomyelitis treatment. Curr Drug Deliv 10:241–250
123. Nayak AK, Sen KK (2009) Hydroxyapatite-ciprofloxacin implantable minipellets for bone delivery: preparation, characterization, *in vitro* drug adsorption and dissolution studies. Int J Drug Develop Res 1(1):47–59
124. Nayak AK, Laha B, Sen KK (2011) Development of hydroxyapatite-ciprofloxacin bone-implants using >>Quality by Design<< Acta Pharma 61(1):25–36
125. Nayak AK, Bhattacharyya A, Sen KK (2010) Hydroxyapatite-antibiotic implantable minipellets for bacterial bone infections using precipitation technique: preparation, characterization and *in-vitro* antibiotic release studies. J Pharma Res 3(1):53–59
126. Fernandes SCM, Freire CSR, Silvestre AJD, Pascoal Neto C, Gandini A, Berglund LA, Salmén L (2010) Transparent chitosan films reinforced with a highcontent of nanofibrillated cellulose. Carbohydr Polym 81:394–401
127. Iliescu RI, Andronescu E, Ghitulica CD, Voicu G, Ficai A, Hoteteu M (2014) Montmorillonite–alginate nanocomposite as a drug delivery system–incorporation and *in vitro* release of irinotecan. Int J Pharm 463:184–192
128. Iliescu RI, Andronescu E, Ghitulica CD, Berger D, Ficai A (2011) Montmorillonite-alginate nanocomposite beads as drug carrier for oral administration of carboplatin-preparation and characterization. UPB Sci Bull Ser B 73:3–16
129. Azhar FF, Olad A (2014) A study on sustained release formulations for oral delivery of 5-fluorouracil based on alginate–chitosan/montmorillonite nanocomposite systems. Appl Clay Sci 101:288–296
130. Liu K-H, Liu T-Y, Chen S-Y, Liu D-M (2008) Drug release behavior of chitosan–montmorillonite nanocomposite hydrogels following electrostimulation. Acta Biomater 4:1038–1045
131. Justin R, Chen B (2014) Strong and conductive chitosan-reduced graphene oxide nanocomposites for transdermal drug delivery. J Mater Chem B 2:3759–3770
132. Venkatesan P, Puvvada N, Dash R, Prashanth kumar BN, Sarkar D, Azab B, Pathak A, Kundu SC, Fisher PB, Mandal M (2011) The potential of celecoxib-loaded hydroxyapatite-chitosan nanocomposite for the treatment of colon cancer. Biomaterials 32:3794–3806
133. Nanda R, Sasmal A, Nayak PL (2011) Preparation and characterization of chitosan–polylactide composites blended with Cloisite 30B for control release of the anticancer drug paclitaxel. Carbohydr Polym 83:988–994
134. Wu J, Jiang W, Shen Y, Jiang W, Tian R (2017) Synthesis and characterization of mesoporous magnetic nanocomposites wrapped with chitosan gatekeepersfor pH-sensitive controlled release of doxorubicin. Mater Sci Eng C Mater Biol Appl 70:132–140

135. Nayak AK, Sen KK (2016) Bone-targeted drug delivery systems. In: Maiti S, Sen KK (eds) Bio-targets & drug delivery approaches. CRC Press, Boca Raton, FL, USA, pp 207–231
136. Hesaraki S, Moztarzadeh F, Nemati R, Nezafati N (2009) Preparation and characterization of calcium sulfate-biomimetic apatite nanocomposites for controlled release of antibiotics. J Biomed Mater Res B Appl Biomater 91:651–661
137. Aghaei H, Nourbakhsh AA, Karbasi S, Kalbasid R, Rafienia M, Nourbakhsh N, Bonakdar S, Mackenzie KJD (2014) Investigation on bioactivity and cytotoxicity of mesoporous nano-composite MCM-48/hydroxyapatite for ibuprofen drug delivery. Ceramics Int 40:7355–7362
138. Nithya R, Meenakshi Sundaram N (2015) Biodegradation and cytotoxicity of ciprofloxacin-loaded hydroxyapatite-polycaprolactone nanocomposite film for sustainable bone implants. Int J Nanomed 10:119–127
139. Gholamali I, Yadollahi M (2020) Doxorubicin-loaded carboxymethyl cellulose/Starch/ZnO nanocomposite hydrogel beads as an anticancer drug carrier agent [published online ahead of print, 2020 May 30]. Int J Biol Macromol 160:724–735
140. Mallakpour S, Hatami M (2019) Fabrication and characterization of pH-sensitive bio-nanocomposite beads havening folic acid intercalated LDH and chitosan: drug release and mechanism evaluation. Int J Biol Macromol 122:157–167
141. Kodoth AK, Ghate VM, Lewis SA, Badalamoole V (2018) Application of pectin-zinc oxide hybrid nanocomposite in the delivery of a hydrophilic drug and a study of its isotherm, kinetics and release mechanism. Int J Biol Macromol 115:418–430
142. Wang SY, Meng YJ, Li J, Liu JP, Liu ZQ, Li DQ (2020) A novel and simple oral colon-specific drug delivery system based on the pectin/modified nano-carbon sphere nanocomposite gel films. Int J Biol Macromol 157:170–176
143. Taufiq A, Nikmah A, Hidayat A, Sunaryono S, Mufti N, Hidayat N, Susanto H (2020) Synthesis of magnetite/silica nanocomposites from natural sand to create a drug delivery vehicle. Heliyon 6(4):e03784
144. Ding Y, Yin H, Shen S, Sun K, Liu F (2017) Chitosan-based magnetic/fluorescent nanocomposites for cell labelling and controlled drug release. New J Chem 41:1736–1743
145. Raj V, Prabha G (2016) Synthesis, characterization and *in vitro* drug release of cisplatin loaded Cassava starch acetate–PEG/gelatin nanocomposites. J Assoc Arab Univ Basic Appl Sci 21:10–16
146. Hasnain MS, Nayak AK, Singh M, Tabish M, Ansari MT, Ara TJ (2016) Alginate-based bipolymeric-nanobioceramic composite matrices for sustained drug release. Int J Biol Macromol 83:71–77
147. Kajthunyakarn W, Sakloetsakun D, Pongjanyakul T (2018) Sodium caseinate-magnesium aluminum silicate nanocomposite films for modified-release tablets. Mater Sci Eng C Mater Biol Appl 92:827–839
148. Asare-Addo K, Totea AM, Nokhodchi A (2020) Drug release from magnesium aluminium silicate-polyethylene oxide (PEO) nanocomposite matrices: an investigation using the USP III apparatus. Eur J Pharm Sci 153:105474
149. Roul J, Mohapatra R, Sahoo SK (2013) Preparation, characterization and drug delivery behavior of novel biopolymer/hydroxyapatite nanocomposite beads. Asian J Biomed Pharm Sci 3:33–38
150. Kharaziha M, Fathi MH, Edris H, Nourbakhsh N, Talebi A, Salmanizadeh S (2015) PCL-forsterite nanocomposite fibrous membranes for controlled release of dexamethasone. J Mater Sci Mater Med 26:53–64
151. Bala Subramanian S, Francis AP, Devasena T (2014) Chitosan-starch nanocomposite particles as a drug carrier for the delivery of bis-desmethoxy curcumin analog. Carbohydr Polym 114:170–178
152. Wang X, Yang B, Xu X, Su M, Xi M, Yin Z (2020) Dextran sulfate-modified pH-sensitive layered double hydroxide nanocomposites for treatment of rheumatoid arthritis. Drug Deliv Transl Res. https://doi.org/10.1007/s13346-020-00832-2
153. Kolanthai E, Sindu PA, Arul KT, Chandra VS, Manikandan E, Kalkura SN (2017) Agarose encapsulated mesoporous carbonated hydroxyapatite nanocomposites powder for drug delivery. J. Photochem Photobiol B Biol 166:220–231

154. Kevadiya BD, Joshi GV, Patel HA, Ingole PG, Mody HM, Bajaj HC (2010) Montmorillonite-alginate nanocomposites as a drug delivery system: intercalation and *in vitro* release of vitamin B1 and vitamin B6. J Biomater Appl 25(2):161–177
155. Kummari SVKR, Kummara MR, Palem RR, Nagellea SR, Shchipunov Y, Ha C-S (2015) Chitosan-poly(aminopropyl/phenylsilsesquioxane) hybrid nanocomposite membranes for antibacterial and drug delivery applications. Polym Int 64(2):293–302
156. Lin J, Li Y, Li Y, Wu H, Yu F, Zhou S, Xie L, Luo F, Lin C, Hou Z (2015) Drug/Dye-Loaded, multifunctional PEG-Chitosan-Iron oxide nanocomposites for methotraxate synergistically self-Targeted cancer therapy and dual model imaging. ACS Appl Mater Interfaces 7:11908–11920
157. Yadollahi M, Farhoudian S, Namazi H (2015) One-pot synthesis of antibacterial chitosan/silver bio-nanocomposite hydrogel beads as drug delivery systems. Int J Biol Macromol 79:37–43
158. Vedakumari WS, Prabu P, Sastry TP (2015) Chitosan-fibrin nanocomposites as drug delivering and wound healing materials. J Biomed Nanotechnol 11(4):657–667
159. Basu T, Pal B, Singh S (2018) Hollow chitosan nanocomposite as drug carrier system for controlled delivery of ramipril. Chem Phys Lett 706:465–471
160. Taleb MFA, Alkahtani A, Mohamed SK (2015) Radiation synthesis and characterization of sodium alginate/chitosan/hydroxyapatite nanocomposite hydrogels: a drug delivery system for liver cancer. Polym Bullet 72(4):725–742
161. Thomas TJ, Tajmir-Riahi HA, Pillai CKS (2019) Biodegradable polymers for gene delivery. Molecules 24(20):3744
162. Oliveira AV, Rosa da Costa AM, Silva GA (2017) Non-viral strategies for ocular gene delivery. Mater Sci Eng C 77:1275–1289
163. Fernandez-Piñeiro I, Pensado A, Badiola I, Sanchez A (2018) Development and characterisation of chondroitin sulfate- and hyaluronic acid-incorporated sorbitan ester nanoparticles as gene delivery systems. Eur J Pharm Biopharm 125:85–94
164. Kashkouli KI, Torkzadeh-Mahani M, Mosaddegh E (2018) Synthesis and characterization of aminotetrazole-functionalized magnetic chitosan nanocomposite as a novel nanocarrier for targeted gene delivery. Mater Sci Eng, C 89:166–174
165. Xie Y, Qiao H, Su Z, Chen M, Ping Q, Sun M (2014) PEGylated carboxymethyl chitosan/calcium phosphate hybrid anionic nanoparticles mediated hTERT siRNA delivery for anticancer therapy. Biomaterials 35(27):7978–7991
166. Zhu Q, Zhou Y, Guan M, Zhou XF, Yang SD, Liu Y et al (2014) Low-density lipoprotein-coupled N-succinyl chitosan nanoparticles co-delivering siRNA and doxorubicin for hepatocyte-targeted therapy. Biomaterials 35(22):5965–5976
167. Yan C-Y, Gu J-W, Hou D-P, Jing HY, Wang J, Guo YZ et al (2015) Synthesis of tat tagged and folate modified N-succinyl-chitosan self-assembly nanoparticles as a novel gene vector. Int J Biol Macromol 72:751–756
168. Yao L, Wang X, Weng W, Fu Y, Cheng K (2020) Bioactive nanocomposite coatings under visible light illumination promoted surface-mediated gene delivery. Biomater Sci 8(13):3685–3696
169. Wang X, Coradin T, Hélary C (2018) Modulating inflammation in a cutaneous chronic wound model by IL-10 released from collagen-silica nanocomposites via gene delivery. Biomater Sci 6(2):398–406
170. Kim H, Kim WJ (2014) Photothermally controlled gene delivery by reduced graphene oxide-polyethylenimine nanocomposite. Small 10(1):117–126
171. Oyane A, Araki H, Nakamura M, Shimizu Y, Shubhra QTH, Ito A, Tsurushima H (2016) Controlled superficial assembly of DNA-amorphous calcium phosphate nanocomposite spheres for surface-mediated gene delivery. Colloids Surf B Biointerfaces 1(141):519–527
172. Li L, Zhang R, Gu W, Xu ZP (2018) Mannose-conjugated layered double hydroxide nanocomposite for targeted siRNA delivery to enhance cancer therapy. Nanomedicine 14(7):2355–2364

173. Zhou Q, Wang Y, Xiang J, Piao Y, Zhou Z, Tang J, Liu X, Shen Y (2018) Stabilized calcium phosphate hybrid nanocomposite using a benzoxaborole-containing polymer for pH-responsive siRNA delivery. Biomater Sci 6(12):3178–3188
174. Cheang TY, Lei YY, Zhang ZQ, Zhou HY, Ye RY, Lin Y, Wang S (2018) Graphene oxide-hydroxyapatite nanocomposites effectively deliver HSV-TK suicide gene to inhibit human breast cancer growth. J Biomater Appl 33(2):216–226
175. Stavitskaya A, Batasheva S, Vinokurov V, Fakhrullina G, Sangarov V, Lvov Y, Fakhrullin R (2019) Antimicrobial applications of clay nanotube-based composites. Nanomaterials (Basel) 9(5):708
176. Baranwal A, Srivastava A, Kumar P, Bajpai VK, Maurya PK, Chandra P (2018) Prospects of nanostructure materials and their composites as antimicrobial agents. Front Microbiol 9:422
177. Butchosa N, Brown C, Larsson PT, Berglund LA, Bulone V, Zhou Q (2013) Nanocomposites of bacterial cellulose nanofibers and chitin nanocrystals: fabrication, characterization and bactericidal activity. Green Chem 15(12):3404–3413
178. Rinehart SJ, Campbell TD, Burke KJ, Garcia B, Mlynarski A, Brain SJ, Keleher JJ (2016) Synthesis and characterization of a chitosan/PVA antimicrobial hydrogel nanocomposite for responsive wound management materials. J Microb Biochem Technol
179. Elbarbary AM, El-Sawy NM (2016) Radiation synthesis and characterization of polyvinyl alcohol/chitosan/silver nanocomposite membranes: antimicrobial and blood compatibility studies. Polym Bull 1–18
180. Youssef AM, Abdel-Aziz MS, El-Sayed SM (2014) Chitosan nanocomposite films based on Ag-NP and Au-NP biosynthesis by Bacillus subtilis as packaging materials. Int Biol Macromol 69:185–191
181. Mohamed RR, Sabaa MW (2014) Synthesis and characterization of antimicrobial cross-linked carboxymethyl chitosan nanoparticles loaded with silver. Int J Biol Macromol 69:95–99
182. Farhoudian S, Yadollahi M, Namazi H (2016) Facile synthesis of antibacterial chitosan/CuO bio-nanocomposite hydrogel beads. Int J Biol Macromol 82:837–843
183. Khalil WF, El-Sayyad GS, El Rouby WMA, Sadek MA, Farghali AA, El-Batal AI (2020) Graphene oxide-based nanocomposites (GO-chitosan and GO-EDTA) for outstanding antimicrobial potential against some Candida species and pathogenic bacteria. Int J Biol Macromol 164:1370–1383
184. Chen J, Fan L, Yang C, Wang S, Zhang M, Xu J, Luo S (2020) Facile synthesis of Ag nanoparticles-loaded chitosan antibacterial nanocomposite and its application in polypropylene. Int J Biol Macromol S0141–8130(20):33904
185. Malagurski I, Levic S, Pantic M, Matijasevic D, Mitric M, Pavlovic V, Dimitrijevic-Brankovic S (2017) Synthesis and antimicrobial properties of Zn-mineralized alginate nanocomposites. Carbohydr Polym 165:313–321
186. Davoodbasha M, Kim S-C, Lee S-Y, Kim J-W (2016) The facile synthesis of chitosan based silver nano-biocomposites via a solution plasma process and their potential antimicrobial efficacy. Arch Biochem Biophys 605:49–58
187. Sá NMSM, Mattos ALA, Silva LMA, Brito ES, Rosa MF, Azeredo HMC (2020) From cashew byproducts to biodegradable active materials: Bacterial cellulose-lignin-cellulose nanocrystal nanocomposite films. Int J Biol Macromol S0141–8130(20):34032
188. Abu Elella MH, Mohamed RR, Abdel-Aziz MM, Sabaa MW (2018) Green synthesis of antimicrobial and antitumor N, N, N-trimethyl chitosan chloride/poly (acrylic acid)/silver nanocomposites. Int J Biol Macromol 111:706–716
189. Mathew S, Mathew J, Radhakrishnan EK (2019) Polyvinyl alcohol/silver nanocomposite films fabricated under the influence of solar radiation as effective antimicrobial food packaging material. J Polym Res 26:223–233
190. Jabbar AH, Mezan SO, Tuama AN, Hamzah MQ, Ameruddin ASB, Agam MA (2019) Enhanced bioactivity of polystyrene-silver nanocomposite (PS/Ag NCs)-an antimicrobial study. AIP Conf Proc 2151(1):020002

191. Noohpisheh Z, Amiri H, Farhadi S, Mohammadi-Gholami A (2020) Green synthesis of Ag-ZnO nanocomposites using *Trigonella foenum-graecum* leaf extract and their antibacterial, antifungal, antioxidant and photocatalytic properties. Spectrochim Acta A Mol Biomol Spectrosc 240:118595
192. Mei L, Xu Z, Shi Y, Lin C, Jiao S, Zhang L, Li P (2020) Multivalent and synergistic chitosan oligosaccharide-Ag nanocomposites for therapy of bacterial infection. Sci Rep 10(1):10011
193. Siripatrawan U, Kaewklin P (2018) Fabrication and characterization of chitosan-titanium dioxide nanocomposite film as ethylene scavenging and antimicrobial active food packaging. Food Hydrocolloids 84:125–134
194. Guom Q, Lan T, Chen Y, Xu Y, Peng J, Tao L, Shen X (2019) Enhanced of antibacterial activity of antibiotic-functionalized silver nanocomposites with good biocompatibility. J Mater Sci—Mater Med 30:34–43
195. Abdelwahab NA (2015) Shukry N, Synthesis, characterization and antimicrobial properties of grafted sugarcane bagasse/silver nanocomposites. Carbohydr Polym 115:276–284
196. El Zowalaty ME, Al Ali SHH, Husseiny MI, Geilich BM, Webster TJ, Hussein MZ (2015) The ability of streptomycin-loaded chitosan-coated magnetic nanocomposites to possess antimicrobial and antituberculosis activities. Int J Nanomed 10:3269
197. Ahmadi Y, Yadav M, Ahmad S (2019) Oleo-polyurethane-carbon nanocomposites: effects of in-situ polymerization and sustainable precursor on structure, mechanical, thermal, and antimicrobial surface-activity. Compos B 164:683–692
198. Jannatyha N, Shojaee-Aliabadi S, Moslehishad M, Moradi E (2020) Comparing mechanical, barrier and antimicrobial properties of nanocellulose/CMC and nanochitosan/CMC composite films. Int J Biol Macromol S0141–8130(20):34013–34017
199. Nayak AK, Ahmed SA, Tabish M, Hasnain MS (2019) Natural polysaccharides in tissue engineering application. In: Hasnain MS, Nayak AK (eds) Natural polysaccharides in drug delivery and biomedical applications. Academic Press, Elsevier Inc., pp 531–548
200. Hasnain MS, Nayak AK, Singh R, Ahmad F (2010) Emerging trends of natural-based polymeric systems for drug delivery in tissue engineering applications. Science J UBU 1(2):1–13
201. Meskinfam M, Sadjadi MA, Jazdarreh H, Zare K (2011) Biocompatibility evaluation of nano hydroxyapatite-starch biocomposites. J Biomed Nanotechnol 7(3):455–459
202. Sadjadi M, Meskinfam M, Jazdarreh H (2010) Hydroxyapatite—starch nanobiocomposites synthesis and characterization. Int J Nano Dimen 1:57–63
203. Huang Y, Zhang X, Wu A, Xu H (2016) An injectable nano-hydroxyapatite (n-HA)/glycol chitosan (G-CS)/hyaluronic acid (HyA) composite hydrogel for bone tissue engineering. RSC Adv 6(40):33529–33536
204. Chae T, Yang H, Leung V, Ko F, Troczynski T (2013) Novel biomimetic hydroxyapatite/alginate nanocomposite fibrous scaffolds for bone tissue regeneration. J Mater Sci Mater Med. 24:1885–1894
205. Liu M, Dai L, Shi H, Xiong S, Zhou C (2015) *In vitro* evaluation of alginate/halloysite nanotube composite scaffolds for tissue engineering. Mater Sci Eng C 49:700–712
206. Kawaguchi M, Fukushima T, Hayakawa T, Nakashima N, Inoue Y, Takeda S, Okamura K, Taniguchi K (2006) Preparation of carbon nanotube-alginate nanocomposite gel for tissue engineering. Dental Mater J 25(4):719–725
207. Correia CR, Moreira-Teixeira LS, Moroni L, Reis RL, van Blitterswijk CA, Karperien M, Mano JF (2011) Chitosan scaffolds containing hyaluronic acid for cartilage tissue engineering. Tissue Eng Part C Methods 17:717–730
208. Patel DK, Dutta SD, Hexiu J, Ganguly K, Lim KT (2020) Bioactive electrospun nanocomposite scaffolds of poly(lactic acid)/cellulose nanocrystals for bone tissue engineering. Int J Biol Macromol S0141–8130(20):34010–34011
209. Pathmanapan S, Periyathambi P, Anandasadagopan SK (2020) Fibrin hydrogel incorporated with graphene oxide functionalized nanocomposite scaffolds for bone repair—in vitro and in vivo study. Nanomedicine 29:102251

210. Saber-Samandari S, Saber-Samandari S (2017) Biocompatible nanocomposite scaffolds based on copolymer-grafted chitosan for bone tissue engineering with drug delivery capability. Mater Sci Eng C Mater Biol Appl 75:721–732
211. Xie X, Hu K, Fang D, Shang L, Tran SD, Cerruti M (2015) Graphene and hydroxyapatite self-assemble into homogeneous, free standing nanocomposite hydrogels for bone tissue engineering. Nanoscale 7:7992–8002
212. Bhowmick A, Pramanik N, Jana P, Mitra T, Gnanamani A, Das M, Kundu PP (2017) Development of bone-like zirconium oxide nanoceramic modified chitosan based porous nanocomposites for biomedical application. Int J Biol Macromol 95:348–356
213. Bhowmick A, Banerjee SL, Pramanik N, Jana P, Mitra T, Gnanamani A, Das M, Kundu PP (2018) Organically modified clay supported chitosan/hydroxyapatite-zinc oxide nanocomposites with enhanced mechanical and biological properties for the application in bone tissue engineering. Int J Biol Macromol 106:11–19
214. De Santis R, Russo A, Gloria A, D'Amora U, Russo T, Panseri S, Sandri M, Tampieri A, Marcacci M, Dediu VA, Wilde CJ, Ambrosio L (2015) Towards the design of 3D fiber-deposited poly(ε-caprolactone)/iron-doped hydroxyapatite nanocomposite magnetic scaffolds for bone regeneration. J Biomed Nanotechnol 11:1236–1246
215. Patel A, Zaky SH, Schoedel K, Li H, Sant V, Beniash E, Sfeir C, Stolz DB, Sant S (2020) Design and evaluation of collagen-inspired mineral-hydrogel nanocomposites for bone regeneration. Acta Biomater 112:262–273
216. Sharma C, Dinda AK, Potdar PD, Chou CF, Mishra NC (2016) Fabrication and characterization of novel nano-biocomposite scaffold of chitosan-gelatin-alginate-hydroxyapatite for bone tissue engineering. Mater Sci Eng C Mater Biol Appl 64:416–427
217. Liu S, Sun Y, Fu Y, Chang D, Fu C, Wang G, Liu Y, Tay FR, Zhou Y (2016) Bioinspired collagen-apatite nanocomposites for bone regeneration. J Endod 42:1226–1232
218. Shakir M, Jolly R, Khan MS, Rauf A, Kazmi S (2016) Nano-hydroxyapatite/beta-CD/chitosan nanocomposite for potential applications in bone tissue engineering. Int J Biol Macromol 93:276–289
219. Samadikuchaksaraei A, Gholipourmalekabadi M, Ezadyar EE, Azami M, Mozafari M, Johari B, Kargozar S, Jameie SB, Korourian A, Seifalian AM (2016) Fabrication and in vivo evaluation of an osteoblast-conditioned nano-hydroxyapatite/gelatin composite scaffold for bone tissue regeneration. J Biomed Mater Res A 104:2001–2010
220. Fricain JC, Schlaubitz S, Le Visage C, Arnault I, Derkaoui SM, Siadous R, Catros S, Lalande C, Bareille R, Renard M (2013) A nano-hydroxyapatite–pullulan/dextran polysaccharide composite macroporous material for bone tissue engineering. Biomater 34:2947–2959
221. Nazeer MA, Yilgör E, Yilgör I (2017) Intercalated chitosan/hydroxyapatite nanocomposites: promising materials for bone tissue engineering applications. Carbohydr Polym 175:38–46
222. Mirza S, Jolly R, Zia I, Saad Umar M, Owais M, Shakir M (2020) Bioactive gum Arabic/κ-Carrageenan-incorporated nano-hydroxyapatite nanocomposites and their relative biological functionalities in bone tissue engineering. ACS Omega 5(20):11279–11290
223. Griffin M, Kalaskar D, Butler P (2019) Argon plasma modified nanocomposite polyurethane scaffolds provide an alternative strategy for cartilage tissue engineering. J Nanobiotechnol 17(1):51
224. Zhou Y, Liang K, Zhao S, Zhang C, Li J, Yang H, Liu X, Yin X, Chen D, Xu W, Xiao P (2018) Photopolymerized maleilated chitosan/methacrylated silk fibroin micro/nanocomposite hydrogels as potential scaffolds for cartilage tissue engineering. Int J Biol Macromol 108:383–390
225. Mirmusavi MH, Zadehnajar P, Semnani D, Karbasi S, Fekrat F, Heidari F (2019) Evaluation of physical, mechanical and biological properties of poly 3-hydroxybutyrate-chitosan-multiwalled carbon nanotube/silk nano-micro composite scaffold for cartilage tissue engineering applications. Int J Biol Macromol 132:822–835
226. Stocco TD, Antonioli E, Elias CMV, Rodrigues BVM, Siqueira IAWB, Ferretti M, Marciano FR, Lobo AO (2019) Cell viability of porous poly(d, l-lactic acid)/vertically aligned carbon nanotubes/nanohydroxyapatite scaffolds for osteochondral tissue engineering. Materials (Basel) 12(6):849

227. Kim HJ, Lee JS, Park JM, Lee S, Hong SJ, Park JS, Park KH (2020) Fabrication of nanocomposites complexed with gold nanoparticles on polyaniline and application to their nerve regeneration. ACS Appl Mater Interfaces 12(27):30750–30760
228. Homaeigohar S, Tsai TY, Young TH, Yang HJ, Ji YR (2019) An electroactive alginate hydrogel nanocomposite reinforced by functionalized graphite nanofilaments for neural tissue engineering. Carbohydr Polym 224:115112
229. Dolkhani S, Najafpour A, Mohammadi R (2020) Fabrication and transplantation of chitosan-selenium biodegradable nanocomposite conduit on transected sciatic nerve: a novel study in rat model. Neurol Res 42(6):439–450
230. Shokraei N, Asadpour S, Shokraei S, Nasrollahzadeh Sabet M, Faridi-Majidi R, Ghanbari H (2019) Development of electrically conductive hybrid nanofibers based on CNT-polyurethane nanocomposite for cardiac tissue engineering. Microsc Res Tech 82(8):1316–1325
231. Scholz M-S, Blanchfield JP, Bloom LD, Coburn BH, Elkington M, Fuller JD, Gilbert ME, Muflahi SA, Pernice MF, Rae SI, Trevarthen JA, White SC, Weaver PM, Bond IP (2011) The use of composite materials in modern orthopaedic medicine and prosthetic devices: a review. Comp Sci Technol 71:1791–1803
232. Ghanbari H, Radenkovic D, Marashi SM, Parsno S, Roohpour N, Burriesci G, Seifalian AM (2016) Novel heart valve prosthesis with self-endothelialization potential made of modified polyhedral oligomeric silsesquioxane-nanocomposite material. Biointerphases 11(2):029801
233. Dabees S, Kamel BM, Tirth V, Elshalakny AB (2020) Experimental design of Al_2O_3/MWCNT/HDPE hybrid nanocomposites for hip joint replacement. Bioengineered 11(1):679–692
234. Swain SK, Gotman I, Unger R, Gutmanas EY (2017) Bioresorbable β-TCP-FeAg nanocomposites for load bearing bone implants: High pressure processing, properties and cell compatibility. Mater Sci Eng C 78:88–95
235. Schlaubitz S, Derkaoui SM, Marosa L et al (2014) Pullulan/dextran/nHA macroporous composite beads for bone repair in a femoral condyle defect in rats. PLoS ONE 9:e110251
236. Garai S, Sinha A (2014) Biomimetic nanocomposites of carboxymethyl cellulose-hydroxyapatite: novel three dimensional load bearing bone grafts. Colloids Surf B Biointerf 115:182–190
237. Johari B, Kadivar M, Lak S, Gholipourmalekabadi M, Urbanska AM, Mozafari M, Ahmadzadehzarajabad M, Azarnezhad A, Afshari S, Zargan J, Kargozar S (2016) Osteoblast-seeded bioglass/gelatin nanocomposite: a promising bone substitute in critical-size calvarial defect repair in rat. Int J Artif Org 39:524–533
238. Medupin RO, Abubakre OK, Abdulkareem AS, Muriana RA, Abdulrahman AS (2019) Carbon nanotube reinforced natural rubber nanocomposite for anthropomorphic prosthetic foot purpose. Sci Rep 9(1):20146
239. Manju V, Iyer S, Menon D, Nair SV, Nair MB (2019) Evaluation of osseointegration of staged or simultaneously placed dental implants with nanocomposite fibrous scaffolds in rabbit mandibular defect. Mater Sci Eng C Mater Biol Appl 104:109864
240. Liu X, Zhao K, Gong T, Song J, Bao C, Luo E, Weng J, Zhou S (2014) Delivery of growth factors using a smart porous nanocomposite scaffold to repair a mandibular bone defect. Biomacromol 15:1019–1030
241. Hagiwara Y, Nakajima K (2016) Use of ceria-stabilized zirconia/alumina nanocomposite for fabricating the frameworks of removable dental prostheses: a clinical report. J Prosthet Dent 116(2):166–171
242. Brown EE, Hu D, Abu Lail N, Zhang X (2013) Potential of nanocrystalline cellulose-fibrin nanocomposites for artificial vascular graft applications. Biomacromol 14(4):1063–1071
243. Mohammadi H, Boughner D, Millon LE, Wan WK (2009) Design and simulation of a poly(vinyl alcohol)-bacterial cellulose nanocomposite mechanical aortic heart valve prosthesis. Proc Inst Mech Eng H 223:697–711
244. Ang HY, Toong D, Chow WS, Seisilya W, Wu W, Wong P, Venkatraman SS, Foin N, Huang Y (2018) Radiopaque fully degradable nanocomposites for coronary stents. Sci Rep 8(1):17409

245. Pataquiva Mateus AY, Ferraz MP, Monteiro FJ (2007) Nano-hydoxyapatite microspheres for periodontitis treatment: preparation and cytotoxicity studies. Eur Cells Mater 14:85
246. Ferraz MP, Mateus AY, Sousa JC, Monteiro FJ (2007) Nanohydroxyapatite microspheres as delivery system for antibiotics: release kinetics, antimicrobial activity, and interaction with osteoblasts. J Biomed Mater Res Part A 81A:994–1004
247. Madhumathil K, Jeevana Rekha TS, Sampath Kumar TS (2018) Tailoring antibiotic release for the treatment of periodontal infrabony defects using bioactive gelatin-alginate/apatite nanocomposite films. J Drug Deliv Sci Tech 43:57–64
248. Chen M-H (2010) Update on dental nanocomposites. J Dental Res 89:549–560
249. Choi AH, Ben-Nissan B (2016) Applications of hydroxyapatite nanocoatings and nanocomposite coatings in dentistry. JSM Dent Surg 1:1002
250. Cohen BD, Combe EC (1994) Development of new adhesive pulp capping materials. Dent Update 21:57–62
251. Fioretti F, Mendoza-Palomares C, Helms M, Al Alam D, Richert L, Arntz Y, Rinckenbach S, Garnier F, Haïkel Y, Gangloff SC, Benkirane-Jessel N (2010) Nanostructured assemblies for dental application. ACS Nano 4:3277–3287
252. Panahi F, Rabiee SM, Reza Shidpour R (2017) Synergic effect of chitosan and dicalcium phosphate on tricalcium silicate-based nanocomposites for root-end dental application. Mater Sci Eng C 80:631–641
253. Zhang T, Ying D, Qi M, Li X, Fu L, Sun X, Wang L, Zhou Y (2019) Anti-biofilm property of bioactive upconversion nanocomposites containing chlorin e6 against periodontal pathogens. Molecules 24(15):2692
254. Furtos G, Rivero G, Rapuntean S, Abraham GA (2017) Amoxicillin-loaded electrospun nanocomposite membranes for dental applications. J Biomed Mater Res B Appl Biomater 105(5):966–976
255. Sowmya S, Mony U, Jayachandran P, Reshma S, Kumar RA, Arzate H, Nair SV, Jayakumar R (2017) Tri-layered nanocomposite hydrogel scaffold for the concurrent regeneration of cementum, periodontal ligament, and alveolar bone. Adv Healthc Mater 6(7). https://doi.org/10.1002/adhm.201601251
256. Farooq A, Yar M, Khan AS, Shahzadi L, Siddiqi SA, Mahmood N, Rauf A, Qureshi ZU, Manzoor F, Chaudhry AA, ur Rehman I (2015) Synthesis of piroxicam loaded novel electrospun biodegradable nanocomposite scaffolds for periodontal regeneration. Mater Sci Eng C Mater Biol Appl 1(56):104–113
257. Wang L, Xie X, Imazato S, Weir MD, Reynolds MA, Xu HHK (2016) A protein-repellent and antibacterial nanocomposite for Class-V restorations to inhibit periodontitis-related pathogens. Mater Sci Eng C Mater Biol Appl 67:702–710
258. Sancilio S, Gallorini M, Di Nisio C, Marsich E, Di Pietro R, Schweikl H, Cataldi A (2018) Alginate/hydroxyapatite-based nanocomposite scaffolds for bone tissue engineering improve dental pulp biomineralization and differentiation. Stem Cells Int 2:9643721
259. Aramwit P (2016) Introduction to biomaterials for wound healing. In: Agren M (ed) Wound healing biomaterials, vol 2. Woodhead Publishing, pp 1–38
260. Sezer AD, Cevher E (2011) Biopolymers as wound healing materials: challenges and new strategies. In: Pignatello R (ed) Biomaterials applications for nanomedicine. InTech, Croatia, pp 383–414
261. Shah A, Buabeid MA, Arafa EA, Hussain I, Li L, Murtaza G (2019) The wound healing and antibacterial potential of triple-component nanocomposite (chitosan-silver-sericin) films loaded with moxifloxacin. Int J Pharm 564:22–38
262. Kumar PTS, Laskmanan VK, Anilkumar TV, Ramya C, Reshmi P, Unnikrishnan AG, Nair SV, Jayakumar R (2012) Flexible and microporous chitosan hydrogel/nano ZnO composite bandages for wound dressing: *in vitro* and *in vivo* evaluation. ACS Appl Mater Interfaces 4:2618–2629
263. Sandri G, Aguzzi C, Rossi S, Bonferoni MC, Bruni G, Boselli C, Cornaglia AI, Riva F, Viseras C, Caramella C, Ferrari F (2017) Halloysite and chitosan oligosaccharide nanocomposite for wound healing. Acta Biomater 15(57):216–224

264. Anisha BS, Sankar D, Mohandas A, Chennazhi KP, Nair SV, Jayakumar R (2013) Chitosan-hyaluronan/nano chondroitin sulfate ternary composite sponges for medical use. Carbohydr Polym 92:1470–1476
265. Figueiredo AG, Figueiredo AR, Alonso-Varona A, Fernandes S, Palomares T, Rubio-Azpeitia E et al (2013) Biocompatible bacterial cellulose-poly(2-hydroxyethyl methacrylate) nanocomposite films. BioMedResearch Int, Article ID 698141, 14
266. Gonzaga VAM, Poli AL, Gabriel JS, Tezuka DY, Valdes TA, Leitão A, Rodero CF, Bauab TM, Chorilli M, Schmitt CC (2020) Chitosan-laponite nanocomposite scaffolds for wound dressing application. J Biomed Mater Res B Appl Biomater 108(4):1388–1397
267. Tantiwatcharothai S, Prachayawarakorn J (2019) Characterization of an antibacterial wound dressing from basil seed (*Ocimum basilicum* L.) mucilage-ZnO nanocomposite. Int J Biol Macromol 135:133–140
268. Sadeghianmaryan A, Yazdanpanah Z, Soltani YA, Sardroud HA, Nasirtabrizi MH, Chen X (2020) Curcumin-loaded electrospun polycaprolactone/montmorillonite nanocomposite: wound dressing application with anti-bacterial and low cell toxicity properties. J Biomater Sci Polym Ed 31(2):169–187
269. Khalid A, Khan R, Ul-Islam MR, Khan T, Wahid F (2017) Bacterial cellulose-zinc oxide nanocomposites as a novel dressing system for burn wounds. Carbohydr Polym 164:214–221
270. Kamel NA, El-messieh SLA, Saleh NM (2017) Chitosan/banana peel powder nanocomposites for wound dressing application: preparation and characterization. Mater Sci Eng C 72:543–550
271. García-Villén F, Faccendini A, Aguzzi C, Cerezo P, Bonferoni MC, Rossi S, Grisoli P, Ruggeri M, Ferrari F, Sandri G, Viseras C (2019) Montmorillonite-norfloxacin nanocomposite intended for healing of infected wounds. Int J Nanomed 10(14):5051–5060
272. Gunes S, Tamburaci S, Tihminlioglu F (2019) A novel bilayer zein/MMT nanocomposite incorporated with H. perforatum oil for wound healing. J Mater Sci Mater Med 31(1):7
273. Razali MH, Ismail NA, Mat Amin KA (2020) Titanium dioxide nanotubes incorporated gellan gum bio-nanocomposite film for wound healing: effect of TiO_2 nanotubes concentration. Int J Biol Macromol 153:1117–1135
274. Manuja A, Raguvaran R, Kumar B, Kalia A, Tripathi BN (2020) Accelerated healing of full thickness excised skin wound in rabbits using single application of alginate/acacia based nanocomposites of ZnO nanoparticles. Int J Biol Macromol 155:823–833
275. Prabhakar O, Matta S (2020) Fabrication and characterization of carboxymethyl guar gum nanocomposite for application of wound healing. Int J Biol Macromol S0141-8130(20):34082–34084
276. Yang M, Ward J, Choy KL (2020) Nature-inspired bacterial cellulose/methylglyoxal (BC/MGO) nanocomposite for broad-spectrum antimicrobial wound dressing. Macromol Biosci e2000070
277. Fan Y, Wu W, Lei Y, Gaucher C, Pei S, Zhang J, Xia X (2019) Edaravone-loaded alginate-based nanocomposite hydrogel accelerated chronic wound healing in diabetic mice. Mar Drugs 17(5):285. https://doi.org/10.3390/md17050285
278. Montaser AS, Abdel-Mohsen AM, Ramadan MA, Sleem AA, Sahffie NM, Jancar J, Hebeish A (2016) Preparation and characterization of alginate/silver/nicotinamide nanocomposites for treating diabetic wounds. Int J Biol Macromol 92:739–747
279. Koneru A, Dharmalingam K, Anandalakshmi R (2020) Cellulose based nanocomposite hydrogel films consisting of sodium carboxymethylcellulose-grapefruit seed extract nanoparticles for potential wound healing applications. Int J Biol Macromol 148:833–842
280. Renu S, Shivashangari KS, Ravikumar V (2020) Incorporated plant extract fabricated silver/poly-D, l-lactide-co-glycolide nanocomposites for antimicrobial based wound healing. Spectrochim Acta A Mol Biomol Spectrosc 228:117673
281. Zhai M, Xu Y, Zhou B, Jing W (2018) Keratin-chitosan/n-ZnO nanocomposite hydrogel for antimicrobial treatment of burn wound healing: characterization and biomedical application. J Photochem Photobiol B 180:253–258

282. Li Y, Xu T, Tu Z, Dai W, Xue Y, Tang C, Gao W, Mao C, Lei B, Lin C (2020) Bioactive antibacterial silica-based nanocomposites hydrogel scaffolds with high angiogenesis for promoting diabetic wound healing and skin repair. Theranostics 10(11):4929–4943
283. Najafabadi SAA, Mohammadi A, Kharazi AZ (2020) Polyurethane nanocomposite impregnated with chitosan-modified graphene oxide as a potential antibacterial wound dressing. Mater Sci Eng C Mater Biol Appl 115:110899
284. Shariatinia Z (2019) Pharmaceutical applications of natural polysaccharides. In: Hasnain MS, Nayak AK (eds) Natural polysaccharides in drug delivery and biomedical applications. Academic Press, Elsevier Inc., pp 15–57
285. Lin J, Li Y, Li Y, Wu H, Yu F, Zhou S, Xie L, Luo F, Lin C, Hou Z (2015) Drug/dye-loaded, multifunctional peg-chitosan-iron oxide nanocomposites for methotraxate synergistically self-targeted cancer therapy and dual model imaging. ACS Appl Mater Interfaces 7(22):11908–11920
286. Fu F, Yang B, Hu X, Tang H, Zhang Y, Xu X, Zhang Y, Touhid SSB, Liu X, Zhu Y, Zhou J, Yao J (2020) Biomimetic synthesis of 3D Au-decorated chitosan nanocomposite for sensitive and reliable SERS detection. Chem Eng J 392:123693
287. Liu Y, Wang MK, Zhao F, Xu ZA, Dong SJ (2005) The direct electron transfer of glucose oxidase and glucose biosensor based on carbon nanotubes/chitosan matrix. Biosens Bioelectron 21:984–988
288. Wang Z, Jia T, Sun Q, Kuang Y, Liu B, Xu M, Zhu H, He F, Gai S, Yang P (2020) Construction of Bi/phthalocyanine manganese nanocomposite for trimodal imaging directed photodynamic and photothermal therapy mediated by 808 nm light. Biomaterials 228:119569
289. Zhu T, Ma X, Chen R, Ge Z, Xu J, Shen X, Jia L, Zhou T, Luo Y, Ma T (2017) Using fluorescently-labeled magnetic nanocomposites as a dual contrast agent for optical and magnetic resonance imaging. Biomater Sci 5(6):1090–1100
290. Kim T, Jang H, Kim S, Lee JH, Kim SY, Jeon NL, Song JM, Min DH (2018) Synthesis of fluorescent Au nanocrystals-silica hybrid nanocomposite (FLASH) with enhanced optical features for bioimaging and photodynamic activity. Langmuir 34(1):173–178
291. Wang X, Dai J, Wang X, Hu Q, Huang K, Zhao Z, Lou X, Xia F (2019) MnO_2-DNAzyme-photosensitizer nanocomposite with AIE characteristic for cell imaging and photodynamic-gene therapy. Talanta 1(202):591–599
292. Tan MJH, Ravichandran D, Ang HL, Ong EWY, Lim CQX, Kam GMQ, Kumar AP, Tan ZK (2019) Magneto-fluorescent perovskite nanocomposites for directed cell motion and imaging. Adv Healthc Mater 8(23):e1900859
293. Ruecha N, Rangkupan R, Rodthongkum N, Chailapakul O (2014) Novel paper-based cholesterol biosensor using graphene/polyvinylpyrrolidone/polyaniline nanocomposite. Biosens. Bioelectron 52:13–19
294. Radhapyari K, Kotoky P, Das MR, Khan R (2013) Graphene–polyaniline nanocomposite based biosensor for detection of antimalarial drug artesunate in pharmaceutical formulation and biological fluids. Talanta 111:47–53
295. Chen Z, Liu X, Liu D, Li F, Wang L, Liu S (2020) Ultrasensitive electrochemical DNA biosensor fabrication by coupling an integral multifunctional zirconia-reduced graphene oxide-thionine nanocomposite and exonuclease I-assisted cleavage. Front Chem 8:521
296. Jain R, Tiwari DC, Shrivastava S (2014) Polyaniline–bismuth oxide nanocomposite sensor for quantification of anti-parkinson drug pramipexole in solubilized system. Mater Sci Eng B 185:53–59
297. Yang Y, Zhang S, Kang M, He L, Zhao J, Zhang H et al (2015) Selective detection of silver ions using mushroom-like polyaniline and gold nanoparticle nanocomposite-based electrochemical DNA sensor. Anal Biochem 490:7–13
298. Zhu Q, Gao F, Yang Y, Zhang B, Wang W, Hu Z et al (2015) Electrochemical preparation of polyaniline capped Bi2S3 nanocomposite and its application in impedimetric DNA biosensor. Sens Actuat B 207:819–826
299. Sheta SM, El-Sheikh SM, Osman DI, Salem AM, Ali OI, Harraz FA, Shousha WG, Shoeib MA, Shawky SM, Dionysiou DD (2020) A novel HCV electrochemical biosensor based on a polyaniline@Ni-MOF nanocomposite. Dalton Trans 49(26):8918–8926

300. Wang Z, Li F, Zhang L, Qian J, Cao S (2020) A phase-transfer-assisted synthesis of cysteine-Ag nanoparticles/graphene oxide nanocomposite and its enhanced performance in antibiosis and biosensing. Nanotechnology. https://doi.org/10.1088/1361-6528/aba05c

Antimicrobial Nanocomposites

El-Refaie Kenawy, Mohamed M. Azaam, Syed Anees Ahmed, and Md Saquib Hasnain

1 Introduction

Since last few decades, a huge leap of material researches in particular biomaterial development and applications have been experienced [1–19]. In recent years, a wide variety of advanced materials for biomedical applications have been engineered using biopolymers (such as synthetic, semi-synthetic, and natural biopolymers) [19–36], metals and metallic oxides [37–38], bioceramics [39–40], bioglasses [39], structured carbons [41–50], etc. These engineered biomaterials includes polymer-blends [51–57], gels [2, 58], hydrogels [59–69], particulates (nanoparticles, microparticles, beads, etc.) [70–82], composites and nanocomposites [83–89],scaffolds [90], fibers [91], membranes and films [92–93], pastes [15, 94], emulsions [95–96], suspensions [97–98], nanotubes [99–100], etc. These advanced biomaterials are being used in many important biomedical uses like drug delivery [101–113], cancer therapy [114], protein encapsulation [115], gene delivery [116, 117], antimicrobial [118], tissue engineering [119–120], wound healing [121], dentistry [122], bioimaging [116], biosensor [123], etc. With the fast progress of science, the nanotechnology has been proved as the prosperous field for biomedical applications [124–125]. The significant biomedical nanomaterials include polymeric nanoparticles [126–127], lipid nanoparticles [128], metallic nanoparticles [129], ceramic nanoparticles [130–131],

E.-R. Kenawy (✉) · M. M. Azaam
Polymer Research Group, Department of Chemistry, Faculty of Science, Tanta University, Tanta, Egypt
e-mail: ekenawy@yahoo.com

S. A. Ahmed
Department of Pharmacology, Hygia Institute of Pharmaceutical Education and Research, Lucknow, UP, India

M. S. Hasnain
Department of Pharmacy, Palamau Institute of Pharmacy, Chianki, Daltonganj, Jharkhand, India

A. K. Nayak et al. (eds.), *Biomedical Composites*, Materials Horizons: From Nature to Nanomaterials, https://doi.org/10.1007/978-981-33-4753-3_4

nanogels [132–133], nanofibers [91], nanovesicles [134–136], nanotubes [137–138], nanocomposites [139–145], etc.

Recently, most of the nanocomposites are made of biopolymers (such as synthetic, semi-synthetic, and natural biopolymers), metals and metallic oxides, bioceramics, bioglasses, structured carbons, etc. [83–88, 139–145]. In biomedical field, these nanocomposites are being used for drug delivery [83–87, 141], cancer therapy [146], gene delivery [147], antimicrobial [148], tissue engineering [139, 141], wound healing [149], prosthesis [140], dentistry [142–144], bioimaging [146], biosensor [123, 150], etc. In recent years, numerous antimicrobial biomaterials are being developed for microbial inactivation and characterized by many researchers and scientists [118]. The present chapter presents a widespread assessment of the applications of antimicrobial nanocomposites.

2 Application of Nanocomposites as Antimicrobial Agents

Devi et al. (2019) used a waste material such as jute stick for the sustainable application [151]. Silver loaded activated carbon (Ag-AC) nanocomposite exhibited remarkable antimicrobial activity against gram (−) and (+) bacteria due to the enhancement of the silver nanoparticles' as antibacterial. The antimicrobial activities of (Ag-AC) were better than silver nanoparticles. The nanocomposite also showed an excellent photocatalytic activity [151].

Another manuscript describes an efficient synthesis of Ag nanoparticles immobilized on the magnetic polymeric nanocarriers for applications as antibiotics [152]. Polyethylene glycol coated Fe_3O_4 nanoparticles (Fe_3O_4@PEG) were synthesized, and Ag nanoparticles were supported. The Ag magnetic polymer nanocomposites showed a high antibacterial effect. Besides, magnetic nanocomposites could be readily separated from the solution after the disinfection process by applying an external magnetic field [152].

Hosseini et al. [153] reported the preparation of nanocomposite based on MnS_2, MnS_2/Chitosan-calcium alginate (MnS_2/CS-CaAlg), and MnS_2/Chitosan sodium alginate (MnS_2/CS-NaAlg) [153]. The antibacterial and antifungal activities of the nanocomposites showed good activities as compared to MnS_2. Also, the MnS_2, MnS_2/CS-NaAlg, and MnS_2/CS-CaAlg showed outstanding mechanical and antioxidant properties. The antibacterial effect was observed after 60 min. The reduction of the surviving bacteria under visible light irradiation was observed in the presence of a catalyst in comparison to materials from MnS2. The size of the fungal colonies was reduced by MnS_2/CS-NaAlg and MnS_2/CS-CaAlg nanocomposites that may be due to degradation of the cell wall and structure of the cell membrane [153].

Corn silk as anagricultural waste was introduced on polyester to augment the cell viability of zinc oxide on polyester fabric [154]. Good antimicrobial properties were obtained on the material containing corn silk/zinc oxide nanocomposites close to 100 and 87%, respectively. The advantage of corn silk and ZnO nanoparticles as a polyester fabric is a low cytotoxicity antimicrobial agents [154].

Nanocomposites based on pectin, polyethyleneimine, and Cu nanoparticles were synthesized [155]. The antimicrobial investigations of the prepared nanocomposites revealed they have high antimicrobial activity against *S. aureus and E. coli* strains.

The oxidized nanofibers of cellulose were formulated with cadmium oxide nanostructure from bagasse pulp, and their antibacterial activity was studied [156]. The Minimum Concentration of Inhibition (MIC) highlighted that the formulated nanocomposite has potent antimicrobial activity. MIC values were 1.56 μg/mL for gram (−) bacteria, less than 1.56 for gram (+) bacteria and 0.19 μg/mL for *Candida albicans* [156].

The antimicrobial activity of pure Ag_2S, CdS, and CdS-Ag_2S composites was evaluated [157]. The CdS to Ag_2S ratio was varied to increase antibacterial activity at an optimum level between two compounds. The increase in CdS ratio in CdS-Ag2S nanocomposites improved antibacterial activity.

Alginate-mediated biomineralization with the Zn-mineral phase was prepared as a simple and cost-effective procedure [158]. Both forms of Zn-mineralized nanocomposites had a clear antimicrobial effect on Candida albicans, Escherichia coli, and Staphylococcus aureus. Mineralized samples had higher Zn(II) content and more stable in a biological environment. Also, they released Zn(II) in a controllable and sustained fashion when compared to a non-mineralized sample. Alginate-mediated biomineralization represents promising antimicrobial biomaterials.

Chitosan (CH)/polyvinyl alcohol (PVA) hydrogel and CH/PVA/ZnO nanocomposite was prepared [159]. The antimicrobial activities of the prepared composites were higher than Erythromycin and Metronidazole. Also, the activity was highly improved after ZnO nanocomposites formation. The increase of the ZnO ratio in the nanocomposite increases their antimicrobial activities.

Modified HDPE nanocomposite monofilaments based on MMT were synthesized using nanoparticles made from copper, silver, and zinc oxide [160]. HDPE/modified MMT nanocomposite monofilaments displayed the enhanced activity in bacterial colonies against HDPE/Ag-MMT monofilaments and gram (−) and (+) bacteria. The kinetics of metal ions/NPs dissolution from modified MMT nanostructure reflected its potential as having antimicrobial action. The antimicrobial HDPE monofilaments produced aim to minimize occurrences of these undesired abnormalities. The level of antimicrobial activity for each load (1, 3, and 5 weight percent) of modified MMT is HDPE/Ag-MMT > HDPE/Cu-MMT > HDPE/Zn-MMT [160].

Ghosh et al. (2019) designed a nanohybrid by simultaneous reduction of copper(II) chloride dehydrate and graphene oxide (GO) in the presence of octadecylamine (ODA) [161]. The nanohybrid was then included in the interpenetrating polymer network (IPN) of bio-based PU and polystyrene (PS). The nanocomposites exhibited high antibacterial properties. Consequently, the studied nanocomposites represent a new track in the surface-active advanced thin-film field. Fascinatingly, high hydrophobicity through intrinsic water repellent behavior and the strong antimicrobial property is achieved for the nanocomposites. As a result, the studied nanocomposites are used as protective functional surface materials for biomedical and electrical devices [161].

The graphite (Gr), graphene oxide (GO), and reduced graphene oxide (RGO) reaction formed cross-linked antimicrobial oleo-polyurethane nanocomposites (PUCs) with high performance[162]. The composites prepared showed the biocidal effects of GO and RGO PUCs against all those bacterial classes. Surface-active cross-linked LO-based PUCs utilizing 0.5 wt% of pristine Gr, GO, and RGO was formulated as nano-fillers. GO and RGO PUCs exhibited efficient antimicrobial properties against *S. aureus and E.coli*. The present study may provide new methods and materials for the preparation and designing of advanced contact active nanocomposite films required in various areas, such as the development of medical equipment, functional films, and active packaging [162] (Fig. 1).

Hydroxyapatite nanoparticles (nHAP) with zinc oxide (ZnO) nanoparticle nanocomposite were synthesized [163]. Nanocomposite showed antimicrobial and antibiofilm activity in a ratio of 90:10, 75:25, and 60:40 nHAP and ZnO NPs. For both pathogens, the MIC was found to be 0.2 mg/mL in a ratio of 90:10 nanocomposites. For *S. aureus* and *E. coli*, the minimum bactericidal concentration (MBC) was 0.2 mg/mL of 75:25 and 60:40 nanocomposites. The maximum biofilm inhibition percentage was found at ratio 60 (nHAP): 40 (ZnO NPs). It was against S.

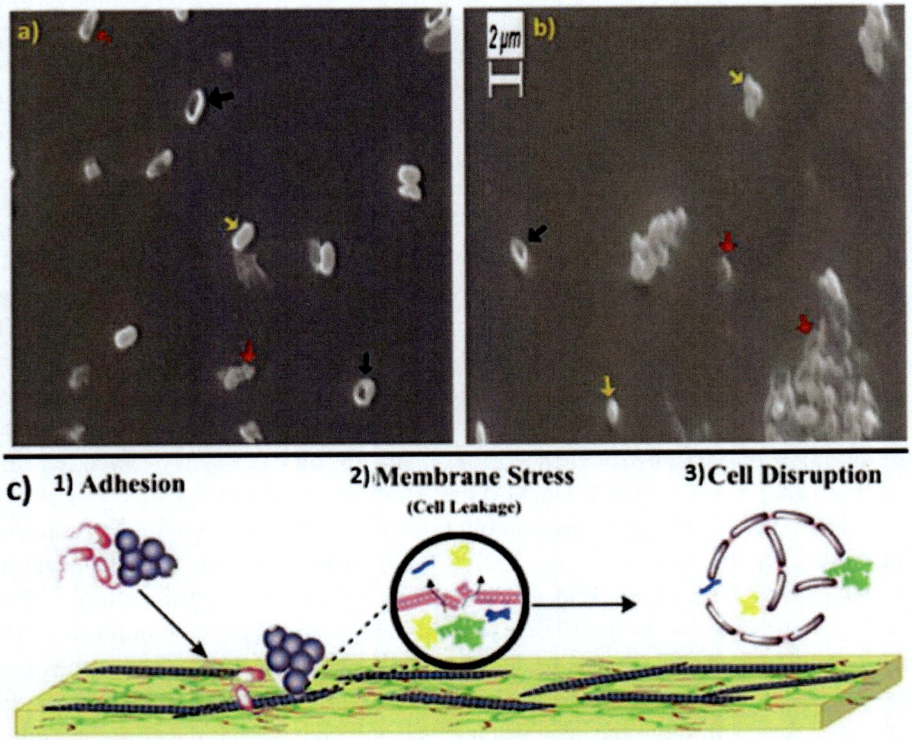

Fig. 1 The permanent antimicrobial effect of GO-PUC against; **a** *E. coli* and **b** *S. aureus*. **c** The three-step antimicrobial action mechanism of Gr-based PUCs [162] (with permission @ 2019 Elsevier Ltd.)

aureus as well as *E.coli* bacteria having inhibition of 52 percent as well as 54 percent and for green synthesized ZnO NPs have inhibition of 51 percent and 52 percent against E.coli and S. aureus, respectively. Biomaterials that involve multiple ratios of nHAP-ZnO NPs could be used in bone regenerative medicine as antibiofilm materials. Increasing the proportion of ZnO NPs in the nanocomposite prevented bacterial infection in orthopedic implants [163].

Silver nanocomposites (AgNPs@Tob&PAGA) was prepared by reaction of AgNPs with tobramycin (Tob) and poly(2-(acrylamide) glucopyranose) (PAGA) [164]. The antibacterial activities of nanocomposites were determined. The introduction of PAGA onto silver nanocomposites enhanced cytocompatibility and antibacterial activity. AgNPs@Tob&PAGA showed more attractive antimicrobial consequence than Tob against *E. coli and S. aureus.* The introduction of (PAGA) into nanocomposites improved the compatibility and the antibacterial effect [164].

The antibacterial activity of different silver sulfide (Ag_2S) nanoparticles cross-linked with chitosan was evaluated [165]. The cross-linked nanopolymers prevent the production of gram (−) antibiotic-resistant bacteria and gram (+) bacteria from 64 up to 32 μg/mL resulting in a positive charged surface. Using glucose as a green capping agent generates nanostructures with acceptable size distribution [165].

Chitosan/silica nanocomposite membrane doped with Al_2O_3 nanoparticles was synthesized [166] and their antimicrobial activities were determined. The modified membranes with a lower concentration of Al_2O_3 ions display better antimicrobial properties due to the high cross-linking between the chitosan matrix and silica-based nanoparticles [166].

Nezami et al. (2019) formulated composites AgNPs, St-A-E/M, and St-A-E/M-Ag nanocomposite beads, and then used as controlled-release drug delivery applications for methyl prednisolone [167]. The release of the nanocomposite drug was significantly high than the composite of free-NPs, as well as the release increased up to 3.3% (V/V) in the polymer matrix by the increase in AgNPs number. The nanocomposite demonstrated more significant antibacterial activity compared to the corresponding composite of free-NPs. The results showed that the amount of drug released from the nanocomposite was around 92 percent compared to free AgNPs composite (68%) for release media having pH 7.4 for 9 h.The nanocomposite with 3.3 (V/V) AgNPs exhibited the highest antibacterial activity against *E. coli* (ZOI = 10.48 mm) and *S. aureus* (ZOI = 8.37 mm) compared to the free AgNPs composite. Therefore, all these observations unanimously illustrated the high potential of the nanocomposite as an oral drug delivery system [167].

An ethnonanocomposite was formulated employing Curcuma zedoaria and a ZnO nanoparticle [168] that was immobilized for effective delivery on the microspheres of poly (ÿ-caprolactone) (PCL). The antimicrobial activity confirmed the notable effectiveness of microspheres loaded with nanocomposite. This study is significant due to the biocompatibility, biodegradability and stability of PCL. The prepared microspheres can be used as efficient antimicrobial agents in applications that require the instant and sustained release of effective antimicrobial agents [168].

$Fe_3O_4/Ag_3PO_4@WO_3$ nanocomposites were prepared by in-situ industrial production and electrospinning to enhance photocatalyst efficiency for methylene blue (MB) deterioration and antibacterial activity [169]. The nanocomposites showed high photocatalytic and antibacterial activity. The $Ag_3PO_4/Fe_3O_4@WO_3$ composite has high antibacterial activities than the other prepared composites. The oleic acid functionalized composite $Fe_3O_4/Ag_3PO_4@WO_3$ also exhibited a high inhibition zone [169].

An oxidation redox coating approach was used to prepare a hybrid $Ag@CeO_2$ nanostructure composite [170]. The synthesized hybrid showed excellent antibacterial activity against gram (−) and (+) bacteria. The hybrid nanostructure bind with the bacterial cell membrane, followed by cell rupturing and finally cell death. The prepared nanocomposites $Ag@CeO_2$ showed more potent bacteriostatic properties compared to other kinds of nanoparticles [170].

3 Application of Antimicrobial Nanocomposites in Food Packaging

Siripatrawan et al. developed antimicrobial packaging nanocomposite based on chitosan and nanosized titanium dioxide (TiO_2) [171]. They reported that 1% TiO_2 was optimal for antimicrobial properties against gram (−) and (+) bacteria and fungi [171] (Fig. 2).

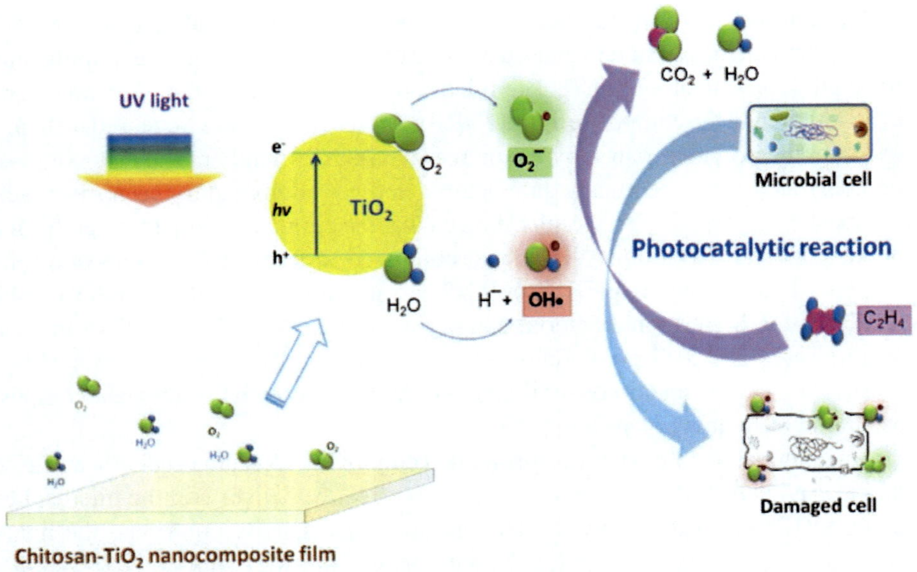

Fig. 2 A schematic diagram of the possible mechanism of photocatalytic degradation of ethylene and antimicrobial activity of the chitosan-TiO_2 nanocomposite film [171] (with permission @ 2018 Elsevier Ltd.)

Certain antimicrobial nanocomposites were synthesized with the root extract of Salvadorapersica L. (SPE) of carboxymethyl cellulose (CMC), cellulose nanofiber (CNF), and miswak [172]. The pure nanocomposites activated by SPE and SPE exhibited strong antibacterial activity. Nanocomposite films based on pullulan/AgNPs and pullulan/pectin/AgNPs films were prepared by Lee et al. (2019) [173]. They showed high antimicrobial activity against food-borne pathogens to maintain food safety [174].

Yu et al. (2019) prepared cellulose nanofibril/silver nanoparticle (CNF/AgNP) as antimicrobial nanocomposite and evaluate their toxicity against human colon cells [175]. The prepared nanocomposite exhibited antimicrobial activities against two important food-borne pathogens, including *Escherichia coli* and *Staphylococcus aureus.*

The nanocomposite CuO/hc-pCUR consisting of copper (II) oxide nanoparticles and strongly cross-linking poly(curcumin) nanospheres (hc-pCUR-NS) was prepared [176]. Relative antimicrobial assays against gram (−) and (+) bacteria were investigated. Hc-pCUR-NS demonstrated more antimicrobial activity than free CUR. Copper nanoparticles added improved the hc-pCUR-NS inhibitory effect.

Polyvinyl alcohol (PVA) reinforced with silver nanoparticles (AgNPs) film was prepared [177]. The presence of AgNPs in the film improved UV and light barrier properties in addition to antibacterial activity against the food-borne pathogen *Salmonella Typhimurium and Staphylococcus aureus.*

The *in-situ* precipitation and a casting method in combination [178] developed nanocomposite films (Zn-carbonate and Zn-phosphate/agar) Reinforcement of the Zn-mineral process enhanced the mechanical properties of the carbonate-mineralized films. Still, they had a marginal effect on the phosphate-mineralized samples. Both nanocomposites exhibited improved optical, thermal and antimicrobial properties. Zn-mineralized nanocomposite agar films have been used as accessible, environmentally safe and active food packaging materials [178].

$CuO/BiVO_4$ (BVO) nanocomposite photocatalyst was immobilized on cotton fabrics through polydopamine (PDA) templating. These nanocomposites showed efficient visible-light-driven photocatalysis, antimicrobial activity and UV protection [179]. The CuO/BVO@ cotton was an antimicrobial agent with a 99.2% reduction in bacterial load against *Escherichia coli.* Also, it was UV protective with the UPF value of 237.17 and UVA and UVB transmittance value of 0.47% and 0.39%, respectively [179].

The dual sputtering method for the mid-range frequency was used to prepare thin films of Ag-plasma polymer fluorocarbon (PPFC) nanocomposite [180]. The thin films of Ag-PPFC nanocomposite, coated on a substratum of polyethylene terephthalate (PET), display higher visible light transmittance than the bare substratum and excellent water repellent. The Ag nanoparticles have had no observed effect on the thin film's optical and surface properties. It has been found that the Ag-PPFC nanocomposite thin films have more potent antimicrobial properties because it reduces bacterial growth by up 92.2% compared to uncoated substrates.This antimicrobial activity is attributable to the decreased humid atmosphere and Ag nanoparticle's antimicrobial characteristics. These coatings have been used in many areas, such as outdoor products, usable fibers, sportswear, medical equipment, touch screen displays, and flexible solar cells [180].

These were manufactured the gelatin-based nanocomposite consisting of chitosan nanofiber (CHINF) and ZnO nanoparticles (ZnONPs) [181] which displayed potent antibacterial activity against pathogenic bacteria raised in food. Chicken fillet and cheese were chosen as food prototypes to evaluate the efficacy of this film for food packaging. Results revealed that coating with nanocomposite film reduced ($p < 0.05$) the growth of inoculation bacteria in chicken fillet and cheese samples significantly.The weight loss of the coated chicken fillet and cheese specimens with nanocomposite at the end of 12-day storage was 18.91 ± 1.96 and $36.11 \pm 3.74\%$, respectively. This study was significant due to: (1) ZnONPs were antimicrobial agents and improved physical properties and barrier properties. (2) They enhance the synergistic effects on antimicrobial properties. (3) biocompatibility of gelatin. G/CHINF/ZnO NPs antibacterial nanocomposite film is suitable for the packaging of poultry meat and cheese, representing a great promise for improving the quality and shelf life of these food products [181].

Nano-sized $Sr1-xAgxTiO_3$ system was prepared using modified Pechini method ($x = 0, 0.02, 0.05$, and 0.07) [182]. Antimicrobial properties have been observed for undoped and Ag-doped nano-sized SrTiO3, while the parent SrTiO3 has no specific region. Different silver concentrations were added to polymer matrices to be used on paper sheets as energetic coating materials. The mechanical and short period behaviors were improved by active nanocomposite coating. Antimicrobial behavior was added to the $SrTiO_3$ parent by doping with silver metal as successfully active packaging coating. Besides, the mechanical properties of Kraft paper were enhanced in tensile, roughness, short span, and water absorption characters [182].

4 Application of Antimicrobial Nanocomposites in Water Treatment

Nanocomposites were prepared, consisting of chitosan and silver nanoparticles (AgNPs), copper nanoparticles (CuNPs), and carbon nanotubes (CNTs) [183]. They have higher antimicrobial activity against gram (−) and (+) bacteria and antifungal activity isolated from the local wastewater sample. The prepared multifunctional nanocomposites act as promising materials in water treatment such as heavy metals removal and water disinfection against different microbes [183] (Figs. 3 and 4).

Nanocomposite based on polyaniline (PANI) and copper nanoparticles solved the problems of easy oxidation and aggregation of copper nanoparticles in addition to their antimicrobial activities [184]. They can control microorganisms in contaminated water before further processing.

A nanocomposite (PPP-TiO_2) was prepared by the reaction of pristine pomegranate peel extract (PPP) and titanium dioxide nanoparticles (TiO_2 NPs) [185].

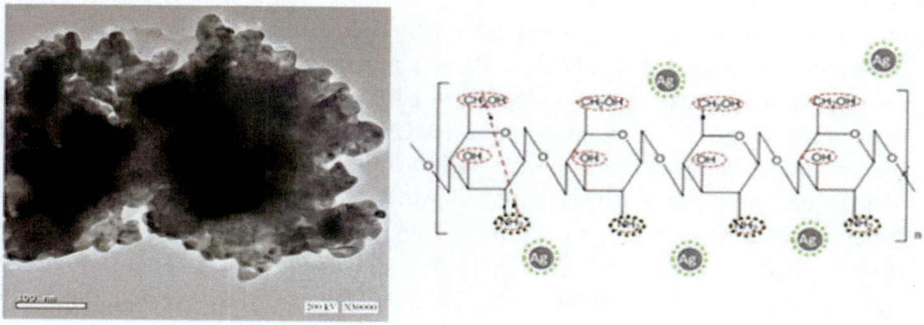

Fig. 3 TEM image and Schematic representations for the suggested structure of chitosan/Ag NPs nanocomposite [183] (with permission @ 2018 Elsevier B.V.)

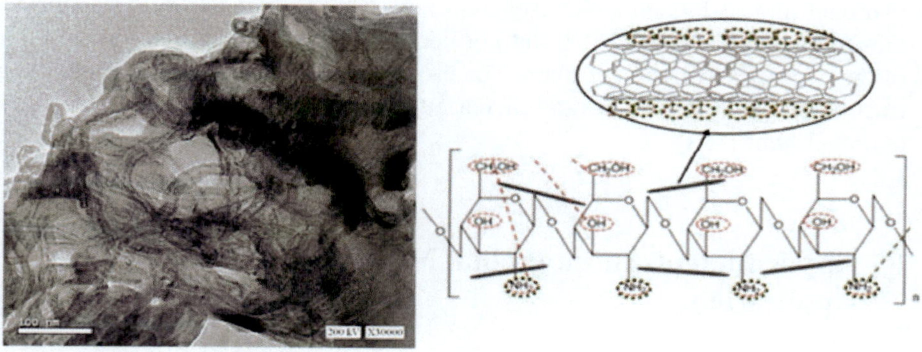

Fig. 4 TEM image and Schematic representations for the suggested structure of chitosan/CNTs nanocomposite [183] (with permission @ 2018 Elsevier B.V.)

PPP-TiO2 antimicrobial activity was 1.5 times that of PPP and TiO2 NPs. The Diameter Inhibition Zone (DIZ) had the highest inhibition of S. aureus as compared to E-coli and P. aeruginosa. The findings investigated lower BOD5 values for PPP-TiO2-containing samples as compared to TiO2 NPs.The approach is simple, feasible, cost-effective, and minimize environmental impact. The samples containing PPP-TiO_2 showed lower BOD5 values and consequently indicated that the sample has lower organic matter and the composites can disinfect water sources without any side effects [185].

5 Antimicrobial Nanocomposites Are Used as Anticancer Agents

Abu Elella et al. synthesized water-soluble polyelectrolyte complex (PEC) with a reaction of N, N, N-trimethyl chitosan chloride (TMC), and poly (acrylic acid) (PAA) in an acidic environment to synthesize AgNPs producing water-soluble nanocomposites (TMC/PAA/Ag) [186]. The formulated nanocomposites had good antimicrobial activity, which increased with an increase of Ag percent. In contrast, the most cytotoxic impact against colon cancer was TMC/PAA/Ag (3%). The nanocomposites have no major harmful effects on the VERO cell lines.

Pristine polyaniline (PANI), PANI-Au nanocomposites, and PANI-Au–Pt nanocomposites were prepared [187]. They demonstrated antibacterial activity against bacterial pathogens with gram (−) and (+) bacteria. They are in vitro anticancer investigations against HepG2 liver cancer cells revealed the highest cytotoxicity for PANI-Au–Pt nanocomposite followed by PANI-Au nanocomposite (32 mg/mL) and pristine PANI (49 mg/mL) [187].

Aminoclays/poly(vinyl alcohol) (AC/PVA) hydrogel films were prepared after three cycles of the freeze-thawing process [188]. Both MgAC/PVA and ZnAC/PVA hydrogel films exhibited antimicrobial effects. The silver nanopaerticles enhanced the antibacterial activity, but it did not decrease the gel fraction or swelling ratio of the MgAC/PVA hydrogel films. The presence of AgNW did not change the gel fraction or swelling ratio but did increase the antibacterial activity of MgAC/PVA hydrogel films [188].

6 Application of Antimicrobial Nanocomposites in Dental Application

Alamgir et al. [189] synthesized poly (methyl methacrylate) and TiO_2 nanocomposites using a twin-screw extruder by melt mixing the compounds and used in dental applications [189]. TiO_2 nanoparticles with different percentages (1 and 2 wt%) were mixed with PMMA through a melt compounding process. The PMMA/2wt%TiO_2 sample exhibited the best mechanical and physical properties [189].

7 Conclusion

Production of various nanocomposites for different biomedical applications has received considerable attention. Numerous antimicrobial biomaterials for microbial inactivation are being developed in recent years and are described by many researchers and scientists. This chapter is focused on the various applications of nanocomposites as antimicrobial agents, in food packaging, wound dressings, water treatment, anticancer and dental applications.

References

1. Guru PR, Bera H, Das M, Hasnain MS, Nayak AK (2018) Aceclofenac-loaded *Plantago ovata* F. husk mucilage-Zn^{+2}-pectinate controlled-release matrices. *Starch—Stärke* 70:1700136
2. Hasnain MS, Rishishwar P, Ali S, Alkahtani S, Tabish M, Milivojevic M, Ansari MT, Nayak AK (2020) Formulation and *ex vivo* skin permeation of lidocaine HCl topical gels using dillenia (*Dillenia indica* L.) fruit gum. Revista Mexicana de Ingeniería Química 19:1465–1476
3. Hasnain MS, Ahmed SA, Alkahtani S, Milivojevic M, Kandar CC, Dhara AK, Nayak AK (2020) Biopolymers for drug delivery. In: Nayak AK, Hasnain MS (eds) Advanced biopolymeric systems for drug delivery. Springer, Cham, pp 1–29
4. Pal D, Nayak AK, Saha S (2019) Cellulose-based hydrogels: present and future. In: Akhtar MS, Swamy MK, Sinniah UR (eds) Natural bio-active compounds. Singapore, Springer, pp 285–332
5. Dey S, Roy S, MS, Nayak AK (2020) Grafted alginates in drug delivery. Nayak AK, Hasnain MS (eds) Alginates in drug delivery. Academic Press, Elsevier Inc., USA, pp 71–100
6. Hasnain MS, Ray P, Nayak AK (2020) Alginate-based interpenetrating polymer networks for sustained drug release. Nayak AK, Hasnain MS (eds) Alginates in drug delivery. Academic Press, Elsevier Inc., USA, pp 101–128
7. Nayak AK, Bera H, Hasnain MS (2020) Particulate matrices of ionotropically gelled alginate- and plant-derived starches for sustained drug release. Nayak AK, Hasnain MS (eds) Alginates in drug delivery. Academic Press, Elsevier Inc., USA, pp 257–295
8. Mohanta BC, Javed MN, Hasnain MS, Nayak AK (2020) Polyelectrolyte complexes of alginate for controlling drug release. Nayak AK, Hasnain MS (eds) Alginates in drug delivery. Academic Press, Elsevier Inc., USA, pp 297–321
9. Bera H, Maiti S, Saha S, Nayak AK (2019) Biopolymers-based gastroretentive buoyant systems for therapeutic management of *Helicobacter pylori* infection. In: Maiti S, Jana S, (eds) Polysaccharide carriers for drug delivery. Woodhead Publishing, Elsevier Inc., pp 713–736
10. Hasnain MS, Nayak AK (2019) Classification of carbon nanotubes. In: Hasnain MS, Nayak AK (eds) Carbon nanotubes for targeted drug delivery. Springer, Singapore, pp 11–15
11. Hasnain MS, Nayak AK (2019) Synthesis of carbon nanotubes. In: Hasnain MS, Nayak AK (eds) Carbon nanotubes for targeted drug delivery. Springer, Singapore, pp 17–20
12. Hasnain MS, Nayak AK (2019) Characterization of carbon nanotubes. In: Hasnain MS, Nayak AK (eds) Carbon nanotubes for targeted drug delivery. Springer, Singapore, pp 29–31
13. Nayak AK, Hasnain MS (2019) Plant polysaccharides in drug delivery applications. In: Nayak AK, Hasnain MS (eds) Plant polysaccharides-based multiple-unit systems for oral drug delivery. Springer, Singapore, pp 19–23
14. Nayak AK, Hasnain MS (2019) Some other plant polysaccharide based multiple units for oral drug delivery. In: Nayak AK, Hasnain MS (eds) Plant polysaccharides-based multiple-unit systems for oral drug delivery. Springer, Singapore, pp 123–1128
15. Hasnain MS, Rishishwar P, Rishishwar S, Ali S, Nayak AK (2018) Extraction and characterization of cashew tree (*Anacardium occidentale*) gum; use in aceclofenac dental pastes. Int J Biol Macromol 116:1074–1081
16. Hasnain MS, Rishishwar P, Rishishwar S, Ali S, Nayak AK (2018) Isolation and characterization of *Linum usitatisimum* polysaccharide to prepare mucoadhesive beads of diclofenac sodium. Int J Biol Macromol 116:162–172
17. Nayak AK, Pal D (2017) Tamarind seed polysaccharide: an emerging excipient for pharmaceutical use. Indian J Pharmaceut Edu Res 51:S136–S146
18. Hasnain MS, Nayak AK (2018) Chitosan as responsive polymer for drug delivery applications. In: Makhlouf ASH, Abu-Thabit NY (eds) Stimuli responsive polymeric nanocarriers for drug delivery applications, vol 1, Types and triggers. Woodhead Publishing Series in Biomaterials, Elsevier Ltd., pp 581–605
19. Pal D, Nayak AK, Saha S (2019) Interpenetrating polymer network hydrogels of chitosan: Applications in controlling drug release. In: Mondal MIH (ed) Cellulose-based superabsorbent hydrogels. Springer, Cham, pp 1727–1768

20. Hasnain MS, Jameel E, Mohanta B, Dhara AK, Alkahtani S, Nayak AK (2020) Alginates: sources, structure, and properties, Nayak AK, Hasnain MS (eds) Alginates in drug delivery. Academic Press, Elsevier Inc., USA, pp 1–17
21. Nayak AK, Ansari MT, Sami F, Balvir Singh HK, Hasnain MS (2020) Alginates as drug delivery excipients. Nayak AK, Hasnain MS (eds) Alginates in drug delivery. Academic Press, Elsevier Inc., USA, pp 19–39
22. Nayak AK, Hasnain MS (2019) Gum Arabic based multiple units for oral drug delivery. In: Nayak AK, Hasnain MS (eds) Plant polysaccharides-based multiple-unit systems for oral drug delivery. Springer, Singapore, pp 25–30
23. Nayak AK, Hasnain MS (2019) Tamarind polysaccharide based multiple units for oral drug delivery. In: Nayak AK, Hasnain MS (eds) Plant polysaccharides-based multiple-unit systems for oral drug delivery. Springer, Singapore, pp 31–59
24. Dey S, Nandy BC, De JN, Hasnain MS, Nayak AK (2019) Tamarind gum in drug delivery applications. In: Hasnain MS, Nayak AK (eds) Natural polysaccharides in drug delivery and biomedical applications. Academic Press, Elsevier Inc., pp 285–306
25. Samanta A, De A, Hasnain MS, Bera H, Nayak AK (2019) Gum odina as pharmaceutical excipient. In: Hasnain MS, Nayak AK (eds) Natural polysaccharides in drug delivery and biomedical applications. Academic Press, Elsevier Inc., pp 327–337
26. Nayak AK, Hasnain MS, Pal D (2018) Gelled microparticles/beads of sterculia gum and tamarind gum for sustained drug release, In: Thakur VK, Thakur MK (eds) Handbook of Springer on polymeric gel. Springer International Publishing, Switzerland, pp 361–414
27. Verma A, Dubey J, Verma N, Nayak AK (2017) Chitosan-hydroxypropyl methylcellulose matrices as carriers for hydrodynamically balanced capsules of moxifloxacin HCl. Curr Drug Deliv 14:83–90
28. Jana S, Samanta A, Nayak AK, Sen KK, Jana S (2015) Novel alginate hydrogel core–shell systems for combination delivery of ranitidine HCl and aceclofenac. Int J Biol Macromol 74:85–92
29. Jana S, Ali SA, Nayak AK, Sen KK, Basu SK (2014) Development and optimization of topical gel containing aceclofenac-crospovidone solid dispersion by "Quality by Design" approach. Chem Eng Res Des 92:2095–2105
30. Nayak AK, Kalia S, Hasnain MS (2013) Optimization of aceclofenac-loaded pectinate-poly (vinyl pyrrolidone) beads by response surface methodology. Int J Biol Macromol 62:194–202
31. Nayak AK, Pal D (2013) Ionotropically-gelled mucoadhesive beads for oral metformin HCl delivery: formulation, optimization and antidiabetic evaluation. J Sci Ind Res 72:15–22
32. Nayak AK, Das B, Maji R (2013) Gastroretentive hydrodynamically balanced system of ofloxacin: formulation and *in vitro* evaluation. Saudi Pharm J 21:113–117
33. Das B, Nayak AK, Nanda U (2013) Topical gels of lidocaine HCl using cashew gum and Carbopol 940: preparation and *in vitro* skin permeation. Int J Biol Macromol 62:514–517
34. Malakar J, Nayak AK, Das A (2013) Modified starch (cationized)-alginate beads containing aceclofenac: formulation optimization using central composite design. Starch-Stärke 65:603–612
35. Pal D, Nayak AK (2011) Development, optimization and anti-diabetic activity of gliclazide-loaded alginate-methyl cellulose mucoadhesive microcapsules. AAPS PharmSciTech 12(4):1431–1441
36. Hasnain MS, Nayak AK (2012) Solubility and dissolution enhancement of ibuprofen by solid dispersion technique using PEG 6000-PVP K 30 combination carrier. Chem: Bulg J Sci Educ 21:118–132
37. Alam MS, Javed MN, Pottoo FH, Waziri A, Almalki FA, Hasnain MS, Garg A, Saifullah MK (2019) QbD approached comparison of reaction mechanism in microwave synthesized gold nanoparticles and their superior catalytic role against hazardous nirto-dye. Appl Organomet Chem 33(9):e5071
38. Majeed S, Shamsuddin AF, Hasnain MS (2018) Novel green approach for synthesis of metallic nanoparticles and its biomedical application. Current Nanomedicine 8(3):177–183

39. Ray P, Hasnain MS, Koley A, Nayak AK (2019) Bone-implantable devices for drug delivery applications. In: Pal K, Kraatz H-H, Li C, Khasnobish A, Bag S, Banerjee I, Kuruganti U (eds) Bioelectronics and medical devices, from materials to devices—fabrication, applications and reliability. Woodhead Publishing Series in Electronic and Optical Materials, Elsevier Inc., pp 333–392
40. Nayak AK, Hasnain MS, Malakar J (2013) Development and optimization of hydroxyapatite-ofloxacin implants for possible bone-implantable delivery in osteomyelitis treatment. Curr Drug Deliv 10:241–250
41. Hasnain MS, Nayak AK (2019) Applications of carbon nanotubes. In: Hasnain MS, Nayak AK (eds) Carbon nanotubes for targeted drug delivery. Springer, Singapore, pp 33–36
42. Hasnain MS, Nayak AK (2019) Targeted delivery with carbon nanotubes. In: Hasnain MS, Nayak AK (eds) Carbon nanotubes for targeted drug delivery. Springer, Singapore, pp 37–50
43. Hasnain MS, Nayak AK (2019) Carbon nanotubes in controlled drug delivery. In: Hasnain MS, Nayak AK (eds) Carbon nanotubes for targeted drug delivery. Springer, Singapore, pp 51–54
44. Hasnain MS, Nayak AK (2019) CNTs in solubility enhancement. In: Hasnain MS, Nayak AK (eds) Carbon nanotubes for targeted drug delivery. Springer, Singapore, pp 55–57
45. Hasnain MS, Nayak AK (2019) Carbon nanotubes as quantum dots for therapeutic purpose. In: Hasnain MS, Nayak AK (eds) Carbon nanotubes for targeted drug delivery. Springer, Singapore, pp 59–64
46. Hasnain MS, Nayak AK (2019) Absorption and transportation of carbon nanotubes. In: Hasnain MS, Nayak AK (eds) Carbon nanotubes for targeted drug delivery. Springer, Singapore, pp 65–68
47. Hasnain MS, Nayak AK (2019) Carbon nanotubes in vaccine delivery. In: Hasnain MS, Nayak AK (eds) Carbon nanotubes for targeted drug delivery. Springer, Singapore, pp 69–73
48. Hasnain MS, Nayak AK (2019) Carbon nanotubes in gene delivery. In: Hasnain MS, Nayak AK (eds) Carbon nanotubes for targeted drug delivery. Springer, Singapore, pp 75–87
49. Hasnain MS, Nayak AK (2019) Toxicity consideration of carbon nanotubes. In: Hasnain MS, Nayak AK (eds) Carbon nanotubes for targeted drug delivery. Springer, Singapore, pp 89–101
50. Hasnain MS, Nayak AK (2019) Regulatory considerations of carbon nanotubes. In: Hasnain MS, Nayak AK (eds) Carbon nanotubes for targeted drug delivery. Springer, Singapore, pp 103–106
51. Nayak AK, Pal D (2017) Natural starches-blended ionotropically-gelled micrparticles/beads for sustained drug release. In: Thakur VK, Thakur MK, Kessler MR (eds) Handbook of composites from renewable materials, vol 8, Nanocomposites: Advanced Applications. WILEY-Scrivener, USA, pp 527–560
52. Pal D, Nayak AK (2017) Plant polysaccharides-blended ionotropically-gelled alginate multiple-unit systems for sustained drug release. In: Thakur VK, Thakur MK, Kessler MR (eds) Handbook of composites from renewable materials, vol 6, polymeric composites. WILEY-Scrivener, USA, pp 399–400
53. Pal D, Nayak AK (2015) Interpenetrating polymer networks (IPNs): natural polymeric blends for drug delivery. In: Mishra M (ed) Encyclopedia of biomedical polymers and polymeric biomaterials. Taylor & Francis Group, USA, vol VI, pp 4120–4130
54. Pal D, Nayak AK (2015) Alginates, blends and microspheres: controlled drug delivery. In: Mishra M (ed) Encyclopedia of biomedical polymers and polymeric biomaterials. Taylor & Francis Group, USA, vol I, pp 89–98
55. Nayak AK, Pal D, Hasnain MS (2013) Development, optimization and *in vitro-in vivo* evaluation of pioglitazone-loaded jackfruit seed starch-alginate beads. Curr Drug Deliv 10:608–619
56. Nayak AK, Pal D, Pradhan J, Hasnain MS (2013) Fenugreek seed mucilage-alginate mucoadhesive beads of metformin HCl: design, optimization and evaluation. Int J Biol Macromol 54:144–154
57. Bera H, Kandukuri SG, Nayak AK, Boddupalli S (2015) Alginate-sterculia gum gel-coated oil-entrapped alginate beads for gastroretentive risperidone delivery. Carbohyd Polym 120:74–84

58. Nayak AK, Das B (2018) Introduction to polymeric gels. In: Pal K, Bannerjee I (eds) Polymeric gels, characterization, properties and biomedical applications. Woodhead Publishing Series in Biomaterials, Elsevier Ltd., pp 3–27
59. Ray P, Maity M, Barik H, Sahoo GS, Hasnain MS, Hoda MN, Nayak AK (2020) Alginate-based hydrogels for drug delivery applications. Nayak AK, Hasnain MS (eds) Alginates in drug delivery. Academic Press, Elsevier Inc., USA, pp 41–70
60. Nayak AK, Hasnain MS, Pal K, Banerjee I, Pal D. Gum-based hydrogels in drug delivery. In: Pal K, Banerjee I, Sarkar P, Kim D, Deng W-P, Dubey NK, Majumder K (eds) Biopolymer-based formulations. Biomedical and Food Applications, Elsevier Inc., pp 605–645
61. Nayak AK, Hasnain MS (2019) Sterculia gum based multiple units for oral drug delivery. In: Nayak AK, Hasnain MS (eds) Plant polysaccharides-based multiple-unit systems for oral drug delivery. Springer, Singapore, pp 67–82
62. Milivojevic M, Pajic-Lijakovic I, Bugarski B, Nayak AK, Hasnain MS (2019) Gellan gum in drug delivery applications. In: Hasnain MS, Nayak AK (eds) Natural polysaccharides in drug delivery and biomedical applications. Academic Press, Elsevier Inc., pp 145–186
63. Bera H, Abbasi YF, Hasnain MS, Nayak AK (2019) Sterculia gum in drug delivery applications. In: Hasnain MS, Nayak AK (eds) Natural polysaccharides in drug delivery and biomedical applications. Academic Press, Elsevier Inc., pp 223–247
64. Nayak AK, Pal D (2018) Functionalization of tamarind gum for drug delivery. In: Thakur VK, Thakur MK (eds) Functional biopolymers. Springer International Publishing, Switzerland, pp 35–56
65. Nayak AK, Pal D (2016) Sterculia gum-based hydrogels for drug delivery applications. In: Kalia S (ed). Polymeric hydrogels as smart biomaterials, springer series on polymer and composite materials. Springer International Publishing, Switzerland, pp 105–151
66. Nayak AK (2016) Tamarind seed polysaccharide-based multiple-unit systems for sustained drug release. In: Kalia S, Averous L (eds) Biodegradable and bio-based polymers: environmental and biomedical applications. USA, WILEY-Scrivener, pp 471–494
67. Nayak AK, Pal D (2016) Plant-derived polymers: ionically gelled sustained drug release systems. In: M. Mishra M (ed) Encyclopedia of biomedical polymers and polymeric biomaterials. Taylor & Francis Group, USA, vol VIII, pp 6002–6017
68. Nayak AK, Pal D (2015) Chitosan-based interpenetrating polymeric network systems for sustained drug release. In: Tiwari A, Patra HK, Choi J-W (eds) Advanced theranostics materials. USA, WILEY-Scrivener, pp 183–208
69. Nayak AK, Beg S, Hasnain MS, Malakar J, Pal D (2018) Soluble starch-blended Ca^{2+}-Zn^{2+}-alginate composites-based microparticles of aceclofenac: formulation development and *in vitro* characterization. Futur J Pharm Sci 4:63–70
70. Rapalli VK, Singhvi G, Gorantla S, Waghule T, Dubey SK, Saha RN, Hasnain MS, Nayak AK (2019) Stability indicating liquid chromatographic method for simultaneous quantification of betamethasone valerate and tazarotene in *in-vitro* and *ex-vivo* studies of complex nanoformulation. J Sep Sci 42(22):3413–3420
71. Nayak AK, Pal D, Santra K (2016) Swelling and drug release behavior of metformin HCl-loaded tamarind seed polysaccharide-alginate beads. Int J Biol Macromol 82:1023–1027
72. Sinha P, Ubaidulla U, Hasnain MS, Nayak AK, Rama B (2015) Alginate-okra gum blend beads of diclofenac sodium from aqueous template using $ZnSO_4$ as a cross-linker. Int J Biol Macromol 79:555–563
73. Nayak AK, Pal D (2014) *Trigonella foenum-graecum* L. seed mucilage-gellan mucoadhesive beads for controlled release of metformin HCl. Carbohydr Polym 107:31–40
74. Nayak AK, Pal D, Santra K (2014) *Artocarpus heterophyllus* L. seed starch-blended gellan gum mucoadhesive beads of metformin HCl. Int J Biol Macromol 65:329–339
75. Das B, Dutta S, Nayak AK, Nanda U (2014) Zinc alginate-carboxymethyl cashew gum microbeads for prolonged drug release: development and optimization. Int J Biol Macromol 70:505–515
76. Nayak AK, Pal D, Das S (2013) Calcium pectinate-fenugreek seed mucilage mucoadhesive beads for controlled delivery of metformin HCl. Carbohyd Polym 96:349–357

77. Nayak AK, Pal D (2013) Blends of jackfruit seed starch-pectin in the development of mucoadhesive beads containing metformin HCl. Int J Biol Macromol 62:137–145
78. Nayak AK, Pal D (2013) Formulation optimization of jackfruit seed starch-alginate mucoadhesive beads of metformin HCl. Int J Biol Macromol 59:264–272
79. Nayak AK, Pal D, Malakar J (2013) Development, optimization and evaluation of emulsion-gelled floating beads using natural polysaccharide-blend for controlled drug release. Polym Eng Sci 53:338–350
80. Guru PR, Nayak AK, Sahu RK (2013) Oil-entrapped sterculia gum-alginate buoyant systems of aceclofenac: development and *in vitro* evaluation. Colloids Surf B: Biointerfaces 104:268–275
81. Jana S, Das A, Nayak AK, Sen KK, Basu SK (2013) Aceclofenac-loaded unsaturated esterified alginate/gellan gum microspheres: *in vitro* and *in vivo* assessment. Int J Biol Macromol 57:129–137
82. Nayak AK, Das B, Maji R (2012) Calcium alginate/gum arabic beads containing glibenclamide: development and *in vitro* characterization. Int J Biol Macromol 51:1070–1078
83. Chowdhury S, Chakraborty S, Maity M, Hasnain MS, Nayak AK (2020) Biocomposites of alginates in drug delivery. In: Nayak AK, Hasnain MS (eds) Alginates in drug delivery. Academic Press, Elsevier Inc., USA, pp 153–185
84. Das S, Pattanayak D, Nayak AK, Yi DK, Nanda SS, Ansari MT, Hasnain MS (2020) Alginate–montmorillonite composite systems as sustained drug delivery carriers. In: Nayak AK, Hasnain MS (eds) Alginates in drug delivery. Academic Press, Elsevier Inc., USA, pp 187–201
85. Hasnain MS, Ahmed SA, Behera A, Alkahtani S, Nayak AK. Inorganic materials–alginate composites in drug delivery. Nayak AK, Hasnain MS (eds) Alginates in drug delivery. Academic Press, Elsevier Inc., USA, pp 231–256
86. Nanda SS, Yi DK, Hasnain MS, Nayak AK (2019) Hydroxyapatite-alginate composites in drug delivery. In: Hasnain MS, Nayak AK (eds) Alginate: versatile polymer in biomedical applications and therapeutics. Apple Academic Press, USA, pp 483–503
87. Hasnain MS, Nayak AK (2018) Alginate-inorganic composite particles as sustained drug delivery matrices. In: Inamuddin, Asiri AM, Mohammad A (eds) Applications of nanocomposite materials in drug delivery. Elsevier Inc., pp 39–74
88. Nayak AK, Ara TJ, Hasnain MS, Hoda N (2018) Okra gum-alginate composites for controlled releasing drug delivery. In: Inamuddin, Asiri AM, Mohammad A (eds) Applications of nanocomposite materials in drug delivery. Woodhead Publishing Series in Biomaterials, Elsevier Inc., pp 761–785
89. Nayak AK, Hasnain MS, Nanda SS, Yi DK (2019) Hydroxyapatite-alginate based matrices for drug delivery. Curr Pharm Des 25:3406–3416
90. Nayak AK, Mohanta BC, Hasnain MS, Hoda MN, Tripathi G (2020) Alginate-based scaffolds for drug delivery in tissue engineering. In: Nayak AK, Hasnain MS (eds) Alginates in drug delivery. Academic Press, Elsevier Inc., USA, pp 359–386
91. Kenawy ER (2020) MS Hasnain nanofibers for filtration applications. In: Nayak AK, Hasnain MS (eds) Advanced biopolymeric systems for drug delivery. Springer, Cham, pp 361–371
92. Hasnain MS, Guru PR, Rishishwar P, Ali S, Ansari MT, Nayak AK (2020) Atenolol-releasing buccal patches made of *Dillenia indica* L. fruit gum: preparation and *ex vivo* evaluations. SN Appl Sci 2(1):57
93. Rath Adhikari SN, Nayak BS, Nayak AK, Mohanty B (2010) Formulation and evaluation of buccal patches for delivery of atenolol. AAPS PharmSciTech. 11(3):1034–1044
94. Hasnain MS, Rishishwar P, Ali S, Nayak AK (2020) Preparation and evaluation of aceclofenac dental pastes using dillenia fruit gum for periodontitis treatment. SN Appl Sci 2(3):1–8
95. Jena AK, Nayak AK, De A, Mitra D, Samanta A (2018) Development of lamivudine containing multiple emulsions stabilized by gum odina. Futur J Pharm Sci 4:71–79
96. Malakar J, Basu A, Nayak AK (2014) Candesartan cilexetil microemulsions for transdermal delivery: formulation, *in-vitro* skin permeation and stability assessment. Curr Drug Deliv 11:313–321

97. Nayak AK, Pal D, Pradhan J, Ghorai T (2012) The potential of *Trigonella foenum-graecum* L. seed mucilage as suspending agent. Indian J Pharm Educ Res 46:312–317
98. Nayak AK, Pal D, Pany DR, Mohanty B (2010) Evaluation of *Spinacia oleracea* L. leaves mucilage as innovative suspending agent. J Adv Pharmceutical Technol Res 1(3):338–341
99. Hasnain MS, Nayak AK (2019) Background: carbon nanotubes for targeted drug delivery. In: Hasnain MS, Nayak AK (eds) Carbon nanotubes for targeted drug delivery. Springer, Singapore, pp 1–9
100. Hasnain MS, Nayak AK (2019) Functionalization of carbon nanotubes. In: Hasnain MS, Nayak AK (eds) Carbon nanotubes for targeted drug delivery. Springer, Singapore, pp 21–28
101. Nayak AK, Hasnain MS (2020) Ionotropically gelled alginate particles in sustained drug release. In: Nayak AK, Hasnain MS (eds) Alginates in drug delivery. Academic Press, Elsevier Inc., USA, pp 203–230
102. Nayak AK, Hasnain MS (2019) Locust bean gum based multiple units for oral drug delivery. In: Nayak AK, Hasnain MS (eds) Plant polysaccharides-based multiple-unit systems for oral drug delivery. Springer, Singapore, pp 61–66
103. Nayak AK, Hasnain MS (2019) Okra gum based multiple units for oral drug delivery. In: Nayak AK, Hasnain MS (eds) Plant polysaccharides-based multiple-unit systems for oral drug delivery. Springer, Singapore, pp 83–92
104. Nayak AK, Hasnain MS (2019) Fenugreek seed mucilage based multiple units for oral drug delivery. In: Nayak AK, Hasnain MS (eds) Plant polysaccharides-based multiple-unit systems for oral drug delivery. Springer, Singapore, pp 93–112
105. Nayak AK, Hasnain MS (2019) Potato starch based multiple units for oral drug delivery. In: Nayak AK, Hasnain MS (eds) Plant polysaccharides-based multiple-unit systems for oral drug delivery. Springer, Singapore, pp 113–116
106. Nayak AK, Pal D, Santra K (2014) Ispaghula mucilage-gellan mucoadhesive beads of metformin HCl: development by response surface methodology. Carbohyd Polym 107:41–50
107. Nayak AK, Hasnain MS (2019) Linseed polysaccharide based multiple units for oral drug delivery. In: Nayak AK, Hasnain MS (eds) Plant polysaccharides-based multiple-unit systems for oral drug delivery. Springer, Singapore, pp 117–121
108. Nayak AK, Pal D, Santra K (2014) Tamarind seed polysaccharide-gellan mucoadhesive beads for controlled release of metformin HCl. Carbohyd Polym 103:154–163
109. Nayak AK, Ansari MT, Sami F, Bera H, Hasnain MS (2019) Cashew gum in drug delivery applications. In: Hasnain MS, Nayak AK (eds) Natural polysaccharides in drug delivery and biomedical applications. Academic Press, Elsevier Inc., pp 263–283
110. Nayak AK, Pal D, Santra K (2014) Development of pectinate-ispagula mucilage mucoadhesive beads of metformin HCl by central composite design. Int J Biol Macromol 66:203–221
111. Nayak AK, Bera H, Hasnain MS, Pal D (2018) Synthesis and characterization of graft-copolymers of plant polysaccharides. In: Thakur VK (ed) Biopolymer grafting. Synthesis and Properties, Elsevier Inc., pp 1–62
112. Nayak AK, Ahmad SA, Beg S, Ara TJ, Hasnain MS (2018) Drug delivery: present, past and future of medicine. In: Inamuddin, Asiri AM, Mohammad A (eds) Applications of nanocomposite materials in drug delivery. Woodhead Publishing Series in Biomaterials, Elsevier Inc., pp 255–282
113. Nayak AK, Pal D, Santra K (2014) Development of calcium pectinate-tamarind seed polysaccharide mucoadhesive beads containing metformin HCl. Carbohyd Polym 101:220–230
114. Das SS, Alkahtani S, Bharadwaj P, Ansari MT, ALKahtani MDF, Pang Z, Hasnain MS, Nayak AK, Aminabhavi TM (2020) Molecular insights and novel approaches for targeting tumor metastasis. Int J Pharm 585:119556
115. Nayak AK (2010) Advances in therapeutic protein production and delivery. Int J Pharm Pharm Sci 2(2):1–5
116. Shariatinia Z (2019) Pharmaceutical applications of natural polysaccharides. In: Hasnain MS, Nayak AK (eds) Natural polysaccharides in drug delivery and biomedical applications. Academic Press, Elsevier Inc., pp 15–57

117. Thomas TJ, Tajmir-Riahi HA, Pillai CKS (2019) Biodegradable polymers for gene delivery. Molecules 24(20):3744
118. Baranwal A, Srivastava A, Kumar P, Bajpai VK, Maurya PK, Chandra P (2018) Prospects of nanostructure materials and their composites as antimicrobial agents. Front Microbiol 9:422
119. Hasnain MS, Nayak AK, Singh R, Ahmad F (2010) Emerging trends of natural-based polymeric systems for drug delivery in tissue engineering applications. Sci J UBU 1(2):1–13
120. Nayak AK, Ahmed SA, Tabish M, Hasnain MS (2019) Natural polysaccharides in tissue engineering application. In: Hasnain MS, Nayak AK (eds) Natural polysaccharides in drug delivery and biomedical applications. Academic Press, Elsevier Inc., pp 531–548
121. Kurakula M, Rao GSNK, Kiran V, Hasnain MS, Nayak AK (2020) Alginate-based hydrogel systems for drug releasing in wound healing. In: Nayak AK, Hasnain MS (eds) Alginates in drug delivery. Academic Press, Elsevier Inc., USA, pp 323–358
122. Hasnain MS, Kiran V, Kurakula M, Rao GSNK, Tabish M, Nayak AK (2020) Use of alginates for drug delivery in dentistry. In: Nayak AK, Hasnain MS (eds) Alginates in drug delivery. Academic Press, Elsevier Inc., USA, pp 387–404
123. Ruecha N, Rangkupan R, Rodthongkum N, Chailapakul O (2014) Novel paper-based cholesterol biosensor using graphene/polyvinylpyrrolidone/polyaniline nanocomposite. Biosens. Bioelectron 52:13–19
124. Hasnain MS, Nayak AK, Kurakula M, Hoda MN (2020) Alginate nanoparticles in drug delivery. In: Nayak AK, Hasnain MS (eds) Alginates in drug delivery. Academic Press, Elsevier Inc., USA, pp 129–152
125. Pal D, Nayak AK (2010) Nanotechnology for targeted delivery in cancer therapeutics. Int J Pharm Sci Rev Res 1(1):1–7
126. Ray S, Sinha P, Laha B, Maiti S, Bhattacharyya UK, Nayak AK (2018) Polysorbate 80 coated crosslinked chitosan nanoparticles of ropinirole hydrochloride for brain targeting. J Drug Deliv Sci Technol 48:21–29
127. Jana S, Gangopadhaya A, Bhowmik BB, Nayak AK, Mukhrjee A (2015) Pharmacokinetic evaluation of testosterone-loaded nanocapsules in rats. Int J Biol Macromol 72:28–30
128. Waghule T, Rapalli VK, Singhvi G, Manchanda P, Hans N, Dubey SK, Hasnain MS, Nayak AK (2019) Voriconazole loaded nanostructured lipid carriers based topical delivery system: QbD based designing, characterization, *in-vitro* and *ex-vivo* evaluation. J Drug Deliv Sci Technol 52:303–315
129. Hasnain MS, Javed MN, Alam MS, Rishishwar P, Rishishwar S, Ali S, Nayak AK, Beg S (2019) Purple heart plant leaves extract-mediated silver nanoparticle synthesis: optimization by Box-Behnken design. Mater Sci Eng, C 99:1105–1114
130. Nayak AK (2010) Hydroxyapatite synthesis methodologies: an overview. Int J ChemTech Res 2(2):903–907
131. Nayak AK, Laha B, Sen KK (2011) Development of hydroxyapatite-ciprofloxacin bone-implants using >>Quality by Design<< Acta Pharmaceutica 61(1):25–36
132. Nayak AK, Bera H (2019) *In situ* polysaccharide-based gels for topical drug delivery applications. In: Maiti S, Jana S (eds) Polysaccharide carriers for drug delivery. Woodhead Publishing, Elsevier Inc., pp 615–638
133. Hasnain MS, Nayak AK (2019) Recent progress in responsive polymer-based drug delivery systems. In: Makhlouf ASH, Abu-Thabit NY (eds) Stimuli responsive polymeric nanocarriers for drug delivery applications: vol 2. Advanced nanocarriers for therapeutics. Woodhead Publishing Series in Biomaterials, Elsevier Ltd., pp 569–595
134. Bhattacharjee A, Das PJ, Dey S, Nayak AK, Roy PK, Chakrabarti S, Marbaniang D, Das SK, Ray S, Chattopadhyay P, Mazumder B (2020) Development and optimization of besifloxacin hydrochloride loaded liposomal gel prepared by thin film hydration method using 3^2 full factorial design. Colloids Surf, A 585:124071
135. Das B, Sen SO, Maji R, Nayak AK, Sen KK (2017) Transferosomal gel for transdermal delivery of risperidone: formulation optimization and *ex vivo* permeation. J Drug Deliv Sci Technol 38:59–71

136. Malakar J, Sen SO, Nayak AK, Sen KK (2012) Preparation, optimization and evaluation of transferosomal gel for transdermal insulin delivery. Saudi Pharm J 20:355–363
137. Panigrahi BK, Nayak AK (2020) Carbon nanotubes: an emerging drug delivery carrier in cancer therapeutics. Curr Drug Deliv 17, In Press
138. Hasnain MS, Ahmad SA, Hoda MN, Rishishwar S, Rishishwar P, Nayak AK (2019) Stimuli-responsive carbon nanotubes for targeted drug delivery. In: Makhlouf ASH, Abu-Thabit NY (eds) Stimuli responsive polymeric nanocarriers for drug delivery applications: vol 2, Advanced nanocarriers for therapeutics. Woodhead Publishing Series in Biomaterials, Elsevier Ltd., pp 321–344
139. Hasnain MS, Ahmad SA, Chaudhary N, Hoda MN, Nayak AK (2019) Biodegradable polymer matrix nanocomposites for bone tissue engineering. In: Inamuddin, Asiri AM, Mohammad A (eds) Applications of nanocomposite materials in orthopedics. Woodhead Publishing Series in Biomaterials, Elsevier Inc., pp 1–37
140. Hasnain MS, Ahmad SA, Minhaj MA, Ara TJ, Nayak AK (2019) Nanocomposite materials for prosthetic devices. In: Inamuddin, Asiri AM, Mohammad A (eds) Applications of nanocomposite materials in orthopedics. Elsevier Inc., pp 127–144
141. Hasnain MS, Nayak AK (2019) Nanocomposites for improved orthopedic and bone tissue engineering applications. In: Inamuddin, Asiri AM, Mohammad A (eds) Applications of nanocomposite materials in orthopedics. Woodhead Publishing Series in Biomaterials, Elsevier Inc., pp 145–177
142. Nayak AK, Mazumder S, Ara TJ, Ansari MT, Hasnain MS (2019) Calcium fluoride-based dental nanocomposites. In: Inamuddin, Asiri AM, Mohammad A (eds) Applications of nanocomposite materials in dentistry. Elsevier Inc., pp 27–45
143. Rani P, Pal D, Hoda MN, Ara TJ, Beg S, Hasnain MS, Nayak AK (2019) Dental pulp capping nanocomposites. In: Inamuddin, Asiri AM, Mohammad A (eds) Applications of nanocomposite materials in dentistry. Elsevier Inc., pp 65–91
144. Hasnain MS, Ahmad SA, Chaudhary N, Minhaj MA, Nayak AK (2019) Degradation and failure of dental composite materials. In: Inamuddin, Asiri AM, Mohammad A (eds) Applications of nanocomposite materials in dentistry. Elsevier Inc., pp 108–121
145. Mazumder S, Nayak AK, Ara TJ, Hasnain MS (2019) Hydroxyapatite composites for dentistry. In: Inamuddin, Asiri AM, Mohammad A (eds) Applications of nanocomposite materials in dentistry. Woodhead Publishing Series in Biomaterials, Elsevier Inc., pp 108–121
146. Lin J, Li Y, Li Y, Wu H, Yu F, Zhou S, Xie L, Luo F, Lin C, Hou Z (2015) Drug/dye-loaded, multifunctional PEG-chitosan-iron oxide nanocomposites for methotraxate synergistically self-targeted cancer therapy and dual model imaging. ACS Appl Mater Interfaces 7(22):11908–11920
147. Yan C-Y, Gu J-W, Hou D-P, Jing HY, Wang J, Guo YZ et al (2015) Synthesis of tat tagged and folate modified N-succinyl-chitosan self-assembly nanoparticles as a novel gene vector. Int J Biol Macromol 72(2015):751–756
148. Youssef AM, Abdel-Aziz MS, El-Sayed SM (2014) Chitosan nanocomposite films based on Ag-NP and Au-NP biosynthesis by *Bacillus subtilis* as packaging materials. Int J Biol Macromol 69:185–191
149. Tantiwatcharothai S, Prachayawarakorn J (2019) Characterization of an antibacterial wound dressing from basil seed (*Ocimum basilicum* L.) mucilage-ZnO nanocomposite. Int J Biol Macromol 135:133–140
150. Chen Z, Liu X, Liu D, Li F, Wang L, Liu S (2020) Ultrasensitive electrochemical DNA biosensor fabrication by coupling an integral multifunctional zirconia-reduced graphene oxide-thionine nanocomposite and exonuclease i-assisted cleavage. Front Chem 8:521
151. Devi TB, Mohanta D, Ahmaruzzaman M (2019) Biomass derived activated carbon loaded silver nanoparticles: an effective nanocomposites for enhanced solar photocatalysis and antimicrobial activities. J Ind Eng Chem 76(2019):160–172
152. Zarnegar Z, Safari J, Zahraei Z (2019) Design, synthesis and antimicrobial evaluation of silver decorated magnetic polymeric nanocomposites. Nano-Struct Nano-Objects 19:100368–100374

153. Hosseini M, Sarafbidabad M, Fakhri A, NoorMohammadi Z, Tahami S (2018) Preparation and characterization of MnS2/chitosan sodium alginate and calcium alginate nanocomposites for degradation of analgesic drug: photocorrosion, mechanical, antimicrobial and antioxidant properties studies. Int J Biol Macromol 118:1494–1500
154. Amani A, Montazer M, Mahmoudirad M (2019) Synthesis of applicable hydrogel corn silk/ZnO nanocomposites on polyester fabric with antimicrobial properties and low cytotoxicity. Int J Biol Macromol 123:1079–1090
155. Demchenko V, Riabov S, Rybalchenko N, Goncharenko L, Kobylinskyi S, Shtompel V (2017) X-ray study of structural formation, thermomechanical and antimicrobial properties of copper-containing polymer nanocomposites obtained by the thermal reduction method. Eur Polymer J 96:326–336
156. Mwafy EA, Hasanin MS, Mostafa AM (2019) Cadmium oxide/TEMPO-oxidized cellulose nanocomposites produced by pulsed laser ablation in liquid environment: synthesis, characterization, and antimicrobial activity. Opt Laser Technol 120:105744–105750
157. Iqbal T, Ali F, Khalid NR, Tahir MB, Ijaz M (2019) Facile synthesis and antimicrobial activity of CdS-Ag2S nanocomposites. Bioorganic Chem 90(2019):103064–103069
158. Malagurski I, Levic S, Pantic M, Matijasevic D, Mitric M, Pavlovic V, Dimitrijevic-Brankovic S (2017) Synthesis and antimicrobial properties of Zn-mineralized alginate nano composites. Carbohyd Polym 165:313–321
159. Abdeen ZI, El Farargy AF, Negm NA (2018) Nanocomposite framework of chitosan/polyvinyl alcohol/ZnO: preparation, characterization, swelling and antimicrobial evaluation. J Mol Liq 250:335–343
160. Roy A, Joshi M, Butola BS (2019) Antimicrobial performance of polyethylene nanocomposite monofilaments reinforced with metal nanoparticles decorated montmorillonite. Colloids Surf B: Biointerfaces 178(2019):87–93
161. Ghosh T, Bardhan P, Mandal M, Karak N (2019) Interpenetrating polymer network-based nanocomposites reinforced with octadecylamine capped Cu/reduced graphene oxide nanohybrid with hydrophobic, antimicrobial and antistatic attributes. Mater Sci Eng, C 105:110055–110068
162. Ahmadi Y, Yadav M, Ahmad S (2019) Oleo-polyurethane-carbon nanocomposites: Effects of in-situ polymerization and sustainable precursor on structure, mechanical, thermal, and antimicrobial surface-activity. Compos Part. B Eng 164:683–692
163. Beyene Z, Ghosh R (2019) Effect of zinc oxide addition on antimicrobial and antibiofilm activity of hydroxyapatite: a potential nanocomposite for biomedical applications. Mater Today Commun 21:100612–100619
164. Guom Q, Lan T, Chen Yi, Yini Xu, Peng J, Tao L, Shen X (2019) Enhanced of antibacterial activity of antibiotic-functionalized silver nanocomposites with good biocompatibility. J Mater Sci - Mater Med 30:34–43
165. Hasan Moshafi M, Ranjbar M, Hedayatifar N (2019) Preparation and evaluation of cross-linked chitosan/silver sulfide luminescence nanocomposites by using green capping agent against some pathogenic microbial strains. Macromol Res
166. El Nahrawy AM, Abou Hammad AB, Abdel-Aziz MS, Wassel AR (2019) Spectroscopic and antimicrobial activity of hybrid chitosan/silica membranes doped with Al_2O_3 nanoparticles. Silicon 11:1677–1685
167. Nezami S, Sadeghi M (2019) pH-sensitive free AgNPs composite and nanocomposite beads based on starch as drug delivery systems. Polym Bull. https://doi.org/10.1007/s00289-019-02801-3
168. Snigdha S, Rahul M, Kalarikkal N, Thomas S, Radhakrishnan EK (2019) Poly (ε-caprolactone) microsphere decorated with nano-ZnO based phytoformulation: a promising antimicrobial agent. J Inorg Organomet Polym Mater 29:1503–1513
169. Gasmalla HB, Lu X, Shinger MI, Ni L, Chishti AN, Diao G (2019) Novel magnetically separable of $Fe_3O_4/Ag_3PO_4@WO_3$ nanocomposites for enhanced photocatalytic and antibacterial activity against *Staphylococcus aureus* (*S. aureus*). J Nanobiotechnology 17:58–72

170. Wang K, Su P, Li H, Wu Y, Zhang D, Feng H, Fan H (2019) Synthesis, characterization and antimicrobial activity of hybrid-structured Ag@CeO_2 nanoparticles. Chem Pap 73:1279–1286
171. Siripatrawan U, Kaewklin P (2018) Fabrication and characterization of chitosan-titanium dioxide nanocomposite film as ethylene scavenging and antimicrobial active food packaging. Food Hydrocoll 84:125–134
172. Ahmadi R, Ghanbarzadeh B, Ayaseh A, Kafil HS, Özyurt H, Katourani A, Ostadrahimi A (2019) The antimicrobial bio-nanocomposite containing non-hydrolyzed cellulose nanofiber (CNF) and Miswak (Salvadora persica L.) extract. Carbohydr Polym 214(2019):15–25
173. Lee JH, Jeongb D, Kanmani P (2019) Study on physical and mechanical properties of the biopolymer/silver based active nanocomposite films with antimicrobial activity. Carbohyd Polym 224:115–159
174. Nguyen TNL, Do TV, Nguyen TV, Dao PH, Trinh VT, Mac VP, Nguyen AH, Dinh DA, Nguyen TA, Vo TKA, Tran DL, Le TL (2019) Antimicrobial activity of acrylic polyurethane/Fe_3O_4-Ag nanocomposite coating. Prog Org Coat 132(2019):15–20
175. Zhilong Yu, Wang W, Dhital R, Kong F, Lin M, Mustapha A (2019) Antimicrobial effect and toxicity of cellulose nanofibril/silver nanoparticle nanocomposites prepared by an ultraviolet irradiation method. Colloids Surf, B 180:212–220
176. Masoule SF, Pourhajibagher M, Safari J, Khoobi M (2019) Base-free green synthesis of copper(II) oxide nanoparticles using highly cross-linked poly(curcumin) nanospheres: synergistically improved antimicrobial activity. Res Chem Intermed 45:4449–4462
177. Mathew S, Mathew J, Radhakrishnan EK (2019) Polyvinyl alcohol/silver nanocomposite films fabricated under the influence of solar radiation as effective antimicrobial food packaging material. J Polym Res 26:223–233
178. Malagurski I, Levic S, Nesic A, Mitric M, Pavlovic V, Dimitrijevic-Brankovic S (2017) Mineralized agar-based nanocomposite films: potential food packaging materials with antimicrobial properties. Carbohyd Polym 175:55–62
179. Ran J, Chen H, Bai X, Bi S, Jiang H, Cai G, Cheng D, Wang X (2019) Immobilizing CuO/BiVO4 nanocomposite on PDA-templated cotton fabric for visible light photocatalysis, antimicrobial activity and UV protection. Appl Surf Sci 493:1167–1176
180. Cho E, Kim SH, Kim M, Park JS, Lee SJ (2019) Super-hydrophobic and antimicrobial properties of Ag-PPFC nanocomposite thin films fabricated using a ternary carbon nanotube-Ag-PTFE composite sputtering target. Surf Coat Technol 370:18–23
181. Amjadi S, Emaminia S, Nazari M, Davudian SH, Roufegarinejad L, Hamishehkar H (2019) Application of reinforced ZnO nanoparticle-incorporated gelatin bionanocomposite film with chitosan nanofiber for packaging of chicken fillet and cheese as food models. Food Bioprocess Technol 12:1205–1219
182. Ibrahim S, Abbas HA, Sultan M, Aziz MA (2019) Preparation and characterization of nano-sized $Sr_{1-x}Ag_xTiO_3$ system as antimicrobial nanomaterial coating for paper base packaging materials. J Packag Technol Res 3:67–75
183. Morsi RE, Alsabagh AM, Nasr SA, Zaki MM (2017) Multifunctional nanocomposites of chitosan, silver nanoparticles, copper nanoparticles and carbon nanotubes for water treatment: antimicrobial characteristics. Int J Biol Macromol 97:264–269
184. Bogdanović U, Dimitrijević S, Škapin SD, Popović M, Rakočević Z, Leskovac A, Petrović S, Stoiljković M, Vodnik V (2018) Copper-polyaniline nanocomposite: role of physicochemical properties on the antimicrobial activity and genotoxicity evaluation. Mater Sci Eng, C 93:49–60
185. Abu-Dalo M, Jaradat A, Albiss BA, Al-Rawashdeh NA (2019) Green synthesis of TiO_2 NPs/pristine pomegranate peel extract nanocomposite and its antimicrobial activity for water disinfection. J Environ Chem Eng 7:103370–103383
186. Elella MHA, Mohamed RR, Abdel-Aziz MM, Sabaa MW (2018) Green synthesis of antimicrobial and antitumor N,N,N-trimethyl chitosan chloride/poly (acrylic acid)/silver nanocomposites. Int J Biol Macromol 111(2018):706–716

187. Boomi P, Poorani GP, Palanisamy S, Selvam S, Ramanathan G, Ravikumar S, Barabadi H, Prabu HG, Jeyakanthan J, Saravanan M (2019) Evaluation of antibacterial and anti-cancer potential of polyaniline-bimetal nanocomposites synthesized from chemical reduction method. J Clust Sci 30:715–726
188. Bui VKH, Park D, Nguyen MK, Moon JY, Lee YC (2019) Preparation of antimicrobial aminoclays/Poly(vinyl alcohol) nanocomposite hydrogel films by freeze-thawing method. Macromol Res 27(6):523–525
189. Alamgir M, Mallick A, Nayak GC, Tiwari SK (2019) Development of PMMA/TiO_2 nanocomposites as excellent dental materials. J Mech Sci Technol 33(10):1–6

Potential Application of Silver Nanocomposites for Antimicrobial Activity

Shagufta Haque, Mamatha Julappagari, and Chitta Ranjan Patra

Abbreviations

AFM	Atomic force microscopy
ATP	Adenosine triphosphate
BSA	Bovine serum albumin
CAGR	Compound annual growth rate
DLS	Dynamic light scattering
DSC	Differential scanning calorimetry
EDAX	Energy dispersive X-ray analysis
EPS	Extracellular polymeric substance
FESEM	Field emission scanning electron microscope
FTIR	Fourier transformation infrared spectroscopy
GSH	Glutathione peroxidase
HPLC	High performance liquid chromatography
ICP-OES	Inductively coupled plasma optical emission spectroscopy
MBC	Minimum bacterial concentration
MDR	Multidrug resistant
MIC	Minimum inhibitory concentration
PVA	Poly (vinyl alcohol)
PVP	Poly(N-vinyl-2-pyrrolidone)
ROS	Reactive oxygen species
RT-PCR	Reverse transcription polymerase chain reaction
TEM	Transmission electron microscope

S. Haque · M. Julappagari · C. R. Patra (✉)
Academy of Scientific and Innovative Research (AcSIR), Ghaziabad 201002, UP, India
e-mail: patra.chitta@gmail.com; crpatra@iict.res.in

M. Julappagari · C. R. Patra
Department of Applied Biology, CSIR-Indian Institute of Chemical Technology, Uppal Road, Tarnaka, Hyderabad 500007, Telangana, India

A. K. Nayak et al. (eds.), *Biomedical Composites*, Materials Horizons: From Nature to Nanomaterials, https://doi.org/10.1007/978-981-33-4753-3_5

TGA	Thermogravimetric analysis
XRD	X-ray diffraction analysis
XPS	X-ray photoelectron spectroscopy
WHO	World Health Organization

1 Introduction

The microbial infections have resulted as one of the foremost reasons of increased morbidity and mortality worldwide. The infections make the immune system of the individuals weak resulting in a comorbid situation [1–3]. Sometimes, these infections make the individual susceptible to cancer and other highly fatal diseases. The most common pathogens responsible for the infections include Gram-positive and Gram-negative bacteria, fungi, protozoans and viruses [4–7]. The global statistics indicate the loss of huge sum of monetary expenditure, time and health behind the microbial infections cases which affects millions of lives worldwide [8–10]. The global market value of the antibiotic treatments is accountable to billions in terms of USD [11]. The conventional therapies available for the treatment of microbial infections target them through different routes resulting in either killing or inhibiting the growth of the microbes [12, 13]. But there exist certain limitations of the conventional therapies including the antibiotic resistance, side effects arising from the antibiotics, the cost of production, etc. [14]. Thereafter, scientists have been exploiting various technologies for antimicrobial therapies. In this context, nanotechnology is in every field of biomedical applications that is helping mankind to lead a better life [15–17]. The nanomedicines are being used for the therapeutic and diagnostic purposes of several diseases including cancer, diabetes, antimicrobial agents, etc. [18]. Among the several applications, nanostructured composites are increasingly preferred as a long-term substitute for additives and antibiotics in different yield to confer antimicrobial effect [19–22]. Furthermore, nanotechnology plays a significant role in the development for the treatment of microbial infections.

In nanotechnology diverse nanocomposites are used to overcome microbial resistance by exhibiting a potential antimicrobial agent. The nanoparticles/nanocomposites (both organic and inorganic) act well against a broad range of bacterial strains. Among the different types of nanocomposites, scientists are focusing on inorganic silver metal nanocomposites for modern-day antimicrobial applications [22–25]. Silver nanocomposites have emerged as a potential candidate as antimicrobial agent owing to its versatile applications, efficiency and effectiveness to augment the outcome of conventional antibacterial agents toward multidrug resistance bacteria [19–21]. Therefore, the upcoming sections of the book chapter dwell into the microorganisms and their ways of antimicrobial activity along with the effect of silver nanocomposites in curbing them.

1.1 Pathogenic Strains for Infection

Microorganisms though are invisible to our naked eyes, but are one of the major reasons of woes and worries all over the world. Some microorganisms are also within us. Some are beneficial to us while a majority of them lead to morbidity and mortality [1–3]. The infections are generally caused by viruses, fungi, bacteria and parasites. The pathogens can be further classified as intracellular and extracellular depending on the immunopathological point of view [26]. Generally, Gram-negative bacteria are found to be pathogenic, harmful and hence some of the examples are described herewith. Among the Gram-negative bacteria of family enterobacteriaceae *Escherichia coli* and *Klebsiella pneumoniae* are responsible for causing pneumonia, infections of wounds, bloodstream and urinary tract [6]. *Neisseria gonorrhoeae* is another pathogen which causes sexually transmitted disease called gonorrhea. The symptoms of the disease include abdominal pain, pelvic inflammatory disease as well as infertility [27].

Again multidrug resistant (MDR) *Acinetobacter baumannii* causes infection through wound and bloodstream [28]. Another MDR bacterium includes *Pseudomonas aeruginosa* that induces skin rashes and skin infections. *Shigella* bacteria are found in the stool of the infected human beings and cause abdominal pain, diarrhea and fever. Infection can occur by consuming raw uncooked poultry or unpasteurized milk [29]. Another most common pathogen is the tuberculosis-causing bacteria *Mycobacterium tuberculosis* that mostly affects the lungs [30]. Other disease-causing pathogens include *Candida* which is a yeast belongs to fungus family. *Candida* leads to infections of mouth, throat oesophagus along with vagina [7]. Even there are reports of different other pathogenic strains of microbes responsible for microbial infections as published in some previously reported literatures. The readers can consult with the published reports to know more about microbial infections in detail [2, 4, 5].

1.2 Global Statistics and Market Value of Bacterial Infections

To evade microorganisms for leading a healthy life, a large amount of economy, labor and time is being spent for decades. In view of a report, infection and contamination caused by microorganisms have lead to mortality of nearly 3 million people in developing countries [22]. If one goes by the statistical analysis, antibiotic resistance is the reason for 25,000 deaths per year in European Union [8–10]. India is helmed to carry the largest trouble regarding drug resistant pathogens worldwide including the high cases of MDR tuberculosis [31, 32]. In India, with passing of bacterial resistant infection from mothers has lead to death of 58,000 babies [8–10]. It is estimated that nearly two million deaths are estimated to happen in India due to antimicrobial resistance by 2050 [32, 33]. The antibiotic resistance global market value was estimated at USD 7.81 billion in 2017. It is being anticipated to attain a CAGR of 5.6% from 2018 to 2025.

The increased resistance to antibiotic along with the rise of the multidrug resistance pathogens has led to the augmenting of the market. Various companies such as Melinta Therapeutics, Nabriva Therapeutics and Achaogen are in the process of developing new therapeutics for antibiotic resistance. The Gram-positive antibiotic-resistant bacteria is responsible for a vast amount of infections but the availability of large amount of treatment therapies has decreased the overall cost of treatment in comparison with Gram-negative bacteria [12]. The drug market includes lipoglycopeptides, oxazolidinones, tetracyclines, combination therapies and others. Among them it is estimated that lipoglycopeptides will occupy 17% of the overall revenue by 2025. On the other hand, the combination therapies comprising of fixed doses of combination of beta-lactamase and beta-lactams inhibitors are anticipated to have maximum growth within forecast period [34]. The revenue for the antibiotic resistance market has been received maximum from North America in 2017. The U.S. will continue to be a major contributor for the revenue during the forecast period. After U.S., Europe is the second largest contributor of revenue in this market by 2017, the reason being the increased number of antibiotic resistance infections [11].

1.3 Conventional Treatment Strategies for Bacterial Infections

The conventional treatment strategies available for the therapeutic usage of microbial infections are antibiotics that work against microbial infections. They work either through killing the bacteria or inhibiting the growth of bacteria [35]. Antibiotics are either produced naturally by a certain organism in response to another organism (such as penicillin) or synthetically (such as antiseptics and sulphonamides). There are varieties of antibiotics targeting the microbe through various pathways. Even the antibiotics are classified on the type of bacteria they are attacking such as targeting a specific group including the Gram-negative and Gram-positive or targeting a wide range of bacteria classified as broad-spectrum group [36]. Some of them target the bacterial cell wall namely cephalosporins and penicillin, other hinders the bacterial enzymes from working properly such as quinolones, sulphonamides, rifamycins [12, 13]. Also some antibiotics help in the inhibition of the protein synthesis such as lincosamides, macrolides and tetracyclines. These protein synthesis inhibitors are usually bacteriostatic in nature [37]. Along with these, there are certain antibiotics that are semisynthetic in origin such as beta-lactam antibiotics which are a combination of penicillin, carbapenems and cephalosporins. Although the antibiotics hold a major portion of the antimicrobial treatment, there lie major disadvantages in using them. From grave side effects to costly treatments antibiotics are helmed responsible. But the main disadvantage faced by the antibiotics is the growing resistance by the bacterial strains [14]. Thereafter to combat the limitations of the conventional treatment new alternatives are being checked out.

1.4 Limitations to Current Treatment Strategies

The bacteria is attaining resistance to antibiotics which pose as a major threat and initiate the emergence of multidrug resistance (MDR). These bacteria and superbacteria are resistance to all antibiotics owing to the unobstructed uses of antibiotics leading to genetic and morphological changes in bacteria overtime [38]. The main causes of resistance owe to the expression of efflux pumps, expression of some enzymes such as aminoglycosides and modification of cell components. There are even reports of a gene regarded as supergene NMD-1 (New Delhi metallo-β-lactamase) in superbacteria responsible for broad-spectrum antibiotic resistance [39]. Some bacterial species are reported to show microbial aggregation on solid surface by interaction with other bacteria and also secretes extracellular polymeric substance (EPS) on its outside. The bacterial aggregates on the EPS are regarded as biofilm. The bacteria within the mature biofilm produces the superantigens that attack the immune system and pose a serious threat to healthcare and biomedical areas [40–42]. Also depending upon the market, majority of the pharmaceutical companies are less active in the development of antibiotics. The reason for this is that the antibiotic resistance market offers slow growth in marketed drugs revenues which in turn account for low return on investments [43]. Hence a new technology is urgently required to overcome all these limitations. To this end nanotechnology may play significant role in combating bacterial infections and overcome the existing limitations of the conventional therapies.

1.5 Nanotechnology and Its Applications

With the growing backdrops of the conventional therapies toward the treatment of microbial infections, scientists are exploring nanotechnology as an alternative means to combat limitations. Nanoparticles/nanocomposites have the capability to enter into the bacterial cell causing interference in the molecular mechanistic pathways thus exhibiting antimicrobial actions [17, 22–25]. They have the tendency to overcome the commonly faced limitations from antibiotic resistance mainly inactivation of enzymes, changes in the target site followed by more efflux of the efflux pumps [22]. The characteristic features of the nanocomposites also play a major role in imparting the antimicrobial activity such as increased surface to volume ratio.

In nanotechnology different types of nanoparticles are exploited for treating bacterial infections. Among various nanoparticles, metal-based nanocomposites involving gold, silver, zinc copper, magnesium, nickel, selenium, aluminium, palladium and iron are reported to exhibit antibacterial activities as per published literatures [23, 25, 44–46]. For example, gold platinum bimetallic nanocomposites exhibit antibacterial actions against *E. coli* which is a multidrug resistant bacteria. The nanoparticles cause bacterial membrane potential dissipation along with increase in adenosine triphosphate (ATP) levels [23]. In another report, iron oxide nanoparticles have been found to cause damage to the bacterial cell by increasing oxidative stress [22]. Even there are reports of copper nanoparticles exhibiting bactericidal and fungicidal activity through either penetrating the bacterial cell wall and damaging the nucleic acid or directly binding with helical structure of DNA and destroying it [46, 47]. Table 1 depicts the list of various metal nanocomposites and their role as antibacterial agents.

TiO_2 nanoparticles also exhibit antimicrobial activity through ROS generation mainly through OH free radicals formation as shown in Table 1 [22]. Among all these metals, silver nanoparticles as well as nanocomposites have drawn much attention because of several advantages such as easy synthesis procedures, unique properties, characterization and biological property such as antimicrobial activity [16, 19–21]. Silver nanocomposites have been found to be highly toxic toward microbes hence provides efficient antimicrobial activity. The upcoming sections focus on various aspects of silver toward treating microbial infections.

2 Historical Background of Silver

If one traces back to the old time, before antibiotics, silver is utilized in various applications. Silver has been used for making vessels, liquid container for water storage, silver coins, sutures, shaving and lastly medicinal requirements that was first documented in 750 AD [48]. Silver was being used for treating certain diseases and ailments even before its discoveries having antibacterial properties. It has been recorded that several civilizations of Phoenicians, primordial Romans, Egyptians and Greeks used silver for food preservation and in water [49]. It was reported that Macedonians and Hippocrates were the first to use silver as wound healing agent. Gabor in 702–705 AD used silver as silver nitrate. There are reports of Avicenna in 980 AD and Angelo Sala in 1614 utilized silver for the purpose of blood purification, wound healing and other infections [50, 51]. Silver was earlier used in the forms of silver zeolite and silver sulfadiazine in clinical applications such as wound healing [52, 53].

Later the silver was converted to silver nanoparticles and silver nanocomposites with the advent of nanotechnology for several biomedical applications. With the formation of the nanoform of silver, it exhibits different properties like increased surface area to volume ratio leading to elevated bioavailability allowing efficient antimicrobial activity with respect to bulk material [54]. The nanosilver at same molar concentration shows less toxicity in comparison to its ionic form. The antimicrobial action of the silver nanocomposites largely depends on their physicochemical features including shape, size, charge and stability. The features in turn are formed, based upon the process of synthesis conditions such as temperature, nature of solvent, concentration and strength of reducing agent [55, 56].

If one goes by the commercialization of the nanoparticles, then silver in the structure of nanoforms occupies 50% of the worldwide nanomaterial end user products. The annual manufacture market estimation of silver nanoforms accounts for 320 tons of manufacture that is being utilized for healthcare along with food products [57].

Table 1 Role of different metal nanocomposites as antibacterial agents

Nanoparticle	Conjugation	Targeted microbes	Mode of action	References
Gold	Cefotaxime	*Klebsiella pneumoniae, Escherichia coli*	Damage to the cell wall of bacteria followed by DNA disruption	[123]
	Carbapenems (meropenem and imipenem)	*Klebsiella pneumoniae, Acinteobacter baumannii* and *Proteus mirabilis*	Changing the osmotic balance leading to the damage of the cell integrity as well as cell wall	[124]
Aluminium Oxide (Al_2O_3)	–	*Staphyloccocus aureus*	Generation of ROS that damages the bacterial cell membrane	[125]
	–	*Escherichia coli*	Deposition and penetration within the cell wall of bacteria	[126]
Titanium Oxide	–	*Escherichia coli*	Peroxidation and accumulation of the fatty acids of the bacterial cell membrane	[127]
	–	*Escherichia coli*	Production of ROS that disrupts the bacterial cell membrane	[128]
Iron Oxide	–	*Escherichia coli*	Damaging the cell membrane of bacteria with the help of magnetic core shell nanoparticles coupled with radiofrequency	[129]
	–	*Staphyloccocus aureus, Escherichia coli* and *Pseudomonas aeruginosa*	Pentration through the bacterial membrane and interference with the electron transfer	[130]
Zinc Oxide	PVA	*Escherichia coli*	Generation of ROS and damage to bacterial cell membrane	[45]

(continued)

Table 1 (continued)

Nanoparticle	Conjugation	Targeted microbes	Mode of action	References
	–	*Bacillus subtilis*	Inhibition of bacterial cell growth and biofilm formation	[131]
Copper	–	*Pseudomonas aeruginosa*	Penetration through the cell wall followed by destruction of nucleic acid	[46]
	–	*Staphyloccocus aureus*	Release of Cu ions that binds with the helical structure of DNA and disrupts them	[132]
Silver	–	*Escherichia coli, Bacillus subtilis, Klebsiella aerogenes, Pseudomonas aeruginosa*	Bacterial cell membrane damage	[24, 25]

It has been even assumed that the global market will exceed 3 billion by the year 2020. The demand for the silver nanocomposites is rising in a variety of commercial markets of textile, paint industry, food packaging and water treatment all owing to its antibacterial and antifungal properties [13, 58, 59].

3 Synthesis and Characterization of Silver Nanocomposites

The designing, synthesis and assembly of silver nanocomposites occurs through different processes but is majorly classified into two categories: bottom-up and top-down. In this the bottom-up approach involves silver nanocomposites production using photochemical, chemical and biological reactions. The top-down approach on the other hand employs breakdown of solid materials into nanocomposites [60]. Among the different synthesis techniques employed for silver nanocomposites, there are mainly three methods that are: chemical, physical and biological synthesis [25, 61–63]. The synthesis procedures play major role behind imparting antibacterial property to the silver nanocomposites [64]. Although the main objective of this book chapter is to demonstrate the potential antimicrobial activity of the silver nanocomposites hence it is beyond the scope of the chapter in indulging about the synthesis and characterization. Thereafter, we very briefly describe about the synthesis and characterization of silver nanocomposites in this section.

The fabrication of silver nanocomposites by chemical methods involves the reduction process, thermal reduction, microwave assisted synthesis and polyol process. Among the other types include electrochemical synthetic method, microemulsion, photochemical reduction methods and irradiation method [65–68]. Various scientists from all over the world are involved in the process of synthesis of silver nanocomposites by different chemical methods. To this context, our group synthesized various silver nanocomposites including silver nitroprusside (AgNNPs) and silver Prussian blue analogue nanocomplex using chemical methods (SPBANPs) [24, 25]. The formed silver nanocomposites were characterized with the help of several characterization techniques such as DLS, XRD, TEM and FESEM. The overall characteristics exhibit the crystalline nature of the nanocomposites [24, 25]. The significance of the chemical methods is that it needs decreased reaction time, increased yield of product with lesser energy consumption.

Another method of productions of silver nanocomposites are physical methods [61]. The various physical methods include evaporation and condensation, laser ablation, ac power, thermal, arc discharge, etc. [62, 69, 70]. These methods are used because they prevent any sort of contamination occurring from the solvent. The physical methods help in formation of size of silver nanocomposites in the narrow range [64].

Apart from the chemical and physical synthesis methods, biosynthesis is another procedure for silver nanoparticles/nanocomposites formation. The biosynthesis process is an economic, simple and non-toxic method. The extracts of plant, microbial (such as bacteria, fungi, algae) agents are attributed to the biosynthesis process [71–73]. The chemical and physical methods have certain disadvantages such as presence of toxic precursors or use of large amount of energy, which can be easily overcome by biosynthesis measures. There are several reports on biosynthesis of silver nanocomposites using plant and microbial extracts as bioresource that acts as reducing agent [19]. Many scientists all over the world have biosynthesized silver nanocomposites. The phytochemicals present within the leaf extract helped in the reduction of the silver into its nanoforms. Our group synthesized silver nanocomposites with the help of *Olax scandens* plant extracts by the reduction of silver nitrate [74]. Similarly, Tripathi et al. [63] biosynthesized poly (vinyl alcohol)-silver nanocomposites by using *Ficus benghalensis* leaf extract [63].

4 Role of Shape, Size and Charge of Silver Nanocomposites on Antimicrobial Activity

The silver in its nanoform exhibits unique features. There exists a strong connection between the physical characteristics and the morphological properties (shape, size, charge and surface chemistry) of the silver nanocomposites [18]. The morphological features of the silver nanocomposites are responsible for most of the biological activities of the silver nanocomposites including cellular activation, cellular uptake, intracellular distribution as well as antibacterial activity [21]. The morphological features in turn are dependent upon certain criteria including synthesis procedures, reducing and stabilizing agents as described in the above Sect. 3. The production of nanoforms takes place through three defining steps namely, nucleation, coalescence and growth [75]. The following section briefly describes the different morphological features involved in imparting antimicrobial features to the silver nanocomposites.

The shape of the nanoparticles and nanocomposites (for example nanocubes, triangular, spherical, pyramidal, nanorods, nanoprism, nanobars, flower shaped, nanowires, octahedral as well as tetrahedral) plays a major role in determining their antibacterial property [76, 77]. In this context Logaranjan et al. [78] designed silver nanocomposites that were octahedral in shape [78]. The nanocomplex was synthesized using the extracts of aloe vera plant that acted as reducing agent in both conditions of microwave irradiation and room temperature. The range of the particles was from 5 to 50 nm and exhibited excellent antimicrobial effect against various strains of bacteria. The results revealed that the octahedral-shaped particles exhibited more antibacterial activity in comparison with other nanocomposites [78]. In another report Chakraborty et al. [79] synthesized silver nanocomposites with different shapes by employing various proteins [79]. The protein used during synthesis acts as capping agent. The different types of proteins used are ovalbumin, BSA, catalase, hemoglobin, avidin, apo-transferrin, lysozyme and β-lactoglobulin. The silver was reduced to silver cluster with the help of citric acid followed by mixing with various proteins. Also, the silver nanocomposites were synthesized with individual amino acids in order to check the contribution of each amino acid in determining the shape of the silver nanocomposites. The results depicted that the cystine and lysine formed spherical-shaped nanocomposites and the other amino acids glycine, alanine, valine, threonine, isoleucine and serine produced anisotropic nanocomposites. The results also exhibited that only lysine helped in formation of stable particles [79].

The size of the silver nanocomposites plays a major role in imparting antibacterial property. The smaller the size of the nanocomposites the greater is the surface to volume ratio resulting in increased interaction with the microorganisms exhibiting more antibacterial activity. To exhibit the size-dependent antimicrobial activity of silver nanoparticles Dogru et al. [80] synthesized silver nanocomposites (sizes ranging from 70 ± 5, 52 ± 5, and 37 ± 5 nm) with the extracts of *Matricaria chamomilla* [80]. The silver nanocomposites exhibited efficient antibacterial activity against *E. coli* and *S. aureus*. All the three size range particles exhibited antibacterial effect with 37 ± 5 nm particle having the maximum one. This corroborates with the fact that the small size of the nanocomposites has greater surface volume ratio and hence interacts more efficiently with the bacteria [80].

Similar to shape and size of the nanoparticles, charge helps in inferring certain characteristic features to the nanocomposites. The charge of the silver nanocomposites is usually dependent upon the process of synthesis. The capping and the stabilizing agents used during synthesis affects the surface charge of the nanocomposites which in turn confers stability and bioactivity to the nanoparticles. In this context Salvioni et al. [81] synthesized silver nanocomposites using tannic and citric acid as reducing agents [81]. The formed particle was characterized by several characterization techniques which revealed 20 nm size along with negative charge of the formed nanocomposites. The silver nanocomposites exhibited good antibacterial activity against Gram-negative and Gram-positive bacteria in comparison with commercially available silver nanocomposites [81].

5 Silver Nanocomposites and Their Antimicrobial Activity

The silver nanocomposites are produced by the combination of various sources and precursors. They either act as matrix bed or reducing agents for silver nanocomposites formation and some compounds enhance the antimicrobial property of silver nanoforms through synergistic effect [20, 82].

5.1 Graphene Silver Nanocomposites

There are many reports of graphene being used as an antibacterial agent. Thereafter this property of graphene was used in synthesizing graphene associated silver nanocomposites to impart a synergistic antibacterial effect. In one such report Beiranvand et al. [20] produced graphene oxide/hydroxyapatite/silver (rGO/HAP/Ag) nanocomposite with the help of graphene oxide nanosheets, $(NH_4)_2HPO_4$, $Ca(NO_3)_2$ and $AgNO_3$ using a hydrothermal method [20]. The nanocomposites were characterized by various analytical tools. The results revealed the formation of graphene-based silver nanocomposite. The nanocomposite was checked for its biological activity which clearly exhibited antibacterial activity through disk diffusion method against Gram-positive and Gram-negative bacteria. The results were affirmative toward Gram-positive bacteria [20]. In another report Song et al. [83] synthesized graphene oxide-based silver nanocomposites [83]. The nanocomposites with size of 5–15 nm were characterized by several characterization techniques. The graphene oxide-silver nanocomposites were evaluated for bactericidal activity in water through different assays such as SEM analysis, bactericidal dosage, effect of pH and lipid peroxidation assay using strains of *Staphylococcus aureus* and *Escherichia coli*. The bacterial dosage study depicted that the nanocomposites exhibited dose-dependent and time-dependent antibacterial activity against both the strains. Time of 20–30 min is required for 99.99% killing activity of both the strains. Also the silver could permeate easily in Gram-negative *E. coli* in comparison to Gram-positive *S. aureus* due to the presence of thicker cell wall of the latter. The nanocomposites worked better in showing antibacterial activity in acidic pH as compared to basic one. The SEM results showed the damage of the cell wall membrane of both the bacterial strains with *E. coli* being more. The antioxidant enzyme measurement study revealed the increased oxidative stress followed by cell membrane damage is the mechanism behind the antibacterial activity of the graphene silver nanocomposites. Figure 1 shows the mitochondria through its electron transport chain generates ROS causing damage to

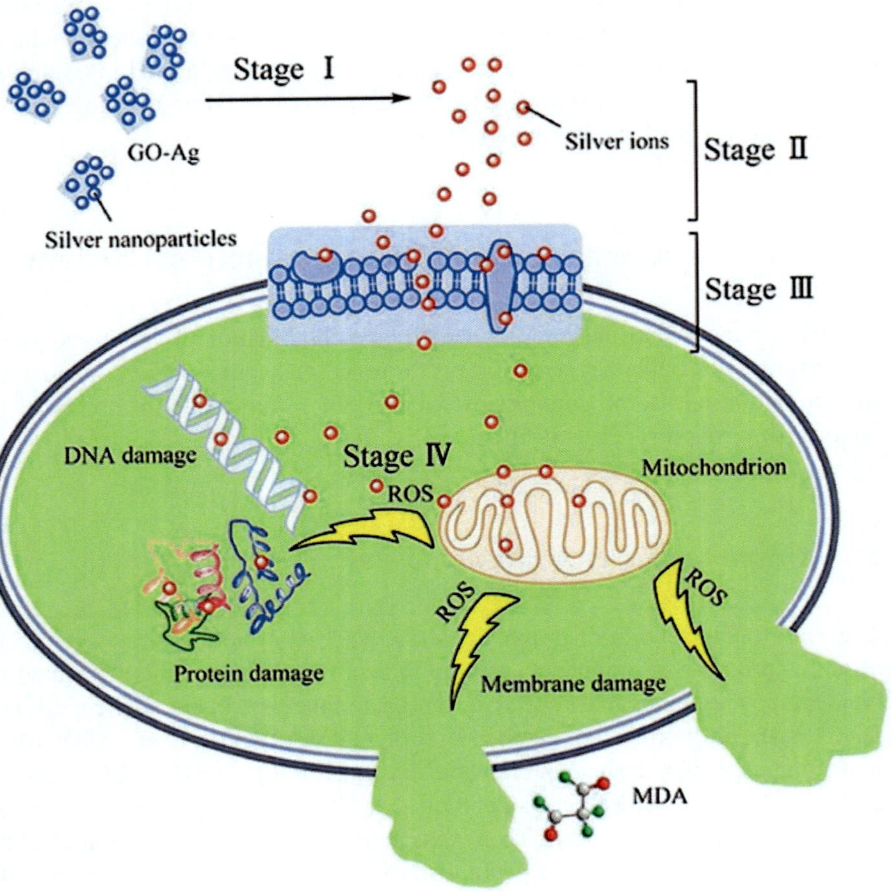

Fig. 1 Schematic mechanisms for antibacterial behaviors of GO-Ag. Blue dots represent silver nanoparticles loaded on GO sheets, and red dots are silver ions. The roman numerals (I, II, III and IV) refer to different stages of the bactericidal process [92]. Reprinted with permission from Song et al. [83]. Copyright © 2016 **Elsevier**

lipids, proteins and DNA. This in turn leads to damage of cellular membrane with the release of cell contents. The oxidation of the biomolecules paves way for the destruction of the bacterial cell [83]. Similarly Jaworski et al. [84] coated graphene oxide with silver nanoparticles leading to the formation of silver nanocomposites [84]. The nanocomposites were checked for the biological activity through presto blue assay, ROS analysis and lactate dehydrogenase assay against four bacterial strains as well as pathogenic yeast. The results exhibited excellent antimicrobial activity against both bacterial and fungal strains [84].

5.2 *Chitosan-Based Silver Nanocomposites*

Chitosan is another such component used for the production of silver nanocomposites acting as a reducing agent for the reduction of silver metal into its nanoform [82]. Vimala et al. [85] designed curcumin encapsulated chitosan-polyvinyl alcohol (PVA)-silver nanocomposites through their reduction in acidic medium [85]. The formation of the silver ions into its respective nanocomposites in the acidic solution was due to the reducing ability of polyvinyl alcohol (PVA) and chitosan (C) which contained functional groups such as –OH, $-NH_2$ and –COOH. The formed nanocomposites were characterized with the help of several characterization techniques. The visualization of the SEM images indicated the presence of silver nanocomposites embedded within the chitosan and polyvinyl alcohol (PVA) matrix. The size of the nanocomposites was around 16.5 nm as indicated by TEM. The effect of the nanocomposites on the bacterial strains (*Escherichia coli*, *Staphylococcus*, *Pseudomonas*, *Micrococcus*, *Pseudomonas aeruginosa*) as well on the fungal strain (*Candida albicans*) was checked with the help of various assays. The results clearly depicted the antibacterial and antifungal activity of the chitosan–curcumin-based silver nancomposites being more than the individual components [85]. In another work, Kaur et al. [86] synthesized silver nanocomposites with the help of chitosan [86]. The nanocomposites were established to have characteristic features through the characterization techniques such as TEM, FTIR, DSC and XRD. The formed nanocomposites showed excellent antibacterial activity against *Staphylococcus aureus* MTCC 1809, *Salmonella entrica* MTCC 1253 and *Pseudomonas aeruginosa* MTCC 424 through disk diffusion methods [86].

5.3 *Nitroprusside-Based Silver Nanocomposites*

Sodium nitroprusside is the class of drug used for the treatment of high blood pressure. Being FDA approved the sodium nitroprusside is used as precursor molecules for the synthesis of several nanocomposites. Our group synthesized silver nanocomposites with the help of sodium nitroprusside. The nanocomposite produced is silver nitroprusside nanocomplex using silver nitrate and sodium nitroprusside [24]. The nitroprusside helps in providing extra stability to the silver and results in its slow release from the complex. The formed nanocomposites were characterized by the help of DLS, XRD, TGA, FTIR, HRTEM, Raman, ICP-OES and FESEM. The characterization results revealed the optimized features of the nanocomposites such as 20–80 nm of size in HRTEM and XRD showing the characteristic peaks of silver as crystalline nature. The nanocomposites were initially evaluated for its biocompatibility in normal cell lines (EA.hy 926 cells, NIH 3T3) and also through chick embryonic assay. Both the results confirmed the biocompatible nature of the formed silver nanocomposites. They were now checked for the antibacterial activity with the help of different methods including zone of inhibition assay, colony formation assay, bacterial cell damage assay and protein expression assay on *Escherichia coli*, *Staphylococcus epidermidis*, *Bacillus subtilis*, *Klebsiella aerogenes*, *Pseudomonas aeruginosa* bacteria. The results showed excellent antibacterial activity through all the assays. Figure 2 illustrated that on treating the bacteria at log phase with the silver nanocomposites at a concentration of 0.5–2 μg/mL for a period of 24 h showed excellent antibacterial activity in comparison to untreated and positive control (commercially available streptomycin). The bacterial colony was inhibited by the treatment of silver nanocomposites as compared to other groups. Figure 3 shows that when the bacterial cells are damaged they have a tendency to join with one another. The confocal images clearly exhibit that the cell membrane adheres to one another. The SEM images depict the damage of the silver nanocomposites treated bacterial cells. The silver nanocomposites so formed were further reported to have shown accelerated wound healing activity in mice due to macrophage conversion [24].

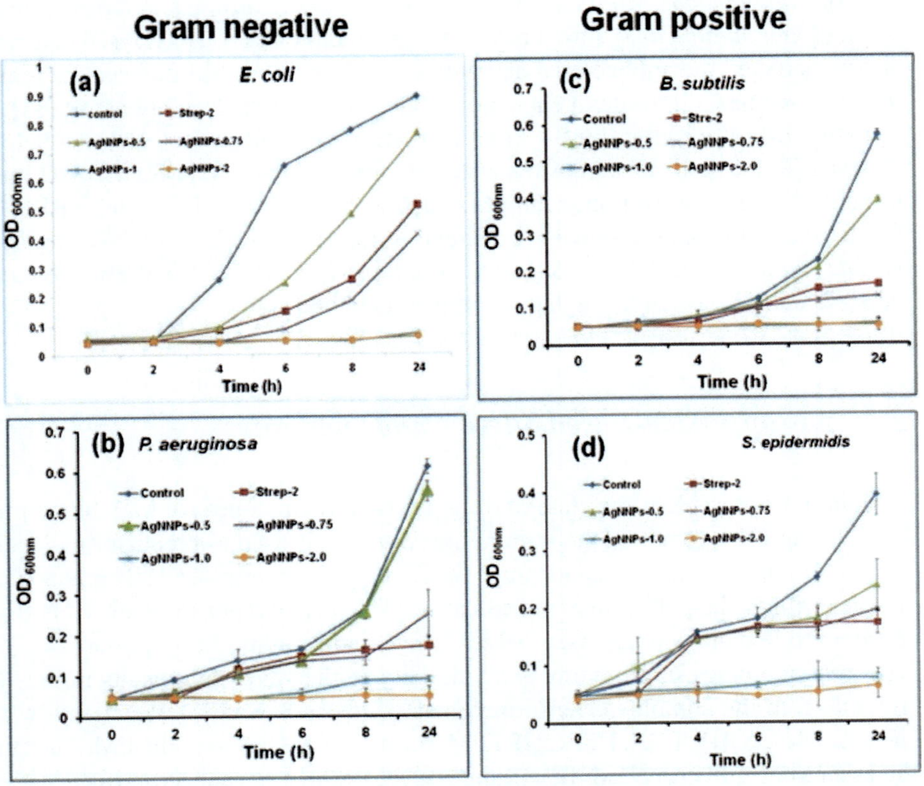

Fig. 2 a–d Calculation of MIC of AgNNPs (0.5–2 μg/mL) on both Gram-positive and Gram-negative bacteria. The bacteria at various time points by measuring OD at 600 nm. Here streptomycin (2 μg/mL) used as positive control [54]. Reprinted with permission from Rao et al. [24]. Copyright © 2018 **American Chemical Society**

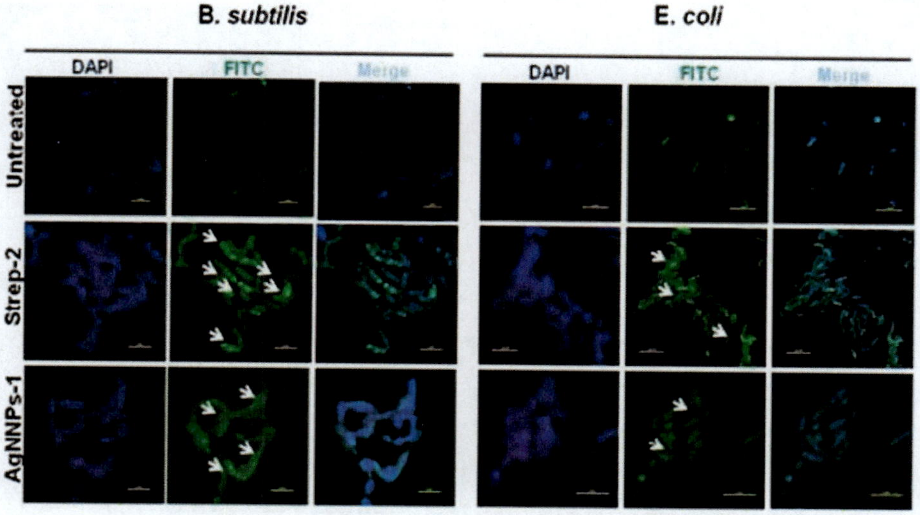

Fig. 3 Confocal microscopic study of bacterial membrane damage. *E. coli* and *B. subtilis* treated with AgNNPs (1 μg/mL), streptomycin (1 μg/mL) for 6 h. AgNNPs treated *E. coli* and *B. subtilis* displayed enhanced membrane damage causing enhanced green fluorescence (FITC:100 μg/mL). Nuclear staining was performed with DAPI (100 μg/mL). First row: untreated control, second row: streptomycin (1 μg/mL) treated bacteria, third row: AgNNPs (1 μg/mL) treated bacteria. Cells were observed under 100X objective, scale bar: 5 μm. The arrows in the corresponding figures represent membrane damage of bacteria [54]. Reprinted with permission from Rao et al. [24]. Copyright © 2018 **American Chemical Society**

5.4 Prussian Blue-Based Silver Nanocomposites

Prussian blue (ferric hexacyanoferrate) is generally used as an oral antidote, also known as Radiogardase for the treatment of poisoning due to radioactive or non-radioactive thallium and radioactive cesium [87]. Being FDA approved it has been utilized for the synthesis of various compounds. Several reports have shown that the Prussian blue and its analogues have been used for the anticancer activity [88]. The Prussian blue prevents the leakage as well as helps in better circulation of silver and forms a stable complex. Therefore, our group used Prussian blue for the synthesis of silver Prussian blue nanocomposites that exhibited excellent antibacterial activity. The silver nanocomposites were designed with silver nitrate and Prussian blue in the presence of PVP [poly(N-vinyl-2-pyrrolidone)] [25]. The process of synthesis is simple, economical and environment friendly. The formed nanocomposites were characterized by different techniques such as UV–visible spectroscopy, TEM, DLS and XRD. The results revealed 2–40 nm size of the nanocomposites through TEM followed by the XRD peaks matching with the earlier published silver nanoparticles. The nanocomposites revealed antibacterial property through various assays such as growth inhibition, zone of inhibition and colony forming assay on both Gram-positive and Gram-negative strains of bacteria. Figure 4a, b depicts the more inhibition of growth of the bacterial strains upon treatment with the 30 μM

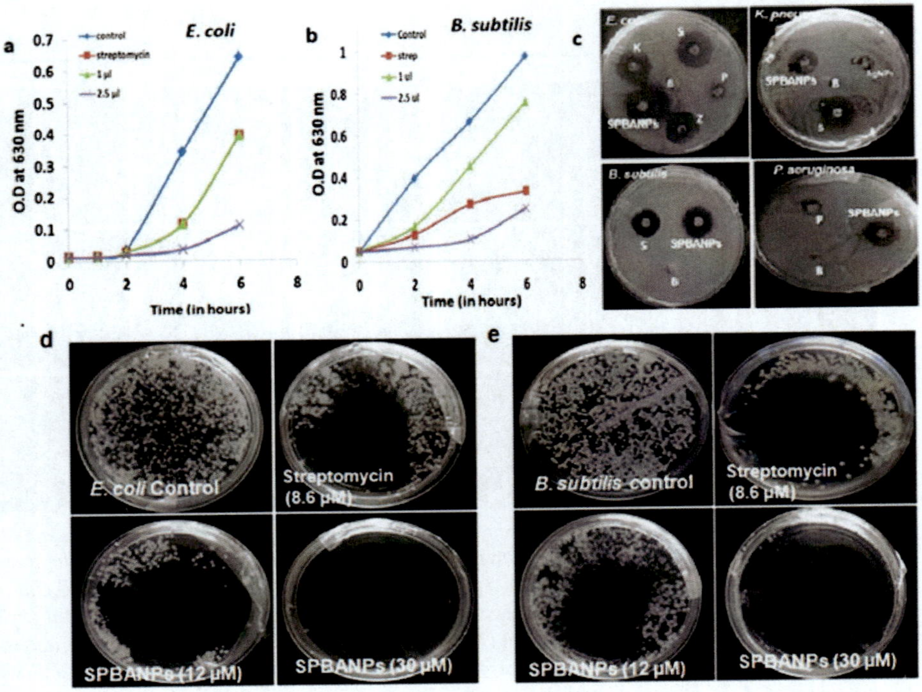

Fig. 4 **a–e**: **a**, **b** Growth inhibition of *E. coli* and *B. subtilis* in presence of SPBANPs. **c** Zone of inhibition of various bacteria treated with SPB(1,1)ANPs-40 and standard antibiotics (1 mg/mL) (P: penicillin; K: kanamycin; S: streptomycin and Z: gentamicin: 0.5 mg/mL. B: blank) in agar plate. **d**, **e** Optical images of bacterial colony formed by *E. coli* and *B. subtilis* after 24 h treatment with SPBANPs [42]. Reprinted with permission from Mukherjee et al. [25]. Copyright © 2020 **American Chemical Society**

concentration of nanocomposites with respect to the control group. Figure 4c shows the results of the zone of inhibition assay where the nanocomposites helped in maximum zone inhibition in comparison with the untreated and four positive controls (treated with kanamycin, gentamycin, streptomycin, penicillin). The colony formation is visible in Fig. 4d, e where the nanocomposites treated bacterial colony of both the strains exhibited maximum inhibition in comparison with untreated and positive control. Figure 5 shows the TEM images of the bacterial cell which upon treatment with the silver nanocomposites shows extensive membrane damage as compared to the untreated ones. Thus the mechanism underlying the antibacterial activity of the silver nanocomposites includes interaction of the bacterial cell with the silver nanocomposites leading to damage of their membrane and leaking of the cellular contents. Also the silver nanocomposites increased the ROS generation resulting in the suppression of the antioxidant enzymes. Oxidative stress generated by the silver Prussian blue analogue complex causes lipid peroxidation leading to plasma membrane damage [25].

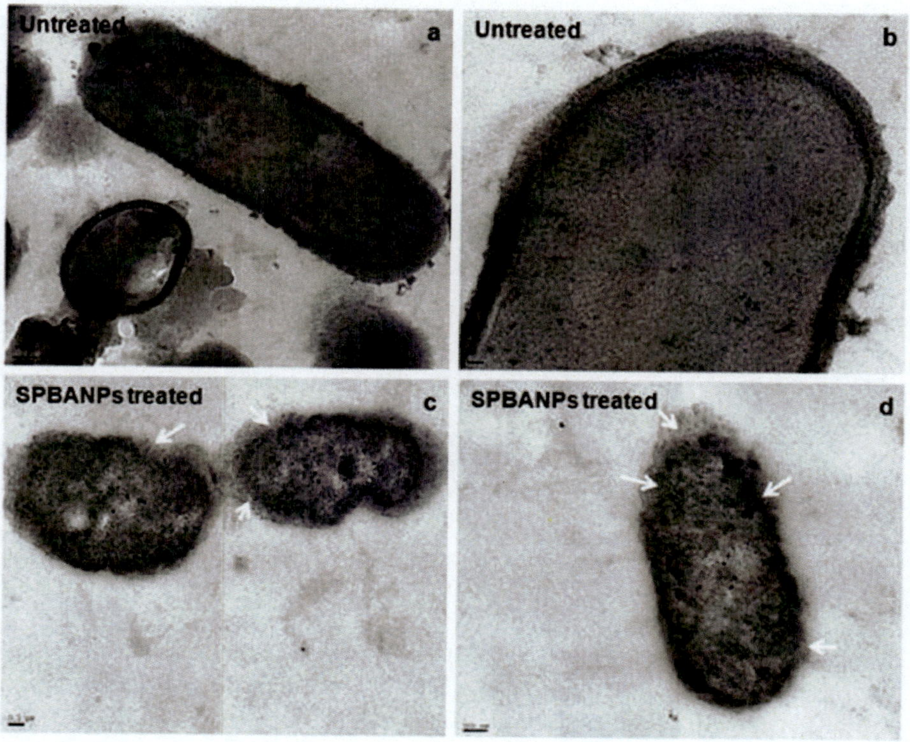

Fig. 5 **a–d**: TEM images of **a**, **b** untreated *E. coli* and **c**, **d** *E. coli* treated with SPBANPs [42]. Reprinted with permission from Mukherjee et al. [25]. Copyright © 2020 **American Chemical Society**

5.5 *Cellulose-Based Silver Nanocomposites*

It has been well established that cellulose polymers are used for the synthesis of silver nanocomposites [61]. The cellulose usually acts in the form of matrix for the formation of silver nanocomposites [89]. To this regards, Li et al. [90] synthesized silver nanocomposites with the help of $AgNO_3$ and cellulose fibers by heating at a temperature of 80 °C for 24 h [90]. The formed nanocomposites were checked for its characteristics features through several characterization tools. Later the antibacterial studies were performed on *Staphylococcus aureus* strain. Figure 6 clearly depicted the enhanced antibacterial property of the silver nanocomposites as compared to the untreated ones. The antibacterial activity augmented with the increase in the concentration of the silver nanocomposites and size of the fibers. The mechanism behind the antibacterial activity of the silver nanocomposites is damage to the bacterial cell wall [90]. In another work Pinto et al. [91] synthesized silver nanocomposites by using different cellulosic substrates (bacterial and vegetables) [91]. The silver nanocomposites exhibited excellent antibacterial activity against three different strains of bacteria.

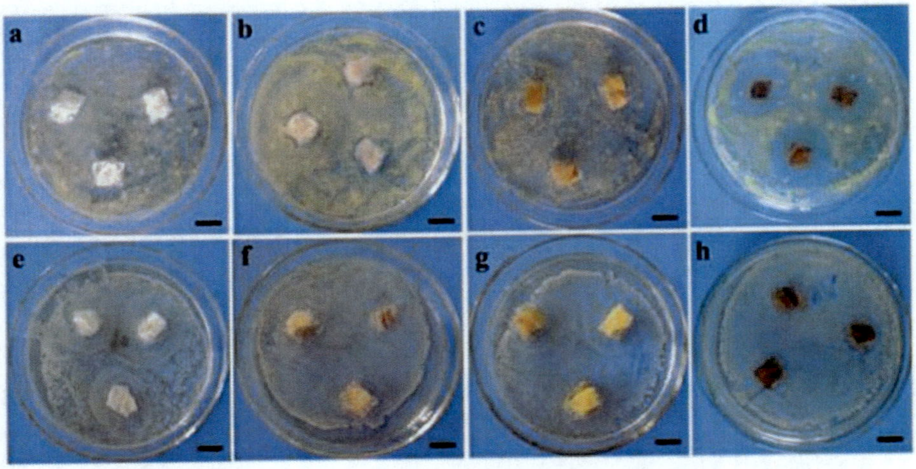

Fig. 6 Antibacterial test of cellulose/silver composite fibers. **a, e**-B1, **b, f**-B3, **c, g**-B4, **d, h**-B5. **a–d**: *Staphyoccocus aureus*; **e–h**: *Echerichia coli*. The scale bar is 1 cm [100]. Reprinted with permission from Li et al. [90]. Copyright © 2014 **Elsevier**

The nanocomposites were effective in showing the antibacterial activity at a concentration even low as 5.0×10^{-4} wt.%. Thus this cellulosic silver nanocomposites exhibits an effective potential antibacterial agent [91].

5.6 *Synthetic Polymer-Based Silver Nanocomposites*

The synthetic polymers are used for wide range of biomedical applications including synthesis of nanocomposites. Many reports suggest silver nanocomposites are produced with the help of synthetic polymers that either acted as reducing agents or matrix bed [89]. To this Palomba et al. [92] used synthetic polymer polystyrene to produce silver polystyrene nanocomposites [92]. The synthesis took place by mixing silver 1,5-cyclooctadiene-hexafluoroacetylacetonate and amorphous polystyrene resulting in a solid solution. During the reaction the silver atoms are decomposed from its initial reactants followed by accumulation in the polymer. The polymer-based silver nanocomposites were checked for its antibacterial activity through several assays such as BacLight fluorescence-based test, turbidity assay on Gram-negative *Escherichia coli* bacteria. The nanocomposites were used in two concentrations (10 and 30% wt) in correspondence with the pure form of silver. Figure 7 depicts the results of BacLight fluorescence-based test where in Fig. 7a shows the green fluorescence of the entire bacterial culture indicating live cell population after 16 h of incubation in liquid medium for the growth of bacteria. Figure 7c, d show the increased red fluorescence indicating dead bacteria upon treatment with silver nanocomposites of 30% wt with mild and strong annealing temperature, respectively. The red fluorescence is due to labeling of propidium iodide with the membrane compromised

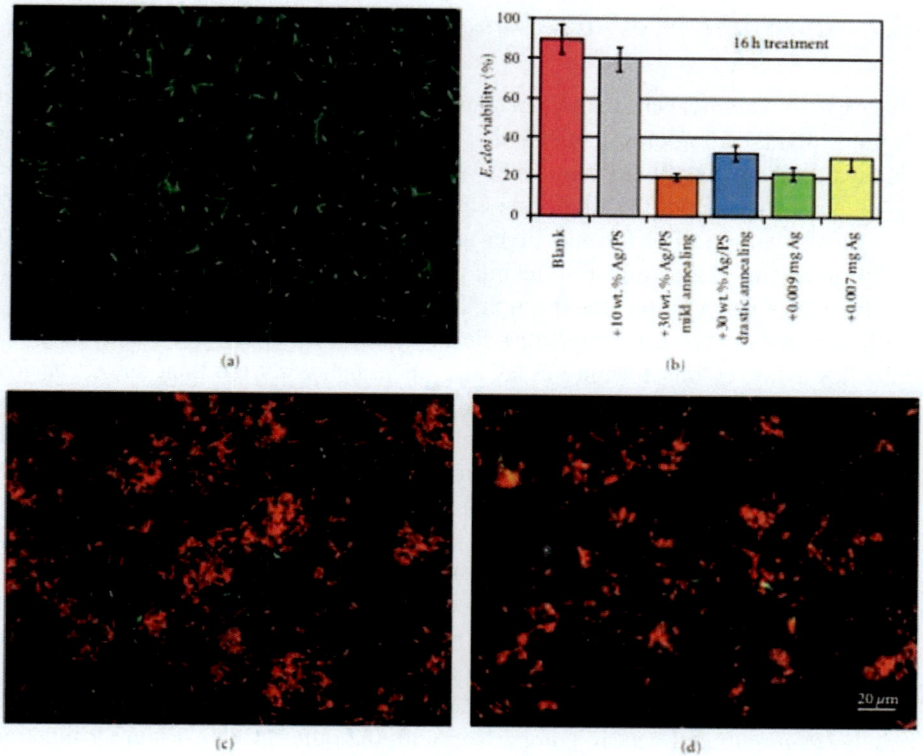

Fig. 7 Live-dead cell staining assay: microscopic analysis and quantitative evaluation. **a** High number of green-labeled living bacteria after 16 h incubation in LB, without Ag/PS. **c**, **d**: high number of red-labeled dead bacteria after 16 h incubation +30 wt.% Ag/PS film (1 cm × 1 cm—0.0177 mg) obtained by using a mild thermal annealing treatment (**c**); similar results were observed in bacteria culture after 16 h of incubation with Ag powder (**d**). Scale bar represents 20 μm. **b** Quantitative analysis of the *E. coli* viability (%) after 16 h incubation at each condition reported in Table 1 (**a**–**e**) [101]. Reprinted with permission from Palomba et al. [92]. Copyright © 2012 **Palomba et al**.

dead bacteria. The results were then quantitatively analyzed as shown in Fig. 7b. The quantitative results revealed decrease in the percentage viability of the *Escherichia coli* is more with the treated samples. The results revealed that the 30% wt of the nanocomposites exhibited the maximum antibacterial activity in comparison with the pure silver [92]. Another polymer-based silver nanocomposite was synthesized by Jabbar et al. [93] where he used silver nanoparticles biosynthesized from *Parthenium* leaf extract doped over polystyrene dissolved in Tetrahydrofuran [93]. The formed nanocomposites hold potential as an antibacterial agent since silver itself imparts antibacterial properties [93].

5.7 Biosynthesized Silver Nanocomposites

Apart from chemical means, silver nanocomposites are also synthesized with the help of plants and microorganisms that impart antimicrobial properties. Our group biosynthesized silver nanocomposites with the help of *Olax scandens* plant extracts [74]. The reduction of the silver ions into its nanoform is achieved with the help of the plants extracts. The formed silver nanocomposites were characterized with the help of several characterization techniques. The results revealed the 2–40 nm size of the silver nanocomposites through TEM images and the XRD shows the crystalline nature of the nanocomposites. Several antibacterial assays such as the growth kinetics assay, colony forming assay and SEM analysis were done on *Escherichia coli* bacterial samples. Figure 8 shows the results of antibacterial studies performed with the help of the biosynthesized silver nanocomposites. Figure 8a shows the inhibition kinetics in graphical manner where the inhibition increased with the dose of the nanocomposites as compared to control and positive control ampicillin. The colony forming assay has been depicted in Fig. 8b–e with control, ampicillin and biosynthesized silver nanocomposites at concentrations of 18 μM and 30 μM, respectively. The 30 μM concentration of the nanocomposites exhibited highest antibacterial activity as compared to the other groups. The SEM images of the bacterial cell are depicted in Fig. 8f–h as control, *Olax* and silver nanocomposites, respectively. Even the SEM images clearly depict damage of the biosynthesized silver nanocomposites treated *Escherichia coli* cell wall in comparison with the other groups. Figure 9 illustrates the overall working antibacterial mechanism of silver nanocomposites. The silver ions released from the nanocomposites result in oxidation of the glutathione which in turn leads to generation of ROS causing toxicity and inhibiting bacterial growth. The glutathione oxidation by silver nanocomposites causes lipid peroxidation that results in membrane damage followed by leakage of cellular components. The silver nanocomposites also disrupt the DNA (both plasmid and chromosomal) by the interaction with the phosphorus groups. Overall the biosynthesized silver nanocomposites act as an excellent antibacterial agent [74].

In another report, Tripathi et al. [63] designed and biosynthesized poly (vinyl alcohol)-silver nanocomposites with the help of *Ficus benghalensis* leaf extract [63]. Initially the silver nanoparticles were made from the plant extracts then the nanocomposites were formed by incorporating nanoparticles into poly (vinyl alcohol) (PVA). The formed nanocomposites were characterized with the help of AFM to exhibit the incorporation of nanoparticles in the film. The silver nanocomposites exhibited good antimicrobial activity against *Salmonella typhimurium* [63]. Scientists are now extensively working upon hydrogel-based silver nanocomposites for several applications. In this context Novientri et al. [94] synthesized sodium alginate-polyvinyl alcohol-g-acrylamide (NaAIg-PVA-g-AAm) nanocomposite with the help of brown algae [94]. The silver ions get trapped in the hydrogel matrix. The various characterization studies revealed the characteristic features of the nanocomposites. The results revealed the size of the silver nanocomposites were of 20 nm that helped them to easily embed in the matrix of the hydrogel. The silver nanocomposites exhibited

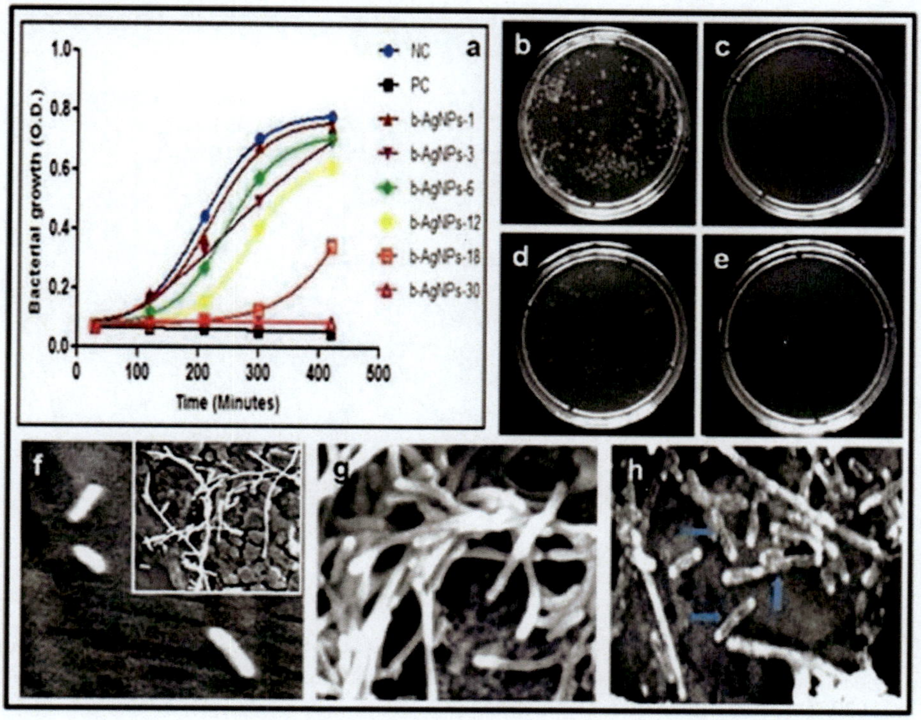

Fig. 8 Study of antibacterial activities: **a** liquid growth inhibition kinetics of *E. coli* using different concentrations of b-AgNPs. b-AgNP-30 (at 30 μM) shows almost 100% growth inhibition. Ampicillin has been used as a positive control (PC) & NC: negative control or untreated *E. coli*. The numerical number indicates the concentration of b-AgNPs in μM, **b–e** optical images of bacterial colonies formed by *E. coli* cells, i.e. colony counting assay (after 24 h): **b**: Control, **c**: Ampicillin (100 μg/ml), **d**: b-AgNPs (18 μM), **e**: b-AgNPs (30 μM) and **f–h** SEM images of *E. coli* cells **f** without being treated (control), **g** treated with Olax for 1 h, **h** treated with b-AgNPs (30 μM) for 1 h. The SEM images show the silver nanoparticles damages the bacterial cell membrane (marked by blue arrow), whereas, the bacterial membranes of untreated and treated *E. coli* with Olax is intact [81]. Reprinted with permission from Mukherjee et al. [74]. Copyright © 2020 **Ivyspring International Publisher**

antibacterial activity against *Staphylococcus aureus* and *Escherichia coli*. The ratio of minimum bacterial concentration (MBC) to minimum inhibitory concentration (MIC) was four when both the bacterial strains were treated with the silver nanocomposite hydrogels. The nanocomposite exhibited prominent results of antibacterial activity against both the Gram-negative and Gram-positive bacteria [94]. Table 2 shows the silver nanocomposites produced by different components as well as their mechanism of antibacterial activity.

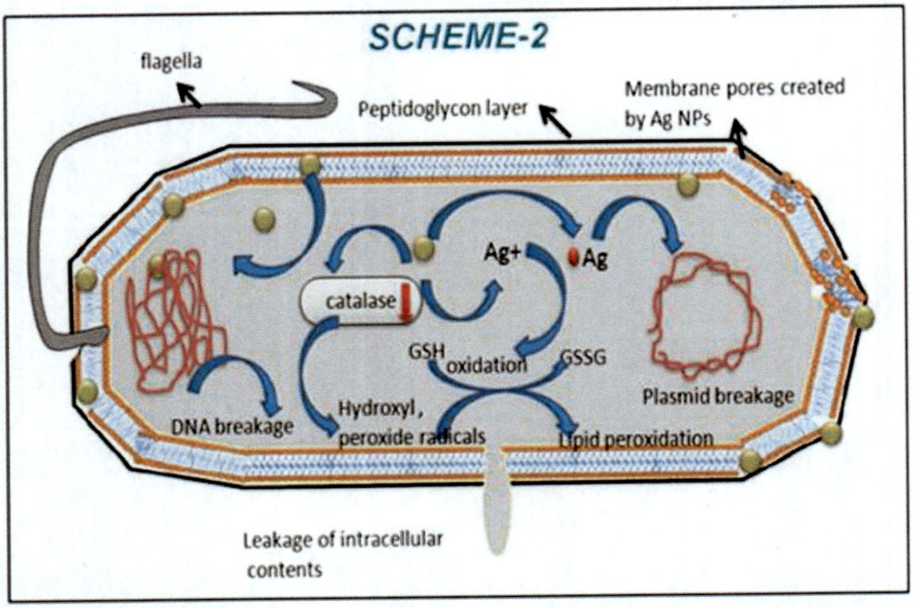

Fig. 9 Schematic presentation of antibacterial activity of b-AgNPs toward *E. coli* using biosynthesized silver nanoparticles [81]. Reprinted with permission from Mukherjee et al. [74]. Copyright © 2020 **Ivyspring International Publisher**

6 Antimicrobial Activity of Silver Nanocomposites: Mechanistic Approach

Silver nanocomposites exhibit excellent activities such as antiviral, antifungal, antibacterial, antiangiogenic and anti-inflammatory because of the characteristic properties. The exact mechanism of action is under research and hence needs further examination for the proof of its antimicrobial activity. There are several reports that demonstrate the probable mechanism of the antibacterial activity of silver nanocomposites. The reports suggest the ability of silver nanocomposites to damage the bacterial membrane leads to release of cellular components and killing the bacterial cell [95]. The effect of silver nanoparticles and nanocomposites depends upon the peptidoglycan layer of the Gram-positive and Gram-negative bacteria. Some reports highlight that the positive charge of the silver nanocomposites helps in easy interaction with the bacterial membrane which is negatively charged. The size, shape and charge of the silver nanocomposites play major role in their internalization into the bacterial cells [96]. Also these properties are responsible for the interaction of the silver nanocomposites with the intracellular components such as nucleic acids, proteins, lipid leading to the inactivation and subsequently death of the bacterial cells. Another mechanism that is being employed by the silver nanocomposites for killing the bacterial cells involves generation of reactive oxygen species (ROS) involving the release of hydrogen peroxide, singlet oxygen and superoxide anion [97]. The augmented

Table 2 Silver nanocomposites from different substances

Precursors	Nanocomposites	Bacterial strain	Mechanism of action	References
Graphene	Graphene Oxide/Hydroxyapatite/silver	*Staphylococcus aureus, Bacillus cereus, Escherichia coli Klebsiella pneumoniae*	Changing properties of bacterial membrane and cell wall and disturbing basic metabolic processes	[20]
	Graphene oxide-based silver nanocomposites	*Staphylococcus aureus, Escherichia coli*	Generation of ROS leading to bacterial cell damage	[83]
Chitosan	curcumin encapsulated chitosan-poly vinyl alcohol (PVA)-silver nanocomposites	*Escherichia coli, Staphylococcus, Pseudomonas, Micrococcus, Candida albicans, Pseudomonas aeruginosa*	Biocidal activity	[85]
	Chitosan-silver nanocomposites	*Staphylococcus aureus, Salmonella entrica, Pseudomonas aeruginosa*	Inhibition of bacterial growth	[86]
Sodium nitroprusside	Silver nitroprusside (AgNNPs)	*Escherichia coli, Staphylococcus epidermidis, Bacillus subtilis, Klebsiella aerogenes, Pseudomonas aeruginosa*	Damaging cell membrane of the bacteria	[24]

(continued)

ROS function results in the decrease of the enzymes such as superoxide dismutase, catalase and glutathione peroxidase (GSH) and also causes DNA damage as well as lipid peroxidation [98]. The toxicity toward the bacterial cell is due to the release of silver ions. But there are also reports that silver ions have less antibacterial activity in comparison with silver nanocomposites, which is an indicative of the fact that the silver in its nanocomposites has different silver ion discharge and different binding pathways [55].

Table 2 (continued)

Precursors	Nanocomposites	Bacterial strain	Mechanism of action	References
Prussian blue	Silver Prussian blue analogue nanoparticles	*Bacillus subtilis, Escherichia coli, Pseudomonas aeruginosa, Klebsiella pneumoniae*	Generation of ROS suppresses the antioxidant activity and oxidative stress results in lipid peroxidation causing cell damage	[25]
Polymers	Silver nanocomposites with cellulosic fibres	*Staphylococcus aureus*	Bacterial cell wall damage	[90]
	Silver polystyrene nanocomposites	*Escherichia coli*	Decrease in viability of bacterial cell along with cell wall damage	[92]
Biosynthesized	*Olax scandens* synthesized silver nanocomposites	*Escherichia coli*	Generation of ROS causing toxicity to bacterial cell, glutathione oxidation causes lipid peroxidation and damages cell membrane	[74]
	Sodium alginate-polyvinyl alcohol-g-acrylamide (NaAIg-PVA-g-AAm) hydrogel with silver nanocomposite	*Staphylococcus aureus, Escherichia coli*	Exhibit bacteriostatic and bacteriocidal activity	[94]

The resistance to antibiotics has emerged as one of the leading causes of increased and stable microbial infections. In context to this, Surwade et al. [99] used a combination therapy by combining antibiotic ampicillin with silver nanocomposites [99]. The combination therapy was used against methicillin-resistant *Staphylococcus aureus* (MRSA) bacteria. The combination was effective at a lower dose of ampicillin and sublethal dose of silver nanoparticles. When combined both acted synergistically [99]. The second most drawbacks to microbial infection after antibiotic resistance are the biofilm formation. The extracellular matrix produced by the bacteria on solid surfaces such as medical devices, pacemakers, catheters, etc., are referred to as biofilms [3]. This is done by the bacteria in order to prevent the action of the antibiotics on their community. To prevent these, Siddique et al. [100] exhibited the role of silver nanocomposites in the inhibition of biofilm formation [100]. The silver nanocomposites were synthesized with $AgNO_3$, polyvinyl pyrrolidone (PVP) and sodium borohydride ($NaBH_4$) where PVP was used for the stabilization of the silver nanocomposites. The silver nanocomposites showed antibacterial activity through broth microdilution and agar diffusion method. When the formed nanocomposites were used upon *K. pneumoniae* strains through tube method and microtiter plate assay, it exhibited decreased biofilm formation. Thus the silver nanocomposites exhibited both antibacterial and antibiofilm formation [100]. In another study Yu et al. [101] exhibited the antibacterial and antibiofilm role of silver nanocomposites on *E.coli* strain obtained from dairy cows affected with mastitis [101]. The silver nanocomposites were formed by the combination of silver nanoparticles with quercitin which is a plant derived drug. The nanocomposites were used to perform experiments such as antibacterial assays, scanning electron microscopy (SEM) experiments, biofilm formation assays as well as real-time reverse transcription PCR (RT-PCR) assays. The results revealed increased antibacterial and antibiofilm activity of the nanocomposites in comparison with its single counterparts. Figure 10 depicts the SEM experiment which clearly shows that the biofilm of the controlled group is normal in comparison with the treatment groups. Moreover the biofilms treated with the silver nanocomposites showed intense damage with respect to the other groups treated with only quercitin and silver nanocomposites. The molecular studies done with the RT-PCR even showed inhibition of transcription biofilm associated genes (bcsA, fliC, csgA, motA, fimA, and wcaF) [101]. Thereafter Fig. 11 illustrates the overall probable mechanism for the antibacterial activity of the silver nanoparticles and nanocomposites that involves adhering to the bacterial cellular membrane, penetrating through the bacterial nucleus, generation of ROS and cell signal modulation leading to the death of the bacterial cells [102].

7 Commercial Applications of Antibacterial Activity of Silver Nanocomposites

Silver nanocomposites due to their antibacterial activity are finding utility in several biomedical applications [16]. From wound healing to medical devices, silver nanocomposites are being used as an antibacterial agent that has been briefly highlighted in the following sections.

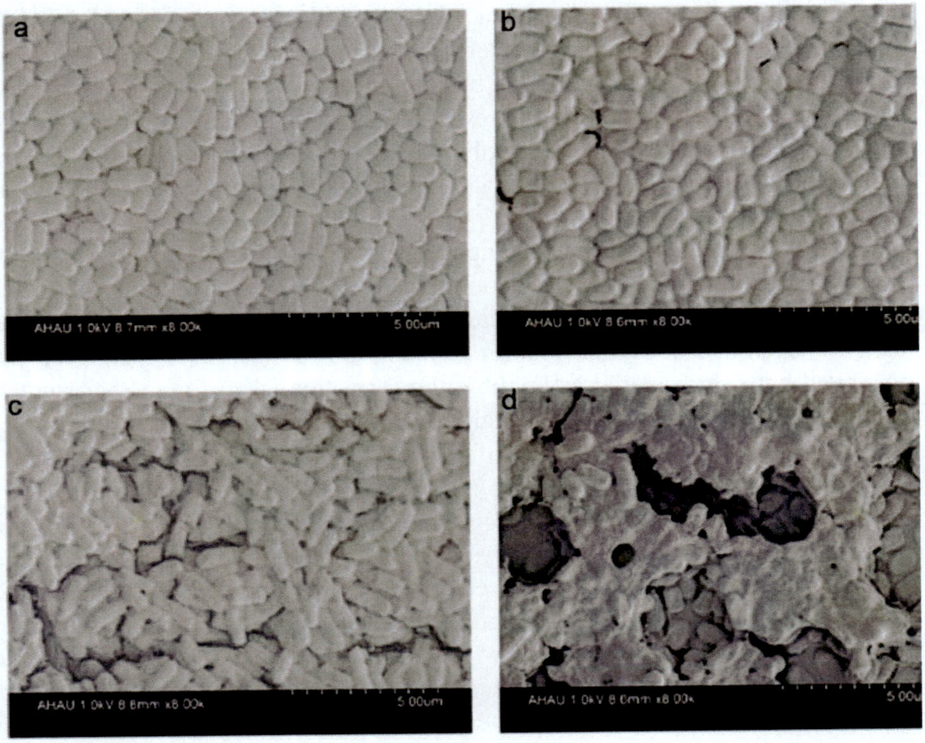

Fig. 10 Effect of Qe, Ag NPs, and QA NPs on biofilm integrity of *E. coli* strain ECDCM1 were monitored by SEM. **a** The control group with no antimicrobial agents; **b** *E. coli* strain ECDCM1 with exposure to Qe; **c** *E. coli* strain ECDCM1 with exposure to Ag NPs; **d** *E. coli* strain ECDCM1 with exposure to QA NPs [109]. Reprinted with permission from Yu et al. [101]. Copyright © 2018 **Yu et al**.

7.1 Wound Healing

Wound healing is one such important aspect of medical field that involves a series of processes. It is a known fact that wound causes an exposure over the skin leading to bacterial infections which in turn causes complications in its healing process [103]. The bacterial and fungal infections cause a delay in wound healing followed by sepsis that sometimes proves fatal to the patient. Since silver is known to be used as a wound healing agent from ancient times, several reports suggest accelerated wound healing. Several groups impregnated wound dressing material with silver nanocomposites for improved wound healing through acting as an antibacterial agent [104]. In this regards, Kim et al. [105] impregnated silver nanocomposites over cotton wool by dipping the cotton in a solution containing alcoholic silver-2 ethylhexyl-carbamates along with thermal reduction at a temperature of 120 °C [105]. The formed cotton fabrics exhibited excellent antibacterial activity against *Escherichia coli* and *Staphyloccocus aureus* along with good anticoagulant action. The silver nanoform impregnated cotton dressings even blocked UV rays in the range of UV-R.

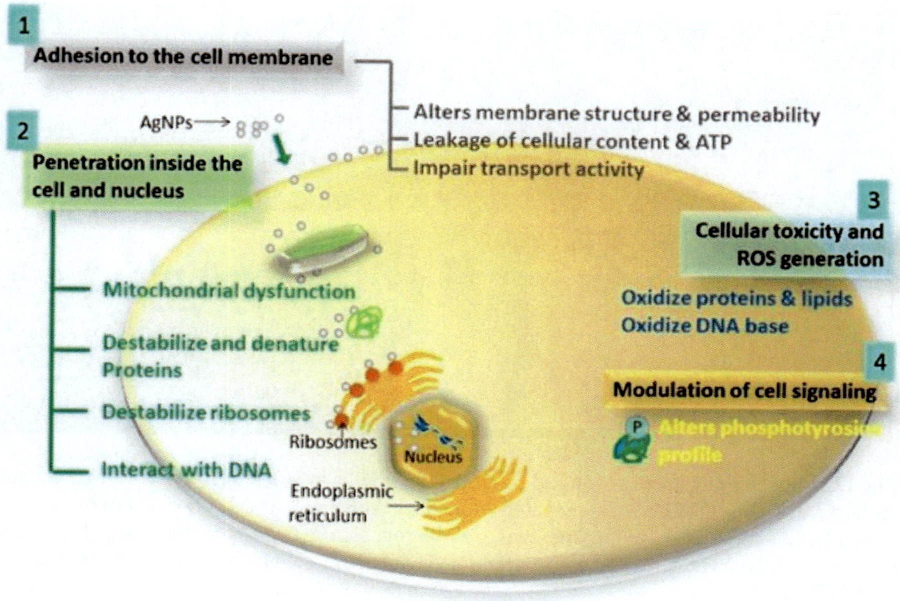

Fig. 11 The four most prominent routes of antimicrobial action of AgNPs. 1, AgNPs adhere to microbial cell surface and results in membrane damage and altered transport activity; 2, AgNPs penetrate inside the microbial cells and interact with cellular organelles and biomolecules, and thereby, affect respective cellular machinery; 3, AgNPs cause increase in ROS inside the microbial cells leading to cell damage and; 4, AgNPs modulate cellular signal system ultimately causing cell death. Reprinted with permission from Dakal et al. [102]. Copyright © 2016 **Dakal et al.**

These properties helped in increased wound healing [105]. Similarly, Adibhesami et al. [58] reported that the antibacterial activity of silver nanocomposites that helped in healing of wounds [58]. For the experiment, firstly the wound was created on the back of mouse then the *Staphyloccocus aureus* solution was applied. The results revealed that the silver nanoforms at a concentration of 0.08 mg/kg of body weight of mice proved to be the most effective one in decreasing the bacterial load along with reduction in wound as compared to untreated and antibiotic gentamicin [58].

7.2 *Medical Devices*

Apart from medicines, medical devices are considered as an inevitable part for the therapeutic and diagnostic purposes. But there remains a major issue regarding infection from the medical devices due to bacterial growth as well as biofilm formation on the exterior of the medical devices [3]. In this regards, application of silver nanocomposites is considered for its antibiomicrobial activity for medical devices. Dybowska-Sarapuk et al. [106] synthesized graphene-based silver nanocomposites for coating of catheters since there exists a serious issue of biofilm formation in the

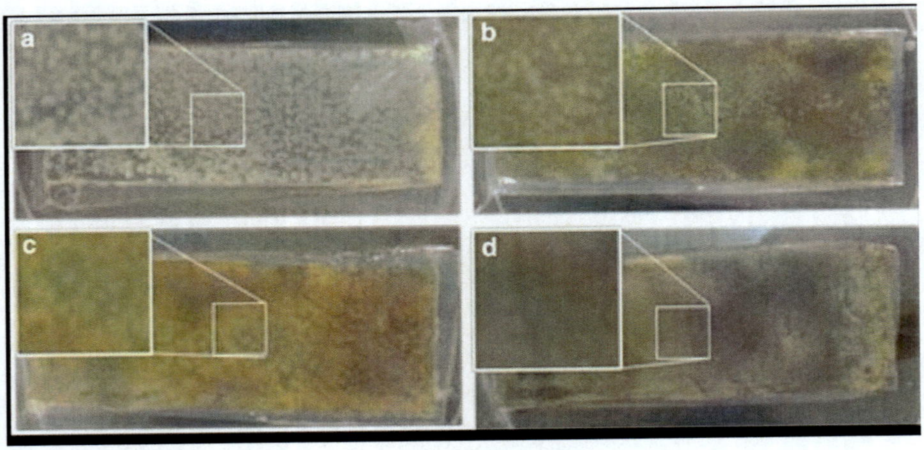

Fig. 12 Photographic images of an AgNP-grown PDMS film at 36 h after inoculation of S. aureus on the surface. **a** Control PDMS was prepared without the AgNP growing reaction. From **b** to **d**, AgNP-grown PDMS films subjected to one, two or three rounds of reactions, respectively. Inset in each figure is an enlarged image of the figure. Reprinted with permission from Kim et al. [107]. Copyright © 2017 **Kim et al**.

catheters of bloodstream and urinary tract infection [106]. The antibiofilm activity was checked in *S. epidermidis*. The results revealed inhibition in the biofilm formation on the catheter surface and the methods were even simple and economical [106]. The other major problem faced by the medical implants is that they are invaded by the bacteria and cause infections. Kim et al. [107] used silver nanocomposites embedded on poly-dimethyl-siloxane (PDMS) for the fabrication of silicon films [107]. The silver nanocomposites formed were permitted to bind electrostatically to the surface of the film through the hydroxyl groups formed on the latter through air plasma treatment. The film when evaluated exhibited excellent antibacterial effect against *Bacillus subtilis* and *Escherichia coli* in comparison with the commercially available silver nanocomposites. Figure 12 clearly shows that the antibacterial activity of the PDMS with silver nanocomposites increased in a concentration-dependent manner in comparison with the naked PDMS without silver nanocomposites. The highest antibacterial activity was shown with three rounds of reaction of silver nanocomposites on PDMS. Thus these silicon films can be effectively used as implanting devices as offering biosafety toward infections caused medical devices [107].

7.3 *Tissue Engineering*

Silver nanocomposites are even being utilized for biomedical engineering designing works. This owes to their unique characteristic features such as antiangiogenic, anti-inflammatory and most importantly antibiofilm as well as antibacterial activity [16]. Silver nanocomposites are extensively used for bone and dental tissue engineering.

Regarding bone tissue implants, Aurore et al. [108] worked to prevent osteoclast infection by *Escherichia coli* and *Staphyloccocus aureus* by synthesizing silver nanoforms [108]. The results clearly depicted that the silver nanocomposites reduced the bacterial load and prevent the infection. The mechanism behind its activity owes to ROS generation followed by damaging of the bacterial cell. These silver nanocomposites can be effectively used for orthopedic and bone implants which can even prevent bacterial infections [108]. For dental care silver nanocomposites are used from centuries as dental amalgams. Due to their antimicrobial property they are highly preferred for dental products. In this context, Inbakandan et al. [109] biosynthesized silver nanocomposites with the extracts of *Haliclona exigua* which is a marine sponge [109]. The shape of the silver nanocomposites was flower like. The silver nanocomplex exhibited excellent oral antibiofilm activity against *S. salivarius*, *S. orulis* and *S. mitis*. The results also revealed that at the lowest concentration (10 μg) the silver nanocomposites inhibit the growth of these bacteria [109].

7.4 Other applications

Apart from the biomedical applications, the silver nanocomplexes and nanocomposites are utilized for textiles, food processing, cosmetics, paints and water management because of its antibacterial activity [110]. The silver nanocomposites are employed for the production of cosmetics as they are effective in removing pathogens from the skin surface without affecting the epidermal lipid layer of the skin [111]. Regarding this Bansod et al. [112] synthesized silver nanocomposites functionalized oil for the production of shampoo, soaps and ointments. The formed compound proved to be antibacterial when topically applied over wounds and other microbial infections [112]. Silver is being widely preferred for the purpose of food packaging since it imparts prevention of the food particles from being destroyed by the microbes. Pulit-Prociak et al. [113] reported to have used silver nanocomposites associated dye for the purpose of dyeing cotton fabrics and imparting antimicrobial protection to the fabric [113]. The silver nanocomposites are being included in paints in order to make them antimicrobial resistance. The silver nanocomposites help the paint from weathering by the microbial particles [114]. The microbes form biofilm over the paint resulting in the retention of spores and dirt. The silver nanocomposites generate ROS and destroy the cell membrane of the bacteria and also interfere with the antioxidant activity of the bacterial cell [115]. All these results exhibit the potential of the silver nanocomposites of being as an efficient antibacterial agent.

8 Toxicity of Silver Nanocomposites

The silver nanocomposites have gained lot of interest among the scientists due to their promising outcomes toward antibacterial activity. The commercialization of the silver nanocomposites requires certain criteria's to be fulfilled where toxicity occupies a major portion. Certain experimental reports exhibit work regarding the toxicity of the silver nanocomposites [57]. In one toxicity report Sujarania et al. [116] studied the toxicity of silver nanocomposites anchored with [2-(2,2-diphenylethylimino) methyl)phenols] [116]. The nanocomposites were characterized using the techniques NMR, IR and UV–Visible. Following that rhodamine B marker was attached to the material. The in vitro and in vivo studies were performed to analyze the toxic effects of the silver nanoparticles anchored 2-[(2,2-dipenylethylimino)methyl] phenols which clearly represented no toxicity. The study demonstrated that the silver nanocomposites along with the Schiff bases is biocompatible in nature [116]. In another work Saha et al. [117] checked for the toxicity of the biosynthesized silver nanocomposites. The silver nanocomposites were produced by coupling with *Euphorbia thymifolia* [117]. For the toxicity studies to be conducted, Swiss albino mice were procured of 6–8 weeks old after in compliance with the rules of Organization for Economic Cooperation Development (OECD). The silver nanocomposites were administered intraperitoneally with a single dose of 2000–5000 mg/kg of body weight to the mice. Even the compounds silver oxides as well as *Euphorbia thymifolia* were provided at low doses to the mice that exhibited biocompatible results. The experiment was performed for 14 days keeping up a constant check on the body and behavioral changes. The end results displayed that the mice were healthy even after the single dose of the silver nanocomposites and showed no toxicity [117].

In a further study, Yu et al. [118] synthesized silver nanocomposites with the help of cellulose fibril [118]. The formed nanocomposites were characterized with different techniques and also exhibited antibacterial property against *Staphylococcus aureus* and *Escherichia coli*. Then the toxicity study was assessed through the effect of the silver nanocomposites on human colon cells. The formed nanocomposites at low concentration exerted no toxicity to the human colon cells but decreased cell viability at higher doses [118]. But some results highlighted the toxic effects of the silver nanocomposites. In one report Ema et al. [119] evaluated the effect of silver nanocomposites on the developmental and reproductive processes of female and male rats [119]. The results revealed that the silver nanocomposites accumulated in the testis of male. In the female rats the formed nanocomposites were found to be accumulated in the visceral yolk sac at time of early gestation followed by deposition in placenta, pre- and postnatal offspring as well as in breast milk in late gestation period. The developmental defects caused by the silver nanoforms include defective cognitive behavior and physical disabilities [119]. The silver nanocomposites when released in the environment can cause some amount of dysfunction. Hence, they must be properly treated before being discharged in nature.

9 Future Perspectives and Challenges

Nanotechnology is paving its way into the everyday life of living beings along with its advancement and improvement. Various diseases including antibacterial disease are diagnosed and cured with the help of nanotechnology [18]. The increased amount of antibiotic resistance toward conventional drugs has led to commercial markets dwell into nanotechnology to formulate new diagnostics and therapeutics. In this aspect silver nanocomposites have emerged as one of the most eligible candidates for antibacterial therapeutics owing to its unique characteristics features [69]. A good number of silver nanocomposites related products are under experimental conditions for clinical trials and FDA approvals.

Although the silver nanocomposites have the potential to establish itself as a strong antibacterial agent, but for enhancement of its activity certain things are to be considered. The primary one is the proper surface functionalization of the silver nanocomposites. This can be done by attaching certain antibiotics, gene, protein and target moieties that will help in the augmentation of the antibacterial activity of the nanocomposites [23]. The second is increase in the yield and proper stability of the silver nanocomposites on a large scale in order to reach a wider population. Usually chemical stability of the silver nanocomposites is preferred for the slow release of silver ions from the complex allowing a longer retention time. Even positively charged silver nanocomposites are preferred for better interaction with the cellular membranes. The shapes of the silver nanocomposites should be given more preferences for proper contact with the bacteria [60]. Since silver is metal, the toxicity experiments should be conducted in a vigorous amount to eliminate any possibility of toxicity through the silver nanocomposites. Also the right amount of dosage as well as accumulation study of the silver nanocomposites possesses a challenge that needs to be met [120]. More amounts of mechanistic studies regarding the antibacterial activity of silver nanocomposites should be conducted. This is due to the fact that the silver nanocomposites is available in various sizes, shapes and charge and each have its own separate mechanism of action [102]. This can even act as a tool to combat antibiotic resistance exhibited by the microbes.

But there remain certain challenges that act as backdrop regarding the usage of silver nanocomposites. The first and the foremost is the resistance of the bacteria toward silver. Certain reports exhibit ineffectiveness of the silver nanocomposites toward some specific bacterial strains [38]. Another challenge involves consumption of the silver nanocomposites on a regular basis does pose a threat of its accumulation in the skin tissues. This even leads to argyrosis following sunlight exposure [121]. Other than these the general difficulties regarding nanocomposites includes cost of production, environmental hazards, immunological effects as well as the pharmacokinetics and pharmacodynamics that needs to be addressed [122]. Thereafter, if all these future perspective and challenges are timely met then it will result in the development of excellent treatments for antibacterial therapy.

10 Conclusions

Since microbial infections are increasing on an alarming rate in day-to-day life even with the availability of commercial products, nanotechnology has paved the way to curb it in its own ways. The book chapter overall focuses on the different pathogenic strains causing the microbial infections, their global statistics in terms of mortality and morbidity followed by the conventional treatment strategies available. But the limitations and backdrops of the commercially available products have led to the rise of nanotechnology especially silver nanocomposites as an alternative treatment for microbial infections. The silver nanocomposites have exhibited excellent antimicrobial activity produced in combination with several precursors and agents through various mechanisms. The silver nanocomposites exhibit its antimicrobial strategy through cell membrane breakage resulting in the leakage of cellular components, damage to DNA, interference in the synthesis of nucleic acids and proteins, generation of oxidative stress, etc. The silver nanocomposites are even utilized for further commercial products. The book chapter further highlights the toxicity related issues of the silver nanocomposites. Finally the future perspective and challenges section discusses the different backdrops of silver nanocomposites such as cost of production, toxicity which when met can prove to produce wonders in the treatment of microbial infections.

Acknowledgements Financial support from Nano Mission-DST (SR/NM/NS-1252/2013; GAP 570), New Delhi, to C.R.P. is duly acknowledged. S.H is thankful to CSIR, New Delhi for supporting Senior Research Fellowship. We thank Director, CSIR-IICT (Ms. No. **IICT/Pubs./2020/184** dated **July 3, 2020**) for providing all the required facilities to carry out the work.

References

1. Zhu X, Radovic-Moreno AF, Wu J, Langer R, Shi J (2014) Nanomedicine in the management of microbial infection—overview and perspectives. Nano Today 9(4):478–498. https://doi.org/10.1016/j.nantod.2014.06.003
2. Rodvold KA, McConeghy KW (2014) Methicillin-resistant Staphylococcus aureus therapy: past, present, and future. Clin Infect Dis 58(suppl_1):S20-S7. https://doi.org/10.1093/cid/cit614
3. Khatoon Z, McTiernan CD, Suuronen EJ, Mah T-F, Alarcon EI (2018) Bacterial biofilm formation on implantable devices and approaches to its treatment and prevention. Heliyon 4(12):e01067. https://doi.org/10.1016/j.heliyon.2018.e01067
4. McMichael AJ (2004) Environmental and social influences on emerging infectious diseases: past, present and future. Philos Trans R Soc Lond B Biol Sci 359(1447):1049–1058. https://doi.org/10.1098/rstb.2004.1480
5. Brachman PS (2003) Infectious diseases—past, present, and future. Int J Epidemiol 32(5):684–686. Oxford University Press. https://doi.org/10.1093/ije/dyg282
6. Peleg AY, Hooper DC (2010) Hospital-acquired infections due to gram-negative bacteria. N Engl J Med 362(19):1804–1813. https://doi.org/10.1056/NEJMra0904124
7. Resistance AD (2020) Antimicrobial resistance threats. National Institute of Allergy and Infectitious Diseases, National Institute of Allergy and Infectitious Diseases

8. Pumart P, Phodha T, Thamlikitkul V, Riewpaiboon A, Prakongsai P, Limwattananon S (2012) Health and economic impacts of antimicrobial resistance in Thailand. J Health Serv Res Policy 6(3):352–360
9. Laxminarayan R, Duse A, Wattal C, Zaidi AK, Wertheim HF, Sumpradit N et al (2013) Antibiotic resistance—the need for global solutions. Lancet Infect Dis 13(12):1057–1098. https://doi.org/10.1016/S1473-3099(13)70318-9
10. Health G (2018) Infographic: antibiotic resistance the global threat. Centers for Disease Control and Prevention: 2018 Centers for Disease Control and Prevention. Report No.
11. Research GV (2018) Antibiotic resistance market size, share & trends analysis report by disease (BSI, CDI, cUTI, cIAI), By Pathogen (E. coli, P. Aeruginosa, K. Pneumoniae), by drug class, and segment forecasts, 2018–2025. 2018 GVR-2-68038-663-9
12. Fair RJ, Tor Y (2014) Antibiotics and bacterial resistance in the 21st century. Perspect Med Chem 6:25–64. https://doi.org/10.4137/PMC.S14459
13. Luthra S, Rominski A, Sander P (2018) The role of antibiotic-target-modifying and antibiotic-modifying enzymes in mycobacterium abscessus drug resistance. Front Microbiol 9(2179). https://doi.org/10.3389/fmicb.2018.02179
14. Opal SM, Pop-Vicas A (2015) 18—molecular mechanisms of antibiotic resistance in bacteria. In: Bennett JE, Dolin R, Blaser MJ (eds) Mandell, Douglas, and Bennett's principles and practice of infectious diseases, 8th edn. Philadelphia, pp 235–251
15. Dowling AP (2004) Development of nanotechnologies. Mater Today 7(12):30–35. https://doi.org/10.1016/S1369-7021(04)00628-5
16. Burdușel A-C, Gherasim O, Grumezescu AM, Mogoantă L, Ficai A, Andronescu E (2018) Biomedical applications of silver nanoparticles: an up-to-date overview. Nanomaterials (Basel) 8(9):681. https://doi.org/10.3390/nano8090681
17. Lee N-Y, Hsueh P-R, Ko W-C (2019) Nanoparticles in the treatment of infections caused by multidrug-resistant organisms. Front Pharmacol 10:1153. https://doi.org/10.3389/fphar.2019.01153
18. Patra JK, Das G, Fraceto LF, Campos EVR, Rodriguez-Torres MDP, Acosta-Torres LS et al (2018) Nano based drug delivery systems: recent developments and future prospects. J Nanobiotechnol 16(1):71. https://doi.org/10.1186/s12951-018-0392-8
19. Siddiqi KS, Husen A, Rao RAK (2018) A review on biosynthesis of silver nanoparticles and their biocidal properties. J Nanobiotechnol 16(1):14. https://doi.org/10.1186/s12951-018-0334-5
20. Beiranvand M, Farhadi S, Mohammadi A (2019) Graphene oxide/hydroxyapatite/silver (rGO/HAP/Ag) nanocomposite: synthesis, characterization, catalytic and antibacterial activity. Int J Nano Dimens 10(2):180–194
21. Arya G, Sharma N, Mankamna R, Nimesh S (2019) Antimicrobial silver nanoparticles: future of nanomaterials. In: Prasad R (ed) Microbial nanobionics: volume 2, basic research and applications. Springer International Publishing, Cham, pp 89–119
22. Baranwal A, Srivastava A, Kumar P, Bajpai VK, Maurya PK, Chandra P (2018) Prospects of nanostructure materials and their composites as antimicrobial agents. Front Microbiol 9:422. https://doi.org/10.3389/fmicb.2018.00422
23. Baptista PV, McCusker MP, Carvalho A, Ferreira DA, Mohan NM, Martins M et al (2018) Nano-strategies to fight multidrug resistant bacteria—"A Battle of the Titans." Front Microbiol 9:1441. https://doi.org/10.3389/fmicb.2018.01441
24. Rao BR, Kotcherlakota R, Nethi SK, Puvvada N, Sreedhar B, Chaudhuri A et al (2018) Ag2[Fe(CN)5NO] nanoparticles exhibit antibacterial activity and wound healing properties. ACS Biomater Sci Eng 4(9):3434–3449. https://doi.org/10.1021/acsbiomaterials.8b00759
25. Mukherjee S, Kotcherlakota R, Haque S, Das S, Nuthi S, Bhattacharya D et al (2020) Silver Prussian blue analogue nanoparticles: rationally designed advanced nanomedicine for multi-functional biomedical applications. ACS Biomater Sci Eng 6(1):690–704. https://doi.org/10.1021/acsbiomaterials.9b01693
26. Thakur A, Mikkelsen H, Jungersen G (2019) Intracellular pathogens: host immunity and microbial persistence strategies. J Immunol Res 2019:1356540. https://doi.org/10.1155/2019/1356540

27. Ng L-K, Martin IE (2005) The laboratory diagnosis of Neisseria gonorrhoeae. Can J Infect Dis Med Microbiol 16(1):15–25. https://doi.org/10.1155/2005/323082
28. Manchanda V, Sanchaita S, Singh N (2010) Multidrug resistant acinetobacter. J Glob Infect Dis 2(3):291–304. https://www.jgid.org/text.asp?2010/2/3/291/68538
29. Lampel KA, Sandlin R (2003) SHIGELLA. In: Caballero B (ed) Encyclopedia of food sciences and nutrition, 2nd edn. Academic Press, Oxford, pp 5261–5268
30. Smith I (2003) Mycobacterium tuberculosis pathogenesis and molecular determinants of virulence. Clin Microbiol Rev 16(3):463–496. https://doi.org/10.1128/cmr.16.3.463-496.2003
31. India T (2017) Revised national tuberculosis program annual status report. Directorate General of Health Services, Ministry of Health and Family Welfare, New Delhi, India
32. Dixit A, Kumar N, Kumar S, Trigun V (2019) Antimicrobial resistance: Progress in the decade since emergence of New Delhi metallo-β-lactamase in India. Indian J Commun Med Off Publ Indian Assoc Prev Soc Med 44(1):4
33. Hellen G, Molly M-P, Suraj P, Sumanth G, Jordan L, Devra B et al (2015) State of the world's antibiotics. Center for Disease Dynamics, Economics & Policy, Washington, DC
34. Dudley MN (1995) Combination beta-lactam and beta-lactamase-inhibitor therapy: pharmacokinetic and pharmacodynamic considerations. Am J Health Syst Pharm AJHP Off J Am Soc Health Syst Pharm 52(6 Suppl 2):S23–S28. https://doi.org/10.1093/ajhp/52.6_Suppl_2.S23
35. Finberg RW, Moellering RC, Tally FP, Craig WA, Pankey GA, Dellinger EP et al (2004) The importance of bactericidal drugs: future directions in infectious disease. Clin Infect Dis 39(9):1314–1320. https://doi.org/10.1086/425009
36. Munita JM, Arias CA (2016) Mechanisms of antibiotic resistance. Microbiol Spectr 4(2). https://doi.org/10.1128/microbiolspec.VMBF-0016-2015
37. Chopra I, Roberts M (2001) Tetracycline antibiotics: mode of action, applications, molecular biology, and epidemiology of bacterial resistance. Microbiol Mol Biol Rev 65(2):232–260. https://doi.org/10.1128/MMBR.65.2.232-260.2001
38. Wang L, Hu C, Shao L (2017) The antimicrobial activity of nanoparticles: present situation and prospects for the future. Int J Nanomed 12:1227–1249. https://doi.org/10.2147/IJN.S121956
39. Hsueh PR (2010) New Delhi metallo-ß-lactamase-1 (NDM-1): an emerging threat among Enterobacteriaceae. J Formos Med Assoc Taiwan yi zhi 109(10):685–687. https://doi.org/10.1016/s0929-6646(10)60111-8
40. Arima S, Ochi H (2018) Staphylococcus pseudintermedius biofilms secrete factors that induce inflammatory reactions in vitro. Lett Appl Microbiol 67(3):214–219. https://doi.org/10.1111/lam.13018
41. Mitsuhashi M, Kibe R, Takahashi K, Kataoka Y (2018) Lett Appl Microbiol. https://doi.org/10.1111/lam.13018
42. Zhu B, Song L, Kong X, Macleod LC, Xu P (2018) A novel regulator modulates glucan production, cell aggregation and biofilm formation in streptococcus sanguinis SK36. Front Microbiol 9:1154. https://doi.org/10.3389/fmicb.2018.01154
43. Cole ST (2014) Who will develop new antibacterial agents? Philos Trans R Soc Lond B Biol Sci 369(1645):20130430. https://doi.org/10.1098/rstb.2013.0430
44. Vimbela GV, Ngo SM, Fraze C, Yang L, Stout DA (2017) Antibacterial properties and toxicity from metallic nanomaterials. Int J Nanomed 12:3941–3965. https://doi.org/10.2147/IJN.S134526
45. Nagvenkar AP, Deokar A, Perelshtein I, Gedanken A (2016) A one-step sonochemical synthesis of stable ZnO–PVA nanocolloid as a potential biocidal agent. J Mater Chem B 4(12):2124–2132. https://doi.org/10.1039/C6TB00033A
46. LewisOscar F, MubarakAli D, Nithya C, Priyanka R, Gopinath V, Alharbi NS et al (2015) One pot synthesis and anti-biofilm potential of copper nanoparticles (CuNPs) against clinical strains of Pseudomonas aeruginosa. Biofouling 31(4):379–391. https://doi.org/10.1080/08927014.2015.1048686

47. Ren G, Hu D, Cheng EW, Vargas-Reus MA, Reip P, Allaker RP (2009) Characterisation of copper oxide nanoparticles for antimicrobial applications. Int J Antimicrob Agents 33(6):587–590. https://doi.org/10.1016/j.ijantimicag.2008.12.004
48. Maillard JY, Hartemann P (2013) Silver as an antimicrobial: facts and gaps in knowledge. Crit Rev Microbiol 39(4):373–383. https://doi.org/10.3109/1040841X.2012.713323
49. Wanhill RJH, Steijaart JPHM, Leenheer R, Koens JFW (1998) Damage assessment and preservation of an Egyptian silver vase (300–200 BC)*. Archaeometry 40(1):123–137. https://doi.org/10.1111/j.1475-4754.1998.tb00828.x
50. Davies RL, Etris SF (1997) The development and functions of silver in water purification and disease control. Catal Today 36(1):107–114. https://doi.org/10.1016/S0920-5861(96)00203-9
51. Levey M (1967) Medieval Arabic minting of gold and silver coins. Chymia 12:3–14. https://doi.org/10.2307/27757271
52. Rai MK, Deshmukh SD, Ingle AP, Gade AK (2012) Silver nanoparticles: the powerful nanoweapon against multidrug-resistant bacteria. J Appl Microbiol 112(5):841–852. https://doi.org/10.1111/j.1365-2672.2012.05253.x
53. Konop M, Damps T, Misicka A, Rudnicka L (2016) Certain aspects of silver and silver nanoparticles in wound care: a minireview. J Nanomater 2016:7614753. https://doi.org/10.1155/2016/7614753
54. Reidy B, Haase A, Luch A, Dawson KA, Lynch I (2013) Mechanisms of silver nanoparticle release, transformation and toxicity: a critical review of current knowledge and recommendations for future studies and applications. Materials 6(6):2295–2350. https://doi.org/10.3390/ma6062295
55. Kędziora A, Speruda M, Krzyżewska E (2018) Similarities and differences between silver ions and silver in nanoforms as antibacterial agents. Int J Mol Sci 19(2). https://doi.org/10.3390/ijms19020444
56. Rybka J, Łukowiak A, Bugla-Płoskońska G (2018) Int J Mol Sci
57. Jeevanandam J, Barhoum A, Chan YS, Dufresne A, Danquah MK (2018) Review on nanoparticles and nanostructured materials: history, sources, toxicity and regulations. Beilstein J Nanotechnol 9:1050–1074. https://doi.org/10.3762/bjnano.9.98
58. Adibhesami M, Ahmadi M, Farshid AA, Sarrafzadeh-Rezaei F, Dalir-Naghadeh B (2017) Effects of silver nanoparticles on Staphylococcus aureus contaminated open wounds healing in mice: an experimental study. Vet Res Forum Int Q J 8(1):23–28
59. Li S, Zhu T, Huang J, Guo Q, Chen G, Lai Y (2017) Durable antibacterial and UV-protective Ag/TiO(2)@ fabrics for sustainable biomedical application. Int J Nanomedicine 12:2593–2606. https://doi.org/10.2147/IJN.S132035
60. Calderón-Jiménez B, Johnson ME, Montoro Bustos AR, Murphy KE, Winchester MR, Vega Baudrit JR (2017) Silver nanoparticles: technological advances, societal impacts, and metrological challenges. Front Chem 5:6. https://europepmc.org/abstract/MED/28271059. https://doi.org/10.3389/fchem.2017.00006. https://europepmc.org/articles/PMC5318410. https://europepmc.org/articles/PMC5318410?pdf=render
61. Iravani S, Korbekandi H, Mirmohammadi SV, Zolfaghari B (2014) Synthesis of silver nanoparticles: chemical, physical and biological methods. Res Pharm Sci 9(6):385–406
62. Alaqad K, Saleh TA (2016) Gold and silver nanoparticles: synthesis methods, characterization routes and applications towards drugs. J Environ Anal Toxicol 6(4):525–2161. https://doi.org/10.4172/2161-0525.1000384
63. Tripathi R, Pudake RN, Shrivastav B, Shrivastav A (2018) Antibacterial activity of poly (vinyl alcohol)—biogenic silver nanocomposite film for food packaging material. Adv Nat Sci Nanosci 9(2):025020. https://doi.org/10.1088/issn.2043-6262
64. Qing Y, Cheng L, Li R, Liu G, Zhang Y, Tang X et al (2018) Potential antibacterial mechanism of silver nanoparticles and the optimization of orthopedic implants by advanced modification technologies. Int J Nanomedicine 13:3311–3327. https://doi.org/10.2147/IJN.S165125
65. Zhang W, Qiao X, Chen J (2007) Synthesis of nanosilver colloidal particles in water/oil microemulsion. Colloid Surf A 299(1):22–28. https://doi.org/10.1016/j.colsurfa.2006.11.012

66. Chen M, Feng Y-G, Wang X, Li T-C, Zhang J-Y, Qian D-J (2007) Silver nanoparticles capped by oleylamine: formation, growth, and self-organization. Langmuir 23(10):5296–5304. https://doi.org/10.1021/la700553d
67. Abid J-P, Wark A, Brevet P-F, Girault H (2002) Preparation of silver nanoparticles in solution from a silver salt by laser irradiation. Chem Commun 7:792–793. https://doi.org/10.1039/B200272H
68. Ma H, Yin B, Wang S, Jiao Y, Pan W, Huang S et al (2004) Synthesis of silver and gold nanoparticles by a novel electrochemical method. ChemPhysChem 5(1):68–75. https://doi.org/10.1002/cphc.200300900
69. Zhang X-F, Liu Z-G, Shen W, Gurunathan S (2016) Silver nanoparticles: synthesis, characterization, properties, applications, and therapeutic approaches. Int J Mol Sci 17(9):1534. https://doi.org/10.3390/ijms17091534
70. Kim K, Kim T (2019) Nanofabrication by thermal plasma jets: from nanoparticles to low-dimensional nanomaterials. J Appl Phys 125(7):070901. https://doi.org/10.1063/1.5060977
71. Sarkar R, Kumbhakar P, Mitra A (2010) Green synthesis of silver nanoparticles and its optical properties. Digit J Nanomater Bios 5
72. Ovais M, Khalil AT, Ayaz M, Ahmad I, Nethi SK, Mukherjee S (2018) Biosynthesis of metal nanoparticles via microbial enzymes: a mechanistic approach. Int J Mol Sci 19(12):4100. https://doi.org/10.3390/ijms19124100
73. Singh A, Jain D, Upadhyay M, Khandelwal N, Verma H (2010) Green synthesis of silver nanoparticles using Argemone Mexicana leaf extract and evaluation of their antimicrobial activities. Digit J Nanomater Bios 5:483–489
74. Mukherjee S, Chowdhury D, Kotcherlakota R, Patra S, Vinothkumar B, Bhadra MP et al (2014) Potential theranostics application of bio-synthesized silver nanoparticles (4-in-1 System). Theranostics 4(3):316–335. https://doi.org/10.7150/thno.7819
75. Thanh NTK, Maclean N, Mahiddine S (2014) Mechanisms of nucleation and growth of nanoparticles in solution. Chem Rev 114(15):7610–7630. https://doi.org/10.1021/cr400544s
76. Chang S, Chen K, Hua Q, Ma Y, Huang W (2011) Evidence for the growth mechanisms of silver nanocubes and nanowires. J Phys Chem C 115(16):7979–7986. https://doi.org/10.1021/jp2010088
77. Xia Y, Xia X, Wang Y, Xie S (2013) Shape-controlled synthesis of metal nanocrystals. MRS Bull 38(4):335–344. https://doi.org/10.1557/mrs.2013.84
78. Logaranjan K, Raiza AJ, Gopinath SC, Chen Y, Pandian K (2016) Shape-and size-controlled synthesis of silver nanoparticles using aloe vera plant extract and their antimicrobial activity. Nanoscale Res Lett 11(1):1–9. https://doi.org/10.1186/s11671-016-1725-x
79. Chakraborty I, Feliu N, Roy S, Dawson K, Parak WJ (2018) Protein-mediated shape control of silver nanoparticles. Bioconjugate Chem 29(4):1261–1265. https://doi.org/10.1021/acs.bioconjchem.8b00034
80. Dogru E, Demirbas A, Altinsoy B, Duman F, Ocsoy I (2017) Formation of Matricaria chamomilla extract-incorporated Ag nanoparticles and size-dependent enhanced antimicrobial property. J Photochem Photobiol B 174:78–83. https://doi.org/10.1016/j.jphotobiol.2017.07.024
81. Salvioni L, Galbiati E, Collico V, Alessio G, Avvakumova S, Corsi F et al (2017) Negatively charged silver nanoparticles with potent antibacterial activity and reduced toxicity for pharmaceutical preparations. Int J Nanomedicine 12:2517–2530. https://doi.org/10.2147/IJN.S127799
82. Mohandas A, Deepthi S, Biswas R, Jayakumar R (2018) Chitosan based metallic nanocomposite scaffolds as antimicrobial wound dressings. Bioact Mater Nanoscale Res Lett 3(3):267–277
83. Song B, Zhang C, Zeng G, Gong J, Chang Y, Jiang Y (2016) Antibacterial properties and mechanism of graphene oxide-silver nanocomposites as bactericidal agents for water disinfection. Arch Biochem Biophys 604:167–176. https://doi.org/10.1016/j.abb.2016.04.018
84. Jaworski S, Wierzbicki M, Sawosz E, Jung A, Gielerak G, Biernat J et al (2018) Graphene oxide-based nanocomposites decorated with silver nanoparticles as an antibacterial agent. 13(1):116. https://doi.org/10.1186/s11671-018-2533

85. Vimala K, Yallapu MM, Varaprasad K, Reddy NN, Ravindra S, Naidu NS et al (2011) Fabrication of curcumin encapsulated chitosan-PVA silver nanocomposite films for improved antimicrobial activity. J Biomater Nanobiotechnol 2(01):55. https://doi.org/10.4236/jbnb.2011.21008
86. Kaur P, Choudhary A, Thakur R (2013) Synthesis of chitosan-silver nanocomposites and their antibacterial activity. Int J Sci Eng Res 4(4):869
87. Crisponi G, Nurchi VM (2016) Chapter 2—chelating agents as therapeutic compounds—basic principles. In: Aaseth J, Crisponi G, Andersen O (eds) Chelation therapy in the treatment of metal intoxication. Academic Press, Boston, pp 35–61
88. Patra CR (2016) Prussian blue nanoparticles and their analogues for application to cancer theranostics. Nanomedicine Nanotechnol Biol Med 11(6):569–572.https://doi.org/10.2217/nnm.16.16
89. Lee SH, Jun B-H (2019) Silver nanoparticles: synthesis and application for nanomedicine. Int J Mol Sci 20(4):865. https://doi.org/10.3390/ijms20040865
90. Li R, He M, Li T, Zhang L (2015) Preparation and properties of cellulose/silver nanocomposite fibers. Carbohydr Polym 115:269–275. https://doi.org/10.1016/j.carbpol.2014.08.046
91. Pinto RJB, Marques PAAP, Neto CP, Trindade T, Daina S, Sadocco P (2009) Antibacterial activity of nanocomposites of silver and bacterial or vegetable cellulosic fibers. Acta Biomater 5(6):2279–2289. https://doi.org/10.1016/j.actbio.2009.02.003
92. Palomba M, Carotenuto G, Cristino L, Di Grazia M, Nicolais F, De Nicola S (2012) Activity of antimicrobial silver polystyrene nanocomposites. J Nanomater 2012. https://doi.org/10.1155/2012/185029
93. Jabbar AH, Mezan SO, Tuama AN, Hamzah MQ, Ameruddin ASB, Agam MA (2019) Enhanced bioactivity of polystyrene-silver nanocomposite (PS/Ag NCs)-an antimicrobial study. AIP Conf Proc 2151(1):020002. https://doi.org/10.1063/1.5124632
94. Novientri G, Abbas GH, Budianto E (2019) Nanocomposite hydrogel-based biopolymer modified with silver nanoparticles as an antibacterial material for wound treatment. J Appl Pharm Sci 9(11):1–9. https://doi.org/10.7324/JAPS.2019.91101
95. Galdiero S, Falanga A, Vitiello M, Cantisani M, Marra V, Galdiero M (2011) Silver nanoparticles as potential antiviral agents. Molecules 16(10):8894–8918. https://doi.org/10.3390/molecules16108894
96. Lei J, Sun L, Huang S, Zhu C, Li P, He J et al (2019) The antimicrobial peptides and their potential clinical applications. Am J Transl Res 11(7):3919–3931
97. Sharma P, Jha AB, Dubey RS, Pessarakli M (2012) Reactive oxygen species, oxidative damage, and antioxidative defense mechanism in plants under stressful conditions. J Bot 2012:217037. https://doi.org/10.1155/2012/217037
98. Park H-J, Kim J, Kim J, Lee J-H, Hahn J-S, Gu M et al (2009) Silver-ion-mediated reactive oxygen species generation affecting bactericidal activity. Water Res 43:1027–1032. https://doi.org/10.1016/j.watres.2008.12.002
99. Surwade P, Ghildyal C, Weikel C, Luxton T, Peloquin D, Fan X et al (2019) Augmented antibacterial activity of ampicillin with silver nanoparticles against methicillin-resistant Staphylococcus aureus (MRSA). J Antibiot 72(1):50–53. https://doi.org/10.1038/s41429-018-0111-6
100. Siddique MH, Aslam B, Imran M, Ashraf A, Nadeem H, Hayat S et al (2020) Effect of silver nanoparticles on biofilm formation and EPS production of multidrug-resistant *Klebsiella pneumoniae*. BioMed Res Int 2020:6398165. https://doi.org/10.1155/2020/6398165
101. Yu L, Shang F, Chen X, Ni J, Yu L, Zhang M et al (2018) The anti-biofilm effect of silver-nanoparticle-decorated quercetin nanoparticles on a multi-drug resistant Escherichia coli strain isolated from a dairy cow with mastitis. PeerJ 6:e5711. https://doi.org/10.7717/peerj.5711
102. Dakal TC, Kumar A, Majumdar RS, Yadav V (2016) Mechanistic basis of antimicrobial actions of silver nanoparticles. Front Microbiol 7:1831. https://doi.org/10.3389/fmicb.2016.01831

103. Guo S, Dipietro LA (2010) Factors affecting wound healing. J Dent Res 89(3):219–229. https://doi.org/10.1177/0022034509359125
104. Gunasekaran T, Nigusse T, Dhanaraju MD (2012) Silver nanoparticles as real topical bullets for wound healing. J Am Coll Clin Wound Spec 3(4):82–96. https://doi.org/10.1016/j.jcws.2012.05.001
105. Kim T-S, Cha J-R, Gong M-S (2018) Investigation of the antimicrobial and wound healing properties of silver nanoparticle-loaded cotton prepared using silver carbamate. Text Res J 88(7):766–776. https://doi.org/10.1177/0040517516688630
106. Dybowska-Sarapuk Ł, Kotela A, Krzemiński J, Wróblewska M, Marchel H, Romaniec M et al (2017) Graphene nanolayers as a new method for bacterial biofilm prevention: preliminary results. J AOAC Int 100(4):900–904. https://doi.org/10.5740/jaoacint.17-0164
107. Kim JH, Park H, Seo SW (2017) In situ synthesis of silver nanoparticles on the surface of PDMS with high antibacterial activity and biosafety toward an implantable medical device. Nano Converg 4(1):33. https://doi.org/10.1186/s40580-017-0126-x
108. Aurore V, Caldana F, Blanchard M, Kharoubi Hess S, Lannes N, Mantel PY et al (2018) Silver-nanoparticles increase bactericidal activity and radical oxygen responses against bacterial pathogens in human osteoclasts. Nanomed Nanotechnol Biol Med 14(2):601–607. https://doi.org/10.1016/j.nano.2017.11.006
109. Inbakandan D, Kumar C, Bavanilatha M, Ravindra DN, Kirubagaran R, Khan SA (2016) Ultrasonic-assisted green synthesis of flower like silver nanocolloids using marine sponge extract and its effect on oral biofilm bacteria and oral cancer cell lines. Microb Pathog 99:135–141. https://doi.org/10.1016/j.micpath.2016.08.018
110. Krishna S, Govender P (2015) Biomedical applications and toxicity of nanosilver: a review. Med Tech SA 29:13–19
111. Rita Fernandes A, Martins-Gomes C, Coutinho T, Durazzo A, Lucarini M, Souto EB et al (2020) Nanomaterials for skin delivery of cosmeceuticals and pharmaceuticals. Appl Sci 10:1594. https://doi.org/10.3390/app10051594
112. Bansod SD, Bawaskar MS, Gade AK, Rai MK (2015) Development of shampoo, soap and ointment formulated by green synthesised silver nanoparticles functionalised with antimicrobial plants oils in veterinary dermatology: treatment and prevention strategies. IET Nanobiotechnol 9(4):165–171. https://doi.org/10.1049/iet-nbt.2014.0042
113. Pulit-Prociak J, Chwastowski J, Kucharski A, Banach M (2016) Functionalization of textiles with silver and zinc oxide nanoparticles. Appl Surf Sci 385:543–553. https://doi.org/10.1016/j.apsusc.2016.05.167
114. Matsumura Y, Yoshikata K, Kunisaki S-i, Tsuchido T (2003) Mode of bactericidal action of silver zeolite and its comparison with that of silver nitrate. Appl Environ Microbiol 69(7):4278–4281. https://doi.org/10.1128/AEM.69.7.4278-4281.2003
115. Gurunathan S, Han JW, Kwon D-N, Kim J-H (2014) Enhanced antibacterial and anti-biofilm activities of silver nanoparticles against Gram-negative and Gram-positive bacteria. Nanoscale Res Lett 9(1):373. https://doi.org/10.1186/1556-276X-9-373
116. Sujarania S, Sironmani TA, Ramu A (2012) Synthesis, characterization and toxicity studies of Schiff bases [2-(2, 2-Diphenylethylimino) Methyl) Phenols] anchored silver nanoparticles. Digit J Nanomater Bios (DJNB) 7(4)
117. Saha S, Bala N, Basu R, Das S (2018) Acute Toxicity Study of Silver Nanoparticle Coupled with Euphorbia thymifolia. J Nanosci Technol 2018:412–414. https://doi.org/10.30799/jnst.125.18040402
118. Yu Z, Wang W, Dhital R, Kong F, Lin M, Mustapha A (2019) Antimicrobial effect and toxicity of cellulose nanofibril/silver nanoparticle nanocomposites prepared by an ultraviolet irradiation method. Colloids Surf B Biointerfaces 180:212–220. https://doi.org/10.1016/j.colsurfb.2019.04.054
119. Ema M, Okuda H, Gamo M, Honda K (2017) A review of reproductive and developmental toxicity of silver nanoparticles in laboratory animals. Reprod Toxicol 67:149–164. https://doi.org/10.1016/j.reprotox.2017.01.005

120. Stensberg MC, Wei Q, McLamore ES, Porterfield DM, Wei A, Sepúlveda MS (2011) Toxicological studies on silver nanoparticles: challenges and opportunities in assessment, monitoring and imaging. Nanomedicine (Lond) 6(5):879–898. https://doi.org/10.2217/nnm.11.78
121. Lansdown ABG (2010) A pharmacological and toxicological profile of silver as an antimicrobial agent in medical devices. Adv Pharmacol Sci 2010:910686. https://doi.org/10.1155/2010/910686
122. Soares S, Sousa J, Pais A, Vitorino C (2018) Nanomedicine: principles, properties, and regulatory issues. Front Chem 6(360). https://doi.org/10.3389/fchem.2018.00360
123. Shaikh S, Rizvi SMD, Shakil S, Hussain T, Alshammari TM, Ahmad W et al (2017) Synthesis and characterization of cefotaxime conjugated gold nanoparticles and their use to target drug-resistant CTX-M-producing bacterial pathogens. J Cell Biochem 118(9):2802–2808. https://doi.org/10.1002/jcb.25929
124. Shaker MA, Shaaban MI (2017) Formulation of carbapenems loaded gold nanoparticles to combat multi-antibiotic bacterial resistance: in vitro antibacterial study. Int J Pharm 525(1):71–84. https://doi.org/10.1016/j.ijpharm.2017.04.019
125. Ansari MA, Khan HM, Khan AA, Pal R, Cameotra SS (2013) Antibacterial potential of Al2O3 nanoparticles against multidrug resistance strains of Staphylococcusaureus isolated from skin exudates. J Nanopart Res 15(10):1970
126. Ansari MA, Khan HM, Khan AA, Cameotra SS, Saquib Q, Musarrat J (2014) Interaction of Al2O3 nanoparticles with E scherichia coli and their cell envelope biomolecules. J Appl Microbiol 116(4):772–783. https://doi.org/10.1111/jam.12423
127. Joost U, Juganson K, Visnapuu M, Mortimer M, Kahru A, Nõmmiste E et al (2015) Photocatalytic antibacterial activity of nano-TiO2 (anatase)-based thin films: effects on Escherichia coli cells and fatty acids. J Photochem Photobiol B 142:178–185. https://doi.org/10.1016/j.jphotobiol.2014.12.010
128. Li Y, Zhang W, Niu J, Chen Y (2012) Mechanism of photogenerated reactive oxygen species and correlation with the antibacterial properties of engineered metal-oxide nanoparticles. ACS Nano 6(6):5164–5173. https://doi.org/10.1021/nn300934k
129. Chaurasia AK, Thorat ND, Tandon A, Kim J-H, Park SH, Kim KK (2016) Coupling of radiofrequency with magnetic nanoparticles treatment as an alternative physical antibacterial strategy against multiple drug resistant bacteria. Sci Rep 6(1):1–13. https://doi.org/10.1038/srep33662
130. El Zowalaty ME, Al Ali SHH, Husseiny MI, Geilich BM, Webster TJ, Hussein MZ (2015) The ability of streptomycin-loaded chitosan-coated magnetic nanocomposites to possess antimicrobial and antituberculosis activities. Int J Nanomed 10:3269. https://doi.org/10.2147/IJN.S74469
131. Hsueh Y-H, Ke W-J, Hsieh C-T, Lin K-S, Tzou D-Y, Chiang C-L (2015) ZnO nanoparticles affect Bacillus subtilis cell growth and biofilm formation. PLoS ONE 10(6):e0128457. https://doi.org/10.1371/journal.pone.0128457
132. Kruk T, Szczepanowicz K, Stefańska J, Socha RP, Warszyński P (2015) Synthesis and antimicrobial activity of monodisperse copper nanoparticles. Colloids Surf B Biointerfaces 128:17–22. https://doi.org/10.1016/j.colsurfb.2015.02.009

Cellulose and Chitin Nanofibers: Potential Applications on Wound Healing

Athira Johnson, M.S. Neelakandan, Jiya Jose, Sabu Thomas, and Nandakumar Kalarikkal

1 Introduction

"Wound" is defined as an injury to the living tissue caused by a scratch, cut, shock, burn, or another impact in which the skin is cut or broken. Wound healing is a step by step process that includes four major stages, hemostasis, inflammation, proliferation, and maturation. Blood factors have a vital role in this biological process [1]. Human skin is composed of three layers, namely, epidermis, dermis, and hypodermis. The outer most layer epidermis is hydrophobic in nature. Connective tissues, hair follicles, and sweat glands are found in dermis layer beneath the epidermis. Inner layer hypodermis consists of fat and connective tissues. The epidermis is rich in keratinocyte and dermis associated with the presence of collagen. Generally, the wound is arisen with swelling, redness, pain, bleeding, and weakening in function. There are two types of wound called acute and chronic. During the past, plant extract and honey are used for healing. Now, these methods are changed by the invention of advanced technologies like nanotechnology and biotechnology. Bionanotechnology knew to be a trend in research field help to plays role in biomedical research. Concentrating onto wound healing, attraction toward biopolymers in nano form has largely increased, which is able to promote cell proliferation, reduce inflammation, and facilitate cell migration. Cellulose and chitin are widely used biopolymers for wound healing due to their availability in nature and attractive properties. Formulation of cellulose nanofibers

A. Johnson · M.S. Neelakandan · J. Jose · S. Thomas (✉)
International and Inter University Centre for Nanoscience and Nanotechnology, Mahatma Gandhi University, Kottayam 686560, Kerala, India
e-mail: sabuthomas@mgu.ac.in

N. Kalarikkal
School of Chemical Sciences, Mahatma Gandhi University, Kottayam 686560, Kerala, India

S. Thomas
School of Pure and Applied Physics, Mahatma Gandhi University, Kottayam 686560, Kerala, India

A. K. Nayak et al. (eds.), *Biomedical Composites*, Materials Horizons: From Nature to Nanomaterials, https://doi.org/10.1007/978-981-33-4753-3_6

and chitin nanofibers helps to overcome the problems associated with ordinary polymers. Glucose monomers connected by a β 1–4 linkage to form cellulose and chitin are derived by elongation of N-acetyl glucosamine units. Cellulose nanofibers and chitin nanofibers were fabricated by electrospinning method with good antimicrobial, anti-inflammatory, and healing properties.

2 Process of Wound Healing and Trends in Wound Dressing

Wound formation by an injurydestroys the integrity of the skin layers. If a wound is formed, the healing process initiates with the formation of a clot. Generally, the healing process is a four-stage process. Removal of microorganisms and facilitation of the activity of growth factors and immune cells like neutrophils, macrophages, endothelial cells, keratinocytes, and fibroblast occurs in the initial stage called hemostasis. New blood vessels were formed in the second stage, angiogenesis [2]. In a wound part, various intracellular and extracellular path way is activated to restore normal structure and functions of the skin. During proliferation, the wound is rebuilt and new tissues are formed to restore the normal structure of the skin. The wound is completely recovered its functions and structure at remodeling phase. Most of the case, the extracellular matrix disorganized and form scar. Depends on the time to heal, the wound is categorized in to acute and chronic. An acute wound is an injury occurs suddenly and heals at an expected rate. A chronic wound develops when any acute wound fails to heal in the expected time, it last for a long term. The process of wound healing is directed by the activity of growth factor called cytokines. Not only the cell growth, migration, matrix formation, differentiation, and enzyme expression but also wound healing activity was controlled by growth factors called cytokines. The process of healing starts when the platelets come in contact with collagen results in the formation of fibrin clot. During this time, platelet produces two important signals called platelet derived growth factor (PDGF), that initiate chemotaxis together with mitogenesis process and transforming growth factor-β (TGF-β) responsible for initiating healing cascade. As a result, connective tissue is deposited at the site of injury and inflammatory stage starts. Within 24 h neutrophils arrives at the wounded area and remove damaged matrix, bacteria, and foreign material. The inflammatory process is accompanied by the signs of redness, heat, swelling, and pain. Macrophage cell is known to be the most important inflammatory cell in wound healing process. Inflammatory phase is followed by proliferative stage. Epithelialisation and angiogenesis were taken place during this stage. Fibroblast begins to produce collagen. Finally, the remodeling of collagen happens and the skin is regenerated [3].

The mechanical distraction of tissue and inflammation are two main issues that arise during an injury formation. Ancient times, human exploits the benefit of natural resources such as honey, animal fat, and plant extract for healing the wound. Honey provides a moist healing environment, rapidly clears the infection, and reduces inflammation, edema, and exudation; promote angiogenesis, granulation, and epithelialization. Some better materials like bandages, cotton wool, lint, and gauze were

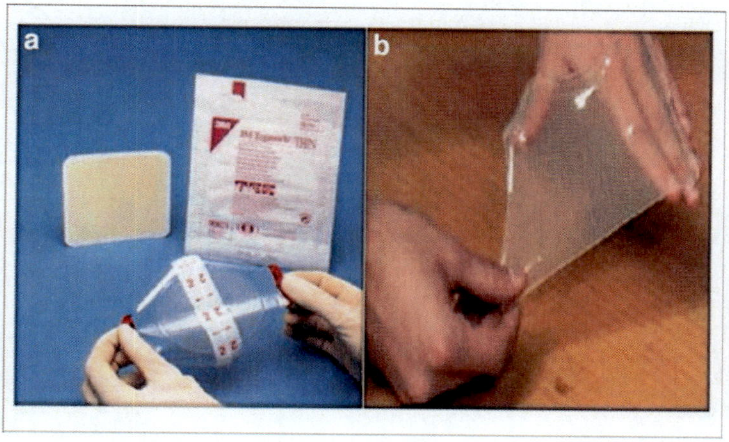

Fig. 1 **a** Hydrocolloid and **b** hydrogel (Boateng et al. 2007)

used for covering the wounded area and thereby preventing inflammation, keeping the wound dry, and allow evaporation in later times [4]. The use of green tea, mushroom, bronze instrument, and other plant powder are used in Chinese medicine which promotes tissue granulation and reduces infection. There is strong evidence for using honey as a wound healing agent from Mesopotamian times. Before dressing, they wash the wound with milk and water. In Greece, the wounded area is washed with vinegar, wine [5]. Adherence is the major problem associated with wound healing process in ancient technique. The first non-adherent dressing material was a double-layered material prepared by using cotton, balsam, and paraffin [6]. In 1947, Blaine developed Calcium alginate extracted from brown seaweed which can be made up of fibers, films, and forms for wound curing [7]. Another type of wound dressing material is called hydrocolloids developed in 1982. It is made up of sodium carboxymethyl cellulose, gelatin, polyisobutylene, and pectin [8]. Hydrogels are another type formed from polymers; they can provide a moisture environment for the dry wounded area. They are able to possess antimicrobial activity. The picture of hydrocolloid and hydrogel are illustrated in Fig. 1.

Polyurethrane foams are also used as dressing materials. The scaffold is the group of dressing material that have rigid nature and promote the infiltration of cells during the healing process [9]. A wound dressing material should possess some properties such as water vapor permeability, sterilization, anti-microbial activity, oxygen permeability, non-toxicity, and biodegradability.

3 Cellulose and Cellulose Nanofibers (CNFs)

Cellulose is known to be the most abundant biopolymer in the world.Anselme Payen, a French Chemist discovered and isolated cellulose from green plants in 1938. It is

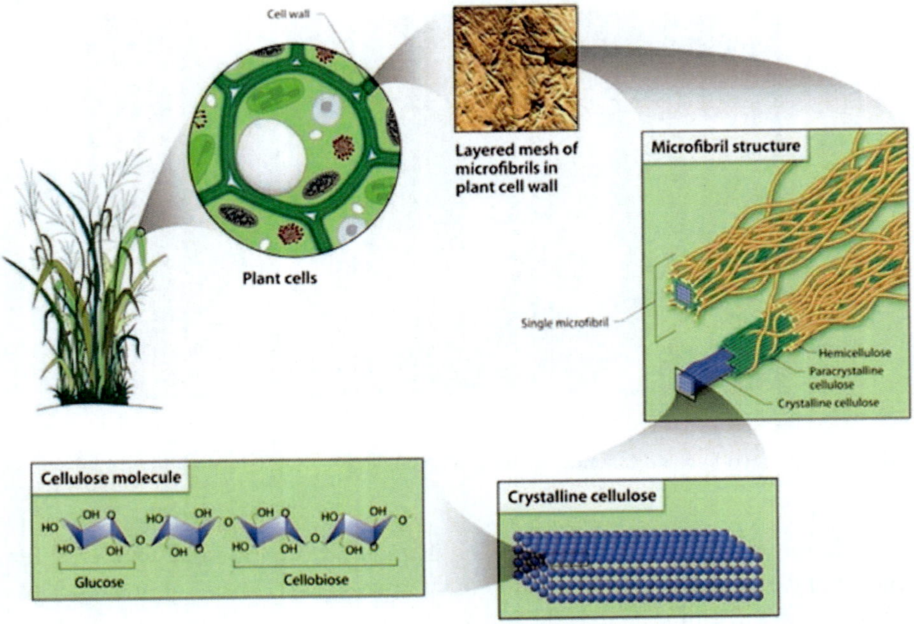

Fig. 2 Illustration of the structure of cellulose from plant cell wall (https://public.ornl.gov/site/gallery/detail.cfm?id=181)

the main constituent of the plant and is also presents in bacteria, fungi, animals, and algae [10]. Cellulose provides protection together with a structural framework for plants and helps to maintain their rigidity. It is insoluble in water and present in almost all part of the plant such as stem, leaves, and stalks. Cellulose $(C_6H_{10}O_5)_n$ is a polysaccharide composed of a linear chain of several hundred to many thousands of $\beta(1 \rightarrow 4)$ linked D-glucose units. Figure 2 illustrates the distribution of cellulose in plants.

Cellulose present in the form of microfibrils with asize between 2 to 50 nm. The crystallize microfibrils may exist in the form of cellulose Iα and Iβ and will be organized further to form the layers, cell walls, and fiber. The number of glucose unit is depended on the location where they found, which means, about 8000 glucose unit per chain (DP = 8000) is present on the primary cell wall, while in the case of the secondary cell wall is seems to be around 15,000 units per chain [11]. Apart from that, cellulose nanofibers (CNFs) have high surface area to volume ratio, high tensile strength, and low-coefficient of thermal expansion. CNFs has size <100 nm in one dimension with a length in micrometer and width in the nanometer range. Nano cellulose fibrils can be divided into two broad categories: amorphous form (cellulose nanofibers) and crystalline form (cellulose nanocrystals). Currently, green nanotechnology is widely spread in almost all part of the world and the attraction toward this area has largely increased especially for fabricating nanocomposite, gas barrier films, and biomedical applications [12]. Commonly CNFs are produced by high-pressure homogenization, refining, and grinding [13].

Quality and standard of dressing materials are important for measuring the rate of healing process. Mechanical parameters like flexibility, tensile strength, and elastic property of the material is considered during dressing material preparation. Stretchy and soft membrane is easy to handle and readily detached from skin. Air permeability of the wound dressing material prevents the activity of bacteria by the open pores present over it. As compared with hydrophobic material, hydrophilic materials are more efficient because they can absorb wound exudates. Controlling of moisture retention is significant to prevent dehydration and exudates formation. Wound infection is reduced by the fluid uptake ability of wound dressing material [15]. Due to the similarities with extra cellular matrix (ECM), biocompatible and biodegradable nature of CNFs is suitable for wound healing.

Ancient methods are replaced by the advancement in technology; the situation is turned into an immense world of research and studies regarding wound healing. Flexibility and porosity of CNFsmake them very attractive material for wound healing. There are certain criteria required for choosing a material as "wound healing material" such as maintaining sterile condition and moisture balance, allow gas exchange, ability to absorb cell exudates, nontoxicity, and non-adherent to wound [14]. Liu et al., 2012 shows that nanofibrous membranes prepared from cellulose acetate (CA) and polyester urethane (PEU) perform good wound healing activity. CA fibers can improve the hydrophilicity and permeability to air and moisture. They prepared by using the blend electrospinning method. Polyhexamethylene biguanide (PHMB) was added to this fibers to prevent clinical infections.

One of the main issue relatedto wound healing material is inflammation, arises due to the activity of microbes which results in the delay of the healing process. The bandage is prepared by the immobilization of cell lytic enzyme called lysostaphin on to the biocompatible fibers {obtained from the solution of cellulose, cellulose-chitosan, and cellulose-poly (methyl methacrylate) (PMMA)} by electrospinning method. Results show that it has an antimicrobial activity against *S. aureus* with low toxicity toward keratinocytes [15]. The fibers prepared by electrospinning are associated with the high surface to volume ratio and length to diameter ratios. Because of these advantages, ultrafine nanofibers exhibit stronger antimicrobial activity than natural fibers. [16] prepared a solution of cellulose acetate ultrafine fibers with silver nanoparticle by an electrospinning method and evaluate the antimicrobial property. The results show that this combination has good activity against microbes [16]. Bacterial cellulose associated in bacterial cell wall, incorporated with silver nanoparticle demonstrated that it performs significant antimicrobial activity against *Staphylococcus aureus*. From this experiment, it is clearly evident that this combination can reduce inflammation and promote wound healing [17]. *Centella asiatica* extracts with cellulose acetate nanofibers are used to heal the wound. In this case, the percentage of water uptake and diameter of the fibers are increased but the contact angle between fiber and water droplet is decreased with increase in solution content. Considering the percentage of water uptake is a remarkable factor for wound healing process. The combination of nanofibers and *Centella asiatica* extract helps to provide a suitable condition for the healing process [18]. Hydrogels are polymer network extensively swollen in water with a high rate of absorption. Currently, there are lots of research focused on the preparation and modification of hydrogel for wound healing application. *Acetobacter xylinum* is a bacteria modified to form 2, 3 dialdehyde cellulose (DABC) by an oxidation process using sodium metaperiodate as an oxidizing agent and this DABC loaded with chloramphenicol (CAP) for preparing a hydrogel. Studies display that, this hydrogel has good biocompatibility, biodegradability, and antimicrobial activity [19]. Nanofibrous membrane mimics extra cellular matrix (ECM) because of its high porosity and high surface area makes them proper material for good wound healing. The ECM consists of proteins and polysaccharides. Electrospun cellulose acetate and gelatin membranes promote proliferation during skin regeneration and perform as a scaffold for wound healing applications [20]. BC is excellent for cleaning and penetration of active substance to the wounded area. Nanocomposite films prepared by BC and montmorillonite (MMT) have good antibacterial activity against *Escherichia coli* and *Staphylococcus aureus* with high tissue regeneration capacity. The BC-MMT composite indicates the attachment and penetration of MMT particles in BC matrix; this penetration provides strength to the composite [21].

Aerogels are formed by removing theliquid part from a gel without damaging the network through freeze drying (lyophilization) or critical point drying [22]. CNFs obtained from wood powder are oxidized by sodium periodate to form dialdehyde CNFs and introduction of collagen into this CNFs forms a matrix. Collagen grows in dialdehyde CNFs and forms composite aerogels. They have a high porosity between 90 and 95% and strong water absorption up to 4000 times. They also exhibit biocompatibility and active proliferation. It support wound healing process [23]. In conclusion, Cellulose nanofibers wound materials have good porosity, surface to volume ratio, biocompatibility, biodegradability, low density, and unique morphology. Due to these characteristics, it is widely employed in wound healing applications [24]. Jin et al. [25] prepared cellulose nanofibrous aerogel by the dissolution of cellulose in calcium thiocyanate followed by drying. Aerogels prepared by solvent exchange drying are good because it has high porosity and specific surface area [25] (Fig. 3).

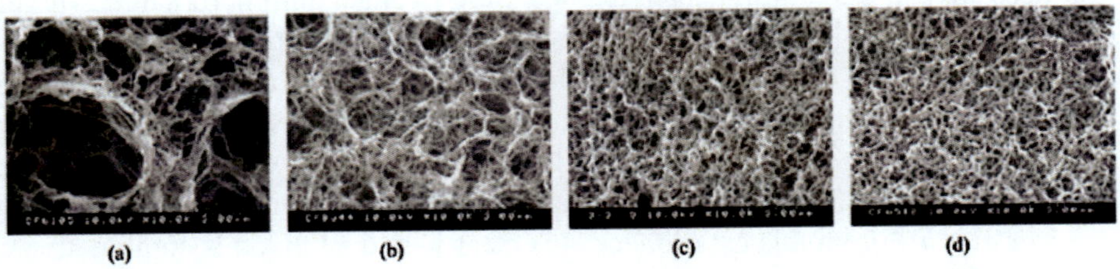

Fig. 3 SEM of CF11 cellulose aerogels by solvent exchange drying **a** 0.5%; **b** 1.0%; **c** 2.0% and **d** 3.0%

4 Chitin and Chitin Nanofibers (CNFs)

Chitin $(C_3H_{13}O_5N)_n$ is a biopolymer formed by the monomer N-acetylglucosamine linked through β 1–4 linkage, which is a primary component of the cell wall in animals, fungi, the exoskeleton of crustaceans, insects, and the scales of fish and lissamphibians. It is a white, hard, nitrogenous polysaccharide with a high percentage of nitrogen (6.89%) [26].

Min et al. studied theformation of chitin nanofibers by an electrospinning method with an average diameter of 110 nm and concluded that they can form a matrix for wound healing in 2004. The solubility of chitin was improved by depolymerizing with gamma irradiation. The chitin matrix gets converted into chitosan matrix by deacetylation [27]. Chitin is known to be the second most abundant biopolymer after cellulose. Compared to cellulose, chitin has low attention from the research field due to its poor solubility and deacetylation of chitin into chitosan. Major investigations associated with chitin in biomedical fields related to wound healing applications. Noh et al. illustrated that these fibers with extracellular matrix proteins like type 1 collagen can promote cell attachment, distribution of keratinocyte and fibroblast during wound healing [28]. The scaffold prepared from chitin has high oxygen permeability, porosity, the morphological similarity with ECM, and promote cell adhesion and cell proliferation. Chitin prepared from shrimp is treated with butyric anhydride results in a water-soluble chitin derivative called Dibutyrylchitin (DBC). This compound elevates glycosaminoglycans (GAG) level and promotes granulation. Charernsriwilaiwat N et al. and his co-workers prepare hydrogels from chitin nanofibers and are also applicable for wound healing process. The wound made on the back of the rat was completely re-epithelialized while using hydrogels [29]. The wound healing activity is evaluated by using lysozyme loaded mixture of chitin fiber-ethylenediaminetetraacetic acid show potential activity during healing [30].

Kang et al. 2012 show that heat-treated polyvinyl alcohol (H-PVA)—chitosan nanofibrous matrix is suitable for wound dressing. Compared to control PVA, chitosan loaded H-PVA shows good collagen regeneration and thick epidermis formation [31]. Chitin nanofibers are prepared by melt blown, self-assembly, spinning, and phase separation [32]. Chen et al. prepared a composite nanofibrous membrane (NFM) with the use of chitosan, type 1 collagen, and polyethylene oxide in 2008, with a fiber diameter of 134 ± 42 nm. It had been reported that Young's modulus and diameter can be increased by adding cross-linker and, water absorption capacity, tensile strength, and strain are decreased. [33] show that 3T3 fibroblast cells are used for measuring cytotoxicity. From the experiment, it was understood that there is no cytotoxicity and good biocompatibility. Animal studies show that NFM is better than normal gauze [33]. The composite composed by using chitosan and silk fibroin show good anti-microbial activity against *Escherichia Coli* and *Staphylococcus aureus* and have good proliferation capacity during wound healing [34]. The composite prepared from β-Chitin, and zinc oxide nanoparticle is used as bandages during wound healing process. In vivo evaluation in Sprague Dawley rat show faster healing and high collagen deposition [35]. During inflammation, the attack of microorganism is increased. The scaffold prepared from chitin and nanosilver is a good candidate for biocompatibility, antimicrobial activity, and blood clotting [36].

By radiation-induced cross-linking, hydrogels are prepared from carboxymethyl chitosan (CM-Chitosan) and gelatin solution. When chitosan content is increased, the swelling ability is also increased but the compressive modulus decreased. After lyophilisation, there is a homogenous pore structure seen. It promotes cell attachment and fibroblast formation [37].

Park et al. [38] formulatedPolyglycolic acid/Chitin blend nanofibers for wound healing applications. They have an average diameter of 140 nm. As can be seen from Fig. 4 their study concluded that these nanofibers coated with bovine serum albumin

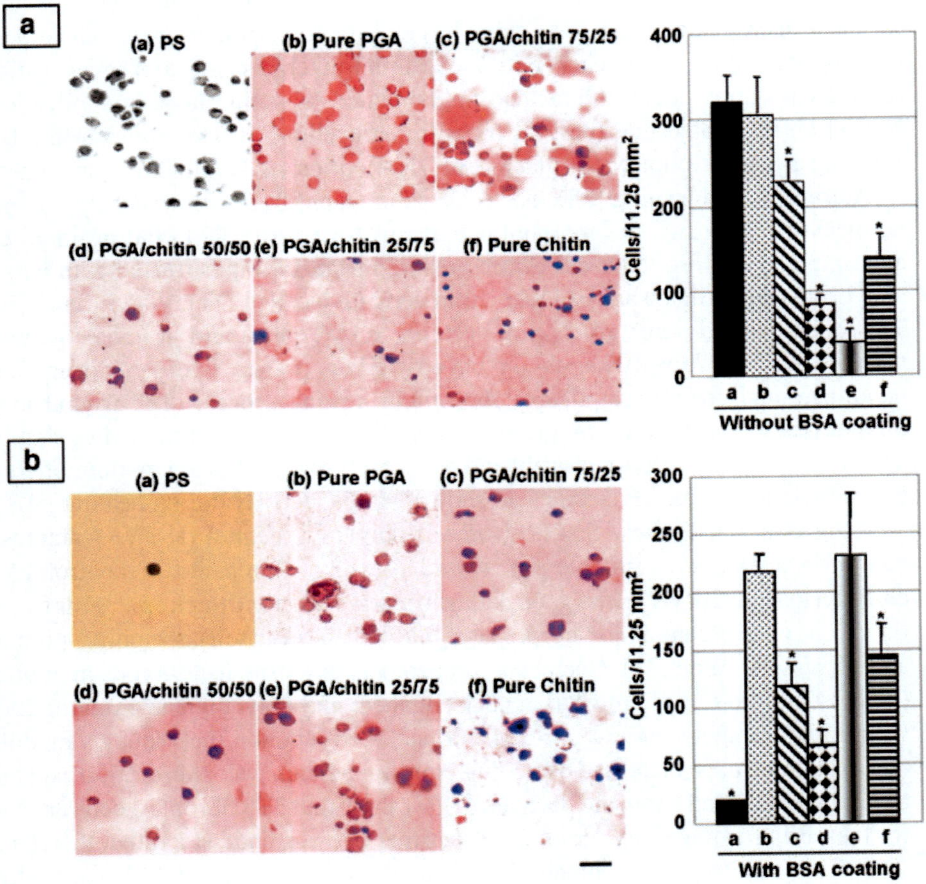

Fig. 4 Cell attachment of proliferating NHEF plated onto PGA/chitin blend nanofibrous matrixes. Microphotographs and level of cell attachment of cultured cells onto PGA/chitin blend nanofibrous matrixes without **a** or with **b** BSA coating. Data are expressed as mean (SD (n) 4). ANOVA: P < 0.05. Pairwise comparisons: *P < 0.05 versus pure PGA nanofibers. **a** polystyrene surface, **b** pure PGA, **c** PGA/chitin (75/25), **d** PGA/chitin (50/50), **e** PGA/chitin (25/75), and **f** pure chitin. Bar 100 ím

promote cell attachment and migration [38]. Requirements for wound dressing are increased the time to time according to the development of new technologies. Prevention of infection and rapid curing are the two major conditions for every dressing material. The materials which provide growth factors and antimicrobial agents get a good impact on the clinical area. Plant and animal derived product are also employed for this purpose.

5 Conclusion

Cellulose and chitin are known to be the most abundant polymers in nature. Cellulose is obtained from plant, fungi, algae, and animal-like tunicates. Chitin is obtained primarily from the exoskeleton of crustaceans and insects. Cellulose and chitin nanofibers are capable of the wound healing process. They are manipulated mainly by electrospinning method. Studies show that when transformed into nano range the properties of fibers are changed and possess advantages than the ordinary one. This fiber can be framed into composite, hydrogel, aerogel, and thin films, etc. They are non-toxic and biocompatible in nature. They are able to promote cell migration, cell proliferation, and prevent infection by the activity microorganism.

References

1. Broughton G, Janis JE, Attinger CE (2006) The basic science of wound healing. Plast Reconstr Surg 117(Suppl.):12S
2. Gurtner GC, Werner S, Barrandon Y, Longaker MT (2008) Wound repair and regeneration. Nature 453:314–319
3. Diegelmann RF, Evans MC (2004) Wound healing: an overview of acute, fibrotic and delayed healing. Front Biosci 9:283–289
4. Boateng JS, Matthew KH, Steven HN, Eccleston GM (2008) Wound healing dressings and drug delivery systems: a review. J Pharm Sci 97(8):2892–2923
5. Majno G (1991) The healing hand: man and wound in the ancient world, 1st paperback. Harvard University Press, Cambridge, Mass
6. Jones AM, San ML (2006) Are modern wound dressings a clinical and cost-effective alternative to the use of gauze? J Wound Care 15(2):65–69
7. Blaine G (1947) Experimental observations on absorbable alginate products in surgery: gel, film. Gauze and Foam Ann Surg 125(1):102–114
8. Dumville JC, O'Meara S, Deshpande S, Speak K (2011) Hydrogel dressings for healing diabetic foot ulcers. Cochrane Database Syst Rev 9:CD009101
9. Juris S, Mueller A, Smith B, Johnston S, Walker R, Kross R (2011) Biodegradable polysaccharide gels for skin scaffolds. J Biomater Nanobiotechnol 2(3):216–25
10. O' Sullivan A.C. (1997) Cellulose: the structure slowly unravels. Cellulose 4(3):173–207
11. Brown RM (2003) Cellulose structure and biosynthesis: what is in store for the 21st century? J Polym Sci, Part a: Polym Chem 42:487–495
12. Chirayil CJ, Mathew L, Thomas S (2014) Review of recent research in nano cellulose preparation from different lignocellulosic fibers. Rev Adv Mater Sci 37:20–28

13. Wang B, Sain MK (2007) Study of structural morphology of hemp fiber from the micro to the nanoscale. Appl Compos Mater 14:89–103
14. Sharma J, Lizu M, Stewart M, Zygula K, Lu Y, Chauhan R, Yan X, Guo Z, Wujcik EK, Wei S (2015) Multifunctional nanofibers towards active biomedical therapeutics. Polymers 7:186–219
15. Miao J, Pangule RC, Paskaleva EE, Hwang EE, Kane RS, Linhardt EJ, Dordick JS (2011) Lysostaphin-functionalized cellulose fibers with antistaphylococcal activity for wound healing applications. Biomaterials 32:9557–9567
16. Son VK, Youk JH, Lee TS, Park WH (2004) Preparation of antimicrobial ultrafine cellulose acetate fibers with silver nanoparticles. Macromol Rapid Commun 25:1632–1637
17. Wu J, Zheng Y, Wen X, Lin Q, Chen X, Wu Z (2014) Silver nanoparticle/bacterial cellulose gel membranes for antibacterial wound dressing: investigation in vitro and in vivo. Biomed Mater 9(3):12
18. Naris B, Anchan M (2014) Preparation of cellulose acetate nanofibers containing centella asiatica extract by electrospinning process as the prototype of wound-healing materials. J Bionanoscience 8(4):313–318(6)
19. Laçin NT (2014) Development of biodegradable antibacterial cellulose based hydrogel membranes for wound healing. Int J Biol Macromol 67:22–27
20. Vatankhah E, PrabhakaranM P, Jin G, Mobarakeh LG, Ramakrishna S (2014) Development of nanofibrous cellulose acetate/gelatin skin substitutes for variety wound treatment applications. J Biomater Appl 28(6):909–921
21. Ul- IM, Khan T, Khatakk VA, Park JK (2013) Bacterial cellulose-MMTs nanoreinforced composite films: novel wound dressing material with antibacterial properties. Cellulose 20:589–596
22. Tan CB, Fung BM, Newman JK, Vu C (2001) Organic aerogels with very high impact strength. Adv Mater 13:644
23. Lu T, Li Q, Chen W, Yu H (2014) Composite aerogels based on dialdehyde nanocellulose and collagen for potential applications as wound dressing and tissue engineering scaffold. Compos Sci Technol 94:132–138
24. Zhang Y, Nypelö T, Salas C, Arboleda J, Hoeger IC, Rojas OJ (2013) Cellulose nanofi brils: from strong materials to bioactive surfaces**. J Renew Mater 1(3):195–211
25. Jin H, Nishiyama Y, Wada M, Kuga S (2004) Nanofibrillar cellulose aerogels: colloids and surfaces a: physicochem. Eng Aspects 240:63–67
26. Majeti NV, Kumar R (2000) A review of chitin and chitosan applications. React Funct Polym 46:1–27
27. Min BM, Lee SW, Lim JN, You Y, Lee TS, Kang PH, Park WH (2004) Chitin and chitosan nanofibers: electrospinning of chitin and deacetylation of chitin nanofibers. Polymer 45(21):7137–7142
28. Noh HK, Lee SW, KimJ M, Oh JE, KimK.W, Chung CP, Choi SC, Park WH, Min BM (2006) Electrospinning of chitin nanofibers: degradation behavior and cellular response to normal human keratinocytes and fibroblasts. Biomaterials 27:3934–3944
29. Jayakumar R, Prabaharan M, Kumar PTS, Nair SV, Tamura H (2011) Biomaterials based on chitin and chitosan in wound dressing applications. Biotechnol Adv 29:322–337
30. Charernsriwilaiwat N, Opanasopit P, Rojanarata T, Ngawhirunpat T (2012) Lysozyme-loaded, electrospun chitosan-based nanofiber mats for wound healing. Int J Pharm 427:379–384
31. Kang YK,Yoon IS, Lee SY, Kim DD, Lee SJ, Park WH, Hudson SM (2009) Chitosan-coated poly(vinyl alcohol) nanofibers for wound dressings. Wiley Periodicals, Inc. J Biomed Mater Res Part B: Appl Biomater 568–576
32. Yao KD, Tu H, Cheng F, Zhang JW, Liu J (1997) pH-sensitivity of the swelling of a chitosan-pectin polyelectrolyte complex. Angew Makromol Chem 245:63–72
33. Chen JP, Chang JY, Chen JK (2008) Electrospun collagen/chitosan nanofibrous membrane as wound dressing: colloids and surfaces a: physicochem. Eng Aspects 313–314:183–188
34. Cai ZX, Mo XM, Zhang KH, Fan LP, Yin AL, He CL, Wang HS (2010) Fabrication of chitosan/silk fibroin composite nanofibers for wound-dressing applications. Int J Mol Sci 11:3529–3539

35. Kumar PTS, Lakshmanan VK, Raj M, Biswas R, Hiroshi T, Nair SV, Jayakumar R (2013) Evaluation of wound healing potential of β-chitin hydrogel/nano zinc oxide composite bandage. Pharm Res 30:523–537
36. Madhumathi K, Sudheesh Kumar PT, Abhilash S, Sreeja V, Tamura H, Manzoor K, Nair SV, Jayakumar R (2010) Development of novel chitin/nanosilver composite scaffolds for wound dressing applications. J Mater Sci: Mater Med 21:807–813
37. Yang C, Xu L, Zhou Y, Zhang X, Huang X, Wang M, Han Y, Zhai M, Wei S, Li J (2010) A green fabrication approach of gelatin/CM-chitosan hybrid hydrogel for wound healing. Carbohyd Polym 82(4):1297–1305
38. Park KE, Kang HK, Lee SJ, Min BM, Park WH (2006) Biomimetic nanofibrous scaffolds: preparation and characterization of PGA/Chitin blend nanofibers. Biomacromol 7(2):635–643

Bionanocomposites for In Situ Drug Delivery in Cancer Therapy: Early and Late Evaluations

Luiza Steffens Reinhardt, Pablo Ricardo Arantes, Jeferson Gustavo Henn, and Dinara Jaqueline Moura

Abbreviations

α-CD	A-cyclodextrin
BBB	Blood brain barrier
CT	Chemotherapy
CTC	Circulation tumor cells
DCA	Sodium dichloroacetate
DOX	Doxorubicin
DPPE	Dipalmitoyl phosphatidylethanoiamine
MWCNT	Multi-walled carbon nanotubes
NF	Nanofiber
NP	Nanoparticle
PCL	Polycaprolactone
PDLLA	Polylactide
PDT	Photodynamic therapy
PEG	Poly (ethylene glycol)
PEI	Poly(ethylene imine)
PEO	Polyethylene glycol
PLA	Polylactic acid
PLGA	Poly (lactic-co-glycolic acid)
PLLA	Poly-l-lactic acid
pNIPAM	Poly(N-isopropylacrylamide)
PS	Photosensitizer
PTX	Paclitaxel
PVA	Polyvinyl alcohol

L. S. Reinhardt (✉) · P. R. Arantes · J. G. Henn · D. J. Moura
Laboratory of Genetic Toxicology, Federal University of Health Sciences of Porto Alegre – UFCSPA, Sarmento Leite 245, Lab.714, Porto Alegre, Rio Grande do Sul, Brazil
e-mail: luizasteffens@live.com

A. K. Nayak et al. (eds.), *Biomedical Composites*, Materials Horizons: From Nature to Nanomaterials, https://doi.org/10.1007/978-981-33-4753-3_7

ROS	Reactive oxygen species
RT	Radiotherapy
SPIONs	Superparamagnetic iron oxide NPs
T1	Two-photon absorption compound
TMZ	Temozolomide
VEGF	Vascular endothelial growth factor

1 Cancer Outline

Cancer is a general word for a complex category of diseases described by unregulated cellular processes such as growth and death [1]. Theoretically, all cancers develop as an outcome of mutations in the DNA sequence, which may be hereditary, provoked by environmental dynamics or a consequence of replication errors of DNA [2]. These explanations are elucidated by the carcinogenesis multistep model (Fig. 1) comprising initiation, promotion, and progression stages [3, 4]. Initiation begins when healthy cells are damaged by contact with physical, biological, or chemical carcinogens resulting in mutations in the DNA [5]. Promotion is described by the clonal growth of initiated cells, inhibiting apoptosis [6]. Lastly, tumor progression is described by a malignant phenotype expression, including characteristics such as uncontrolled growth, genomic instability, metastases development, and modifications in the morphological and biochemical features of cells [6, 7].

Conventional cancer management embraces surgical procedure, systemic chemotherapy (CT), and radiotherapy (RT). The CT therapeutics frequently damage normal cells developing severe side effects [8, 9]. An alternative to the treatment is a nanotechnological intervention since it has transformed the cancer therapy by overcoming existing restrictions of conventional CT, and offering better targets with sustained and controlled drug release also improving both distribution and pharmacokinetics [9–11].

2 Drug Delivery for Cancer: General Reflections

Targeted drug delivery for cancer therapy refers, quantitatively and selectively, to drug accumulation within tumor tissue [12, 13]. Subsequently, targeted medicine or targeted therapy means precise drug efficiency combine with minor side effects and interaction, at the molecular level, between a biomolecule and a drug. As a matter of fact, the concentration of drugs and biomolecules must be high at the tumor site, while its concentration in healthy non-target tissue should be almost undetectable preventing systematic negative reactions [12, 13].

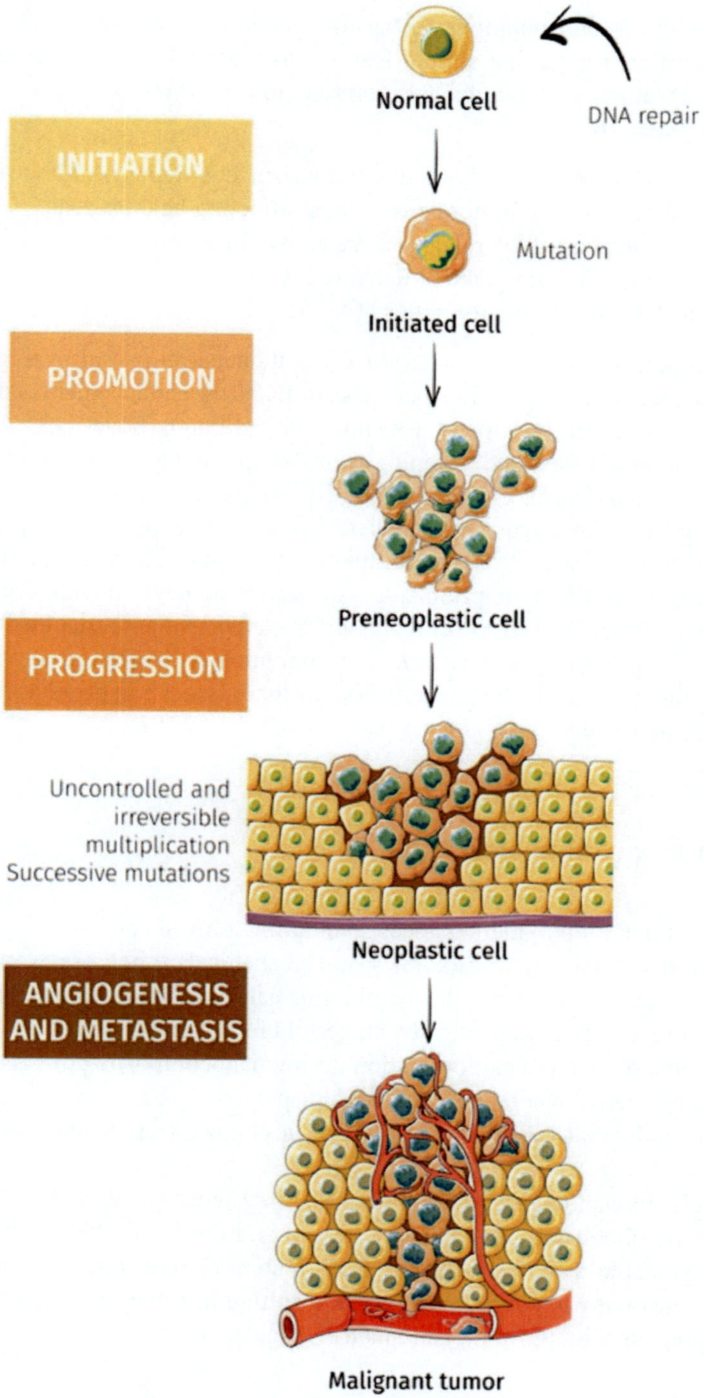

Fig. 1 Carcinogenesis multistep model comprising initiation, promotion, and progression stages

Preferably, for formulations to be efficient in cancer therapy, they should be biofunctionalized including simple features as size and surface charge and multifunctional features as biomolecules combination. Furthermore, they need to fulfill the requirement to:

a. Enhance drug pharmacodynamics and pharmacokinetics profiles.
b. Selectively eradicate tumor cells without affecting healthy cells.
c. Prolong and control the release of the active form of drugs.
d. Improve the cellular uptake of delivered drugs.
e. Diminish dose-limiting toxicities [14, 15].

In all assurance, to reach a successful CT without harming healthy tissues has been the key effort of targeting approaches. It is estimated the enhancement of drug delivery to cancer cells with an improved well-developed nanoproduct [12, 13], however, even with intense research developing in the bionanocomposites field, no significant results with meaningful changes in the life of patients have been accomplished. It is a challenge to bypass drug delivery issues such as early drug degradation, bloodstream circulation time, tissue membranes, and systemic toxicity, therefore, for some cancers, the main promising approach to achieve suitable drug delivery is an in situ platform. In fact, several nanoproducts present an initial burst release that can cause acute toxicity or release the total amount of a therapeutic compound before reaching the tumor cells, fortunately, both of these issues can be addressed by using local drug delivery.

3 Bionanocomposites

It is known that biopolymers possess some limitations in contrast to synthetic polymers in terms of stability, transparency, and mechanical properties nevertheless, these drawbacks can be overcome by combining nanosized materials while producing bionanocomposites. Among all advantages of bionanocomposites, this approach can avoid nanoparticle (NP) agglomeration during nanocomposite preparation by using a polymeric matrix where the NP can be dispersed [16], moreover, nanocomposite biodegradability increases after producing a composite with nanosized materials [17].

By rule, bionanocomposite is a two-phase system, where, at least one part had a nanosized dimension (lower than 100 nm) and the word "bio" indicates the use of biodegradable material [18]. The main feature of bionanocomposites is a large surface area owing to nanosized material resulting in enhanced interaction between its components with the composite matrix (Fig. 2) [19].

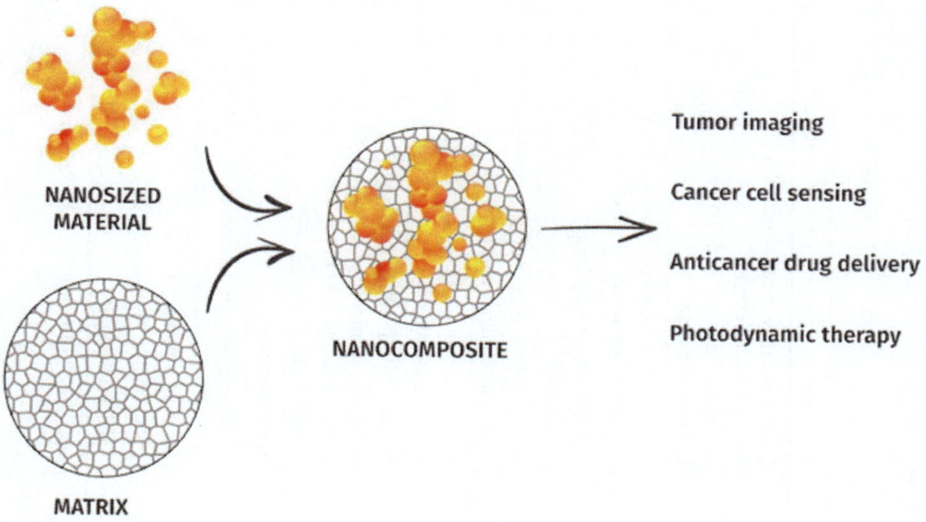

Fig. 2 Features of bionanocomposites

3.1 How to Choose the Right Polymer

In the last decade, the use of synthetic biodegradable polymers as drug delivery systems increased significantly [20], mainly because synthetic polymers made of petrochemicals are toxic and take a long time to degrade. Biodegradable polymers (Table 1) can be synthetic such as polyglycolic acid (PGA), polylactic acid (PLA), polycaprolactone (PCL) and poly-l-lactic acid (PLLA), or natural such as collagen, chitosan, and silk [35, 36].

Features of biocomposite materials such as optical and mechanical properties and conductivity usually are improved from bulk material although matrix composition is the same [37]. Polymers as PCL, poly (ethylene glycol) (PEG), PLA, and poly (lactic-co-glycolic acid) (PLGA) that are synthetic and biodegradable are commonly used as substrates for the production of scaffolds. Moreover, during the past years, other polymers with stimuli-responsive behavior such as poly(*N*-isopropylacrylamide) (pNIPAM) and poly(4-vinylpyridine) have been investigated as thermo and pH-responsive [38].

It is also essential, during the development of biocomposites as drug delivery devices, the solvent selection, which needs to offer adequate solubility for both drug and polymer. The compatibility of polymeric mixture components and their physicochemical properties such as surface tension, conductivity, and viscosity have key effects on the resulting morphology of the carrier [39, 40]. The state of both components, their crystallinity, and their distribution within the composite have significant influences on the drug release kinetics. Furthermore, the establishment of both phases can be influenced by the addition of amphiphilic molecules, which can facilitate the release of hydrophobic drugs to the aqueous environment [41, 42].

Table 1 Features of biodegradable polymers

	Polymer	Biodegradation	Mechanical properties	Physicochemical properties	Key-points	References
Synthetic polymers	PLGA	Hydrolysis	Strong	Amphiphilic	Good solubility, weak toxicity, application in several biomedical devices and fast biofilm formation	[21, 22]
	PLA	Hydrolysis	Strong	Hydrophilic	Weak toxicity and application in several biomedical devices	[23, 24]
	PCL	Hydrolysis	Strong	Hydrophobic	Application in long-term implantable devices	[25]
	PVP	Hydrolysis	Strong	Hydrophilic	Good solubility and application as antimicrobial nanofibers	[26, 27]
	PVA	Enzymatic	Strong	Hydrophilic	Good solubility, nontoxic, and non-carcinogenic	[28, 29]
	PEG	Enzymatic	Weak	Hydrophilic	Good solubility, high range of viscosity, contains toxic impurities, and is nephrotoxic if applied to damaged skin	[30]
Natural polymers	Collagen	Enzymatic	Weak	Hydrophilic	Weak toxicity	[31]
	Chitosan	Enzymatic	Weak	Hydrophilic	Molecular weight dependent solubility, weak toxicity, and several applications in pharmaceutical industry	[32, 33]
	Silk	Enzymatic	Strong	Hydrophilic	Moderate toxicity	[34]

PLGA poly (lactic-co-glycolic acid), *PLA* poly (lactic acid), *PCL* poly (Ɛ-caprolactone), *PVP* polyvinylpyrrolidone, *PVA* polyvinyl alcohol, *PEG* polyethylene glycol

3.2 *How to Investigate Polymer-Drug Interactions*

Polymers present particular characteristics which range from the angstrom level of an individual bond between atoms to nanometers of the polymer chain, micrometers, millimeters, larger in solutions, and polymeric nanofibers. Due to its particular characteristics, the best way to investigate the polymer-drug interactions is through in silico studies [43–47]. The different time scales for each material properties may range from femtoseconds to seconds or even hours. On the literature, there are many examples of the multiscale nature of polymer systems [48–55]. Because of this, several computational methods were developed in order to address these issues [56–64]. These novel methods introduce new possibilities to design, optimize, and predict the structures and properties of polymers. Not only that, these methods allowed the study of these materials with other molecules [43–47]. In the case of this chapter, is the possibility of studying the polymer-drug interactions. One of the main methods to study polymer-drug interactions, nowadays, is the atomistic calculations where all atoms are explicitly represented and treated by a single sphere. The force field, typical interactions in the system, is responsible for the potential energy of the system. These interactions include the bonded interactions that are the bond length, the bond angle, and the dihedral angle potentials between atoms. In addition to the bonded interactions, force fields also contain non-bonded interactions. Non-bonded interactions act between atoms in the same molecule and those in other molecules. Force fields usually divide non-bonded interactions into two: electrostatic interactions and Van der Waals interactions.

Molecular dynamics (MD) simulations are one example used to model molecular processes involving a larger group of atoms, such as drugs, biological compounds [65], carbohydrates [66–69], proteins, membranes [68, 69], nucleic acids, and polymers [43–45]. On the MD simulations studies, usually, to analyze the polymer-drug interactions the root means square deviation (RMSD) [43], solvent accessible surface area (SASA) [43], interaction energy [44, 45], number of contacts [46], and number of hydrogen bonds [46, 47] are the main choices. With these analyses, it is possible to study the polymer behavior in the presence and absence of the drug, verify the drug exposition and interaction to the solvent, verify the drug affinity for the polymer, and calculate the interaction number of contacts between the polymer and drug. The study of polymer-drug interactions requires a comprehensive knowledge of the phenomena. The computational methods, mainly MD simulations allow the study of these systems at the atomistic scale. With this method it is possible to verify some important features on the polymer-drug interaction, solving problems that experimental studies are not able to.

3.3 Current Achievements in Bionanocomposites for In Situ Drug Delivery

Recently, several researchers are developing bionanocomposites for in situ drug delivery aiming to address issues caused by systemic CT (Table 2). A significant drawback in cancer therapy is that 90% of patients with cancer die from cancer metastasis [75, 76]. Metastasis is a complex multistep process comprising tumor cells release and their proliferation at a secondary site [77, 78]. Nowadays, nanotechnology has been addressing this issue with controllable and functionalized structures able to recognize circulating tumor cells (CTCs) [79, 80], nevertheless few products are able to induce in situ CT in blood circulation. Chunmiao Liu and colleagues (2019) developed a multifunctional nanocomposite by conjugating FePt NPs onto the surface of Fe_3O_4 @SiO_2 NPs, which, in tumor pH, was able to release Fe^{2+} catalyzing reactive oxygen species (ROS) formation and inducing ferroptosis in tLyP-1 receptor-positive CTCs such as MCF-7, HeLa, HepG2, and 4T1. Thus, this system could effectively separate and capture CTCs suggesting it as a promising approach for in situ cancer treatment [73].

Angiogenesis displays a critical role in tumor proliferation and development in which, vascular endothelial growth factors regulate the development of new blood vessels and act binding tyrosine kinase receptors including vascular endothelial growth factor (VEGF) receptor 1 and 2 [81, 82]. As expected, these receptors are overexpressed in several tumors and have been studied as a target for treatments [83, 84]. Recently, Carenza et al. developed PLGA nanocapsules containing superparamagnetic iron oxide NPs (SPIONS) and VEGF165 in a polymeric shell able to

Table 2 Bionanocomposites for in situ drug delivery

System	Drug	Outcome	References
Silica-calcium phosphate nanocomposite (SCPC75) discs	Cisplatin	Lower tumor growth in the rats treated with SCPC75-cisplatin discs when compared to systemic cisplatin	[70]
CuS nanoparticles-graphene oxide with polyethylene glycol (PEO-GO/CuS)	DOX	75% cancerous cells death after irradiation at NIR laser	[71]
Albumin, PLGA and magnetic nanoparticles with diphenylhexatriane	5-Fluorouracil	Reduction of tumor size	[72]
FePt nanoparticles	–	Ferroptosis induction in tLyP-1 receptor-positive CTCs	[73]
PLGA nanocapsules with SPIONs	VEGF165	Improved accumulation in the tumor cells	[74]

CuS copper sulfide; *DOX* dooxorubicin, *PLGA* poly (lactic-co-glycolic acid), *FePt* Iron–platinum, *CTCs* circulation tumor cells, *SPIONs* superparamagnetic iron oxide nanoparticles accumulate in the tumor cells [74], showing that magnetic bionanocomposites could be used in targeted and controlled drug delivery.

Multifunctional nanocomposites can be produced as biofunctionalized drug delivery systems that combine the drug transport with biomolecules such as proteins, enzymes, antibodies, nucleic acids, and membrane receptors [85]. Biomolecules are not able to penetrate some body barriers such as the brain-blood barrier (BBB), which creates the necessity of novel drug delivery routes for solid tumors treatment.

3.3.1 Injectable Hydrogels

Photodynamic therapy (PDT) stands for a therapeutic that sensitizes cells to a particular wavelength of light. The radiation and the photosensitizer (PS) are inoffensive when used separately however when combined, they can form several ROS and trigger local inflammation, cell apoptosis, and localized ischemia. Therefore, PDT is very effective as cancer in situ therapy [86]. Recently, Luo and coworkers (2019) established a micellar thermosensitive nanocomposite gel for intra-tumoral two-photon PDT by developing methoxy PEG-polylactide copolymer (mPEG-PDLLA) micelles loaded with PS and two-photon absorption compound (T1). The liquid characteristic of the micelles enabled an in situ gelification after injection maintaining the composite into the breast cancer tumor model and it was shown that the radiation promoted deep tissue penetration and enhanced tumor inhibition in around 50% [87].

Another study aiming PDT developed an injectable nanocomposite hydrogel with chitosan micelles loaded with paclitaxel (PTX) and PEGylated gold nanorods to treat local hepatocellular carcinoma by using xenograft tumor model [88]. After laser irradiation, the photothermal-induced damage was restrained to the tumor without damaging the adjacent healthy tissue and the sustained release of PTX enhanced the antitumor effect of treatment prolonging mice survival compared to PDT alone [88]. These studies indicate that photothermal-CT associated with nanocomposites present improved antitumor effects by suppressing tumor recurrence and prolonging survival.

Glioma is a common central nervous system tumor that presents no significant improvement in therapy in the past years, therefore, the average of patients' survival sits still around 14 months [89]. The challenges in drug delivery for this type of cancer include the existence of the BBB that provides an impediment to standard CT routes of delivery [90]. It was confirmed by Bodell et al. [91] that in situ drug delivery enhances by 113 times the drug concentration when compared to intravenous injection. By exploiting this approach, Ding and colleagues (2017) developed a nanocomposite thermoresponsive hydrogel using mPEG-dipalmitoyl phosphatidylethanoiamine (mPEG-DPPE) calcium phosphate NP and the double emulsion method for PTX and temozolomide (TMZ) delivery. Their results suggest that the autophagy induced by the combination of the drugs regulated tumor cell apoptosis and that the local delivery of the gel on C6 tumor-bearing rats improved significantly the treatment antitumor efficacy [92].

An injectable hydrogel system based on the polypseudorotaxane development between cationic mPEG-b-PCL-b-poly(ethylene imine) (PEI) and α-cyclodextrin (α-CD) was developed for the delivery of plasmid DNA, more specific the Nur77 gene. Given the addition of cyclic α-CD, this biocomposite could form solid hydrogels in situ, owing to the poly pseudorotaxane establishment between α-CDs and PEG. The authors showed that the controlled drug release was able to improve the tumor resistance in vivo [93].

Even that hydrogel studies seem promising for local therapy of cancer, there are some drawbacks to their application. If solid, the samples are usually very stiff and this makes the administration challenging in soft tissues and if liquid, the variable release and distribution could trigger side effects in healthy cells. Thus, different approaches are required for improving biocomposite hydrogels or introducing malleable materials with enhanced release properties such as nanofibrous drug delivery systems.

3.3.2 Electrospinning

The electrospinning method can be used for fibrous bionanocomposites as nanofibers (NF) combined with nanocarriers. The equipment consists of a conductive collector plate, a pump, a high voltage power apparatus, a tip needle, and a syringe containing the polymeric solution [37]. When the solution is loaded in the equipment, the high voltage is applied and the drop in the edge of the needle is transformed into a Taylor cone producing nanofibrous material jetted in the collector plate. By using this method, some features of the nanoproduct can be modulated such as diameter and drug release rate by choosing the adequate properties of the polymeric solution composition as polymeric concentration, viscosity, elasticity, conductivity, and the distance between the collector and the needle [38].

Generally, electrospinning is used for producing NF blends of drugs and polymers, however, it can be used to fabricate fibrous nanocomposites. These systems are constructed by the incorporation of other nanocarriers including nanoparticles (NP), vesicles, and micelles into the NF (Fig. 3). Table 3 shows several recent studies with promising NF for cancer applications.

It is known that the simple incorporation of drugs into nanocarriers commonly leads to burst drug release and to overcome this disadvantage the introduction of these nanocarriers into NF can expand the applicability of nanocarriers as cancer drug delivery, improving their release rate, stability, and acute toxicity. Lately, NF systems aiming at local drug delivery for both tumor therapy and prevention have been significantly studied by researchers worldwide. Table 4 presents an overview of studies focused on the design and evaluation of fibrous bionanocomposite in vivo. Within these studies, the treatment was focused on solid tumors therapy. These NF can increase the specificity of drugs associated with the tumor microenvironment and given their release properties tumor recurrence can be diminished.

One of the most frequent forms of gynecological tumors is cervical cancer, presenting an incidence rate of 25 years in young adults [103]. This type of tumor is

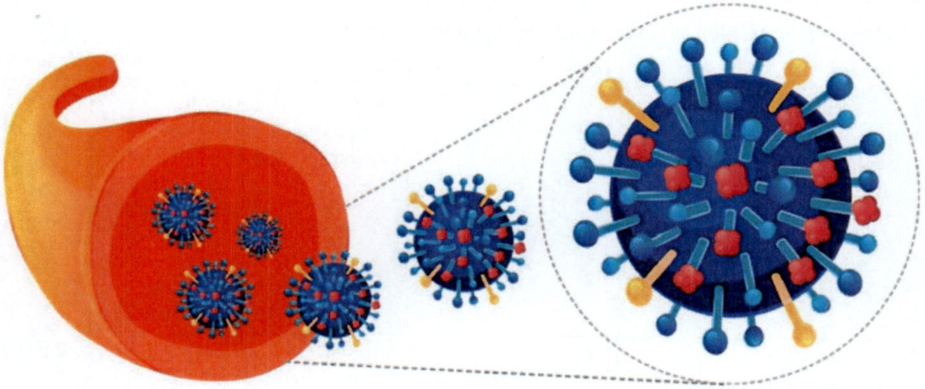

Fig. 3 Fibrous nanocomposites scheme constructed by the incorporation of nanocarriers into nanofibers

developed as a consequence of oncogenic types of human papillomavirus infection [104] and regardless of the mortality and incidence rates are considerably decreasing over the last few years in developed countries, patients with stage IV tumor have a survival rate of only 5 years occasioning in more than 270,000 deaths each year around the world [105]. The cervical cancer therapy depends on the cancer stage varying from surgical resection and RT at stages I to III and CT as a palliative treatment of stage IV (advanced stage or recurrent tumor). Amid the drugs used for the therapy the most common drugs are cisplatin (Pt), ifosfamide, doxorubicin (DOX), and sodium dichloroacetate (DCA) [106]. Albeit platinum compounds present high efficacy against cervical carcinoma, their toxicity on healthy tissues and their systemic toxicity is a drastic therapy problem. To overcome this issue, Zhang and colleagues [107] produced a system by using polyvinyl alcohol (PVA) NF loaded with DCA and Pt(IV)-backboned micelles [107]. Firstly, DCA release induced cell death by apoptosis through the inhibition of pyruvate dehydrogenase kinase and the micelles could highly accumulate in tumor cells via enhanced permeation and retention effect. In tumor pH, Pt(IV) is reduced to Pt(II) that is its active form which can cause DNA crosslinks and consequently inhibition of DNA replication. The nanoproduct presented improved in vitro cytotoxicity against HeLa cells when the two drugs were combined showing their synergistic effect. The in vivo evaluation was performed by using Kumming mice injected with the cervical cancer cell line U14 in the armpit, and the results showed an enhanced drug accumulation on the tumor of mice treated locally with NF when compared to intravenously treated mice. The tumor volume analyses after 16 days of treatment revealed that micelles-DCA-NF reduced tumor volume when compared to single-drug NF and intravenous injection. This study showed the importance of local treatment with synergic drug effects implying that this implantable bionanocomposite could be useful for successful therapy reducing system risks [107].

Table 3 Fibrous bionanocomposites for in situ drug delivery

System	Drug	Cancer	Outcome	References
Core-shell NF Core: iron oxide nanoparticles Shell: PS and collagen NF	–	SKOV-3 ovarian cancer cell line	After the application of an AMF during 10 min to the mats, the cancer cells deposited on the Fe_3O_4-in-fibers were eliminated	[94]
Core-shell NF Core: cobalt ferrite and titanium oxide nanoparticles Shell: DOX-loaded chitosan NF	DOX	B16F10 melanoma cell line	Higher cell death was observed when hyperthermia and chemotherapy were combined, achieving a synergistic effect, which enhances the cytotoxic effect and allows a reduction of side effects	[95]
Core-shell NF Core: vesicles of CTAB and SDBS with 5-Fluorouracil and paenolum Shell: PEO	5-Fluorouracil and paenolum	–	The release investigations concluded that the hydrophilic drug was released in an increased manner when the molar ratio of CTAB/SDBS was higher	[96]
Core-shell NF Core: three different vesicles composed of didodecyldimethylammonium bromide, CTAB/SDBS (7/3) and CTAB/SDBS (3/7) introduced into chitosan and sodium alginate nanocapsules Shell: PEO	5-Fluorouracil	–	The different drug delivery systems showed different release rates and pH-responsive behaviors	[96]
Core-shell NF Core: pluronic (F127) vesicles Shell: PCL NF	Rhodamine-B	L929 murine fibroblast cell line	Reduction in the release rate and increased cell viability	[97]

(continued)

Table 3 (continued)

System	Drug	Cancer	Outcome	References
Electrospun of a mesh using 10% of a hydrophobic PGC-C18 and 90% of PCL loaded with the SN-38	SN-38	HT-29 colorectal cancer cell line	Decreased cell viability	[98]
Core-shell NF Core: colloidal structures formed by a block copolymer composed by mPEG-PLA with 5-Fluorouracil and cefradine Shell: chitosan and PEO	5-Fluorouracil and cefradine	HepG-2human liver cancer cell line	Decreased cell viability	[99]
Core-shell NF Core: silica and hydroxyapatite nanoparticles containing DOX and hydroxycamptothecin Shell: PLGA	DOX and hydroxycamptothecin (topoisomerase inhibitor)	HeLa cell line	Increased thermal stability and inhibited tumor cells growth	[100]
Core-shell NF Core: silica nanoparticles with DOX Shell: PLLA	DOX	HeLa cell line	Sustained and prolonged release and higher antitumor efficacy	[101]
TMZ and polycaprolactone diol combined with gold nanoparticles	TMZ	U-87 MG glioblastoma cell line	The nanoformulation achieves a greater cell death overtime in contrast with free TMZ (25% more)	[102]

NF nanofiber, *PS* polystyrene, *AMF* alternating magnetic field, *DOX* dooxorubicin, *CTAB* cetyl trimethylammonium bromide, *SDBS* sodium dodecylbenenesulfonate, *PEO* polyethylene glycol, *PCL* polycaprolactone, *PGC* poly(glycerol monostearate-co-ε-caprolactone, *mPEG-PLGA* methoxypoly(ethyleneglycol)-block-poly(L-lactide), *PLGA* poly (lactic-co-glycolic acid), *PLLA* poly-l-lactic acid, *TMZ* temozolomide

Another study with cervical cancer focused on CT in combination with hyperthermia by producing PLLA NF loaded with DOX and multi-walled carbon nanotubes (MWCNT) [108]. During the in vitro drug release studies it was found that when near-infrared irradiation was applied, DOX was released in a burst rate in both acid and physiologic pH, however, when the radiation was stopped DOX release returned to slow mode, this behavior was explained by the relation between polymer mobility and the diffusion rate of DOX. By using Kunming mice implanted with U14 cells, this effect was also observed in vivo after 48 h of NF implantation into the tumor and irradiation treatment, moreover, it was detected that DOX accumulated in the tumor tissue more than in other vital organs. Finally, the in vivo antitumor effect of the nanocomposite was assessed by placing the material directly on the

Table 4 Overview of studies focused on the design and evaluation of fibrous bionanocomposite in vivo

System	Drug	Cancer	Outcome	References
Core-shell Core: DCA and Pt(IV)-micelles Shell: PVA NF	DCA, Pt(IV)	Cervical	*In vivo* release showed 58% of Pt released during 30 min, 90% of Pt released during 24 h and the tumor volume on day 15 was found to be half of the untreated control group tumor size	[107]
Core-shell Core: DOX and MWCNTs Shell: PLLA NF	DOX	Cervical	Enhanced DOX release after irradiation and preferable accumulation of DOX in the tumor than in vital organs	[108]
Core-shell Core: PVA micelles loaded with DOX and activefolate groups on the surface cross-linked with gelatin Shell: PCL/PEO	DOX	Breast	Drug concentration increased in the tumor site after 48 h (18 μg/g) when compared to free DOX (2 μg/g) and other organs; NF-treated animals presented a significate increased survival rate	[111]
Thixotropic silk NF hydrogels loaded with DOX	DOX	Breast	The system was capable of solidifying at a specific site and releasing to certain pH conditions. The *in vivo* studies: at the fifth week, showed significant differences in the volume and weight of the tumor treated with DOX-loaded hydrogels compared to free DOX	[112]

DCA sodium dichloroacetate, *Pt(IV)* platinum(IV), *NF* nanofiber, *DOX* dooxorubicin, *MWCNTs* multi-walled carbon nanotubes, *PLLA* poly-l-lactic acid, *PVA* polyvinyl acid, *PEO* polyethylene glycol, *PCL* polycaprolactone

tumor site and after 24 h of implantation the site was irradiated and as expected the MWCNT-DOX NF were effective against cancer cells, nevertheless, cells detected profound inside the tumor presented a certain resistance. Analyzing this study, it was possible to conclude that CT and PDT using MWCNT-DOX-loaded PLLA fibrous composite could have a synergic effect on antitumor activity [108].

A critical type of cancer in relation to mortality worldwide is breast cancer. This type of disease is the leading cause of women's deaths related to cancer, with more than half a million deaths [105] and nearly 2.5 million new cases in 2015 [109]. One important characteristic of this cancer is that the recurrence rate within 10 years can reach 15% even after resection [110]. The standard therapy consists of mastectomy, RT, systemic CT, or hormonal blockade. However, systemic treatment is associated with severe cardiotoxic effects.

Aiming to diminish DOX side effects and reduce the risk of recurrence, Yang and colleagues [111] produced a complex active-targeting system. Firstly, DOX micelles were developed by using PCL-PEO copolymer with active folate groups on their surface. Then, core-shell NF were produced by coaxial electrospinning where the micelles were mixed with PVA to form the core phase and a cross-linked gelatin solution was used as the shell. After NF implantation, the micelles were released from the NF as a consequence of polymeric matrix degradation, then they were able to penetrate into the tumor cells and undergo endocytosis throughout binding to the folate receptors on the cells' membranes. To perform in vivo biodistribution and antitumor activity, 4T1 tumor-bearing Balbc nude mice were used. The distribution results showed a strong DOX accumulation at the tumor when compared to other organs such as heart, lung, kidney, and liver, suggesting that this bionanocomposite is an effective and safe system for DOX delivery. The antitumor activity evaluation compared two different deliveries: (a) intravenous injection of free DOX or micelles and (b) implantation of NF near the tumor. The NF exhibited a pronounced tumor volume reduction after 21 of implantation when compared to the other groups, in addition, the NF-treated mice presented increased body weights and survival rates, and therefore this system presents significant improvements to the breast cancer therapy [111].

It is known that most frequent actions for in vitro release studies embrace:

(a) cumulative release that occurs when a compound is released into the same amount of media volume and;
(b) non-cumulative release that occurs when there is a continuous replacement of the media mimicking a living organism where the drug concentration drops.

Related approaches are used in in vitro biological analysis focus on cytotoxic experiments on different cell lines. Though experimental conditions of in vitro evaluations simulate those of living organisms, for a system as bionanocomposites to be acknowledged as suitable and safe for local antitumor treatment, in vivo investigations are still essential. Unfortunately, only a few studies tested its bionanocomposites products in vivo, therefore the local application is still a challenge given the invasive procedures and the lack of in vivo evaluations.

4 Final Considerations

Bionanocomposites show great properties related to effective therapy of cancer such as increased thermal stability, enhanced tensile strength, reduced side effects and improved targeting. Drug delivery obtained by in situ approaches has the prospective to overcome the CT limitations by their capacity to increase drug concentration in the tumor and in the same instance decreasing the damage to the healthy tissue. Though bionanocomposites may enable the efficacy of individualized drug delivery, issues such as the difficulty of testing the scaffolds and to develop surfaces that are able to maintain an appropriate environment may affect the drug distribution. It is necessary, in the future of bionanocomposites research, for a collaboration among researchers, regulatory agencies, and industry to ensure that novel approaches are safe and effective and will be provided to the society. Further development in bionanocomposite technology may prove potent, safe, controlled, and targeted drug delivery.

References

1. WHO (2017) World Health Organization. What is cancer? www.who.int/cancer/en/
2. Tomasetti C, Li L, Vogelstein B (2017) Stem cell divisions, somatic mutations, cancer etiology, and cancer prevention. Science 355:1330–1334
3. Siddiqui IA, Sanna V, Ahmad N, Sechi M, Mukhatar H (2015) Resveratrol nanoformulation for cancer prevention and therapy. Ann NY Acad Sci 1348:20–31
4. Klaunig JE (2015) Chemical carcinogenesis. In: Roberts SM, James RC, Williams PL (eds) Principles of toxicology: environmental and industrial applications, 3rd edn. Willey, New Jersey, pp 259–282
5. Abel EL, DiGiovanni J (2011) Multistage carcinogenesis. In: Penning TM (ed) Chemical carcinogenesis. Springer, New York, pp 27–52
6. Oliveira PA, Colacao A, Chaves R, Guedes-Pinto H, De-La-Cruz LE, Lopes C (2007) Chemical carcinogenesis. Anais da Academia de Ciencias 79:593–616
7. NCBI (2017) National center for biotechnology information. Multistage carcinogenesis. https://www.ncbi.nlm.nih.gov/books/NBK13982/
8. NCI (2017) National cancer institute. Common cancer types. https://www.cancer.gov/types/common-cancers
9. Muzzalupo R, Tavano L (2016) Advances on magnetic nanocarriers based on natural polymers. Curr Pharm Design 22:3353–3363
10. Wang M, Thanou M (2010) Targeting nanoparticles to cancer. Phamacol Res 62:90–99
11. Zaimy MA, Saffarzadeh N, Mohammadi A, Pourghadamyari H, Izadi P, Sarli A et al (2017) New methods in the diagnosis of cancer and gene therapy of cancer based on nanoparticles. Cancer Gene Ther 24:233–243
12. Torchilin VP (2010) Passive and active drug targeting: drug delivery to tumors as an example. In: Schafer-Korting M (ed) Handbook of experimental pharmacology. Springer, Heidelberg, pp 3–53
13. Bae YH, Park K (2011) Target drug delivery to tumors: myts, reality and possibility. Control Release 153:198–205
14. Yingchoncharoen P, Kalinowski DK, Richardson DR (2016) Lipid-based drug delivery systems in cancer therapy: what is available and what is yet to come. Pharmacol Rev 68:701–787
15. Cho K, Wang XU, Nie S, Shin DM (2008) Therapeutic nanoparticles for drug delivery in cancer. Clin Cancer Res 14:1310–1316
16. Chivrac F, Pollet E, Schmutz M, Averous L (2008) New approach to elaborate exfoliated starch based nanobiocomposite. Biomacromol 9:896–900
17. Sothornvit R, Hong SI, Duck JA, Rhim JW (2010) Effect of clay content on the physical and antimicrobial properties of whey protein isolate/organo-clay composite films. LWT-Food Sci Technol 43:279–284
18. Zhang Z, Yuan L, Wang J, Zhao H, Chen J (2012) Irbesartan drug formulated as nanocomposite particles for the enhancement of the dissolution rate. Particuology 10:462–467

19. Shinde SM, Payghan SA, D'souza JI (2014) Physiochemical assessment of pharmaceutical salt Forms: A quality attribute. Int Res J Invent Pharm Sci 2:46–53
20. Sahoo R, Sahoo S, Nayak P (2011) Controlled release of the drug cefadroxil from polycaprolactone-polylactic acid nanocomposites. Eur J Sci Res 53:154–162
21. Said SS et al (2011) Antimicrobial PLGA ultra-fine fibers: interaction with wound bacteria. Eur J Pharm Biopharm 79(1):108–118
22. Garg T et al (2013) Stimuli-Sensitive hydrogels: an excellent carrier for drug and cell delivery. Crit Rev Therapeut Drug Carrier Syst 30(5):369–409
23. Iwakate A (2008) Cellulose nanofiber-reinfirced polylactic acid. Compos Sci Technol 68(9):2103–2106
24. Holan V et al (2011) Cyclosporine A-loaded and stem cell-seeded electrospun nanofibers for cell-based therapy and local immunosuppression. J Controlled Release 156(3):406–412
25. Cho WJ et al (2009) Hydrophillized polycaprolactone nanofiber mesh embedded poly (glycolic-co-lactic acid) membrane for effective guided bone regeneration. J Biomed Mater Res A. 91(2):400–407
26. Ignatova M et al (2007) Electrospinning of poly(vinyl pyrrolidone)–iodine complex and poly(ethylene oxide)/poly(vinyl pyrrolidone)–iodine complex—a prospective route to antimicrobial wound dressing materials. Eur Polymer J 43(5):1609–1623
27. Zhang K et al (2011) Preparation and characteristics of electrospun multiwalled carbon nanotube/polyvinyl pyrrolidone nano composite nanofiber. J Nanosci Nanotechnol 11(6):5446–5449
28. Briscoe B et al (2000) The effects of hydrogen bonding upon the viscosity of aqueous poly(vinyl alcohol) solutions. Polymer 41(10):3851–3860
29. Cai W, Gupta RB (2000) Hydrogels: Kirk-Othmer encyclopedia of chemical technology. Wiley, Hoboken, pp 1–20
30. Krause TL, Bittner GD (1990) Rapid morphological fusion of severed myelinated axons by polyethylene glycol. Proc Natl Acad Sci USA 87(4):1471–1475
31. Marisca OT et al (2015) Comparison of the *in vitro* uptake and toxicity of collagen and synthetic polymer-coated gold nanoparticles. Nanomaterials 5(3):1418–1430
32. Kean T, Thanou M (2010) Biodegradation, biodistribution and toxicity of chitosan. Adv Drug Deliv Rev 62:3–11
33. Islam S et al (2017) Chitin and chitosan: structure, properties and applications in biomedical engineering. J Polym Environ 25:854–866
34. Naserzadeh P et al (2018) Evaluation of the toxicity effects of silk fibroin on human lymphocytes and monocytes. J Biochem Mol Toxicol 32(6):1–7
35. Nanda R, Sasmal A, Nayak PL (2011) Preparation and characterization of chitosan-polylactide composites blended with cloisite 30B for control release of the anticancer drug paclitaxel. Carbohydr Polym 83:988–994
36. Lau AK, Hussain F, Lafdi K (2010) Biopolymeric nanofibers for tissue engineering. In: Nano- and biocomposites. CRC Press, New York, pp 157–92
37. Patwekar SL, Jamkhande P, Gattani SG, Payghan SA (2016) Nanobiocomposite: a new approach to drug delivery system. Asian J Pharm 10:S646
38. Tzounis L et al (2019) Temperature-controlled catalysis by core-shell-satellite AuAg@pNIPAM@Ag hybrid microgels: a highly efficient catalytic thermoresponsive nanoreactor. ACS Appl Mater Interfaces 11(32):29360–29372
39. Sharma R, Singh H, Joshi M, Sharma A, Garg T, Goyal AK, Rath G (2014) Recent advances in polymeric electrospun nanofibers for drug delivery. Crit Rev Ther Drug Carrier Syst 31:187–217
40. Sirc J, Hobzova R, Kostina N, Munzarova M, Juklickova M, Lhotka M, Kubinova S, Zajicova A, Michalek J (2012) Morphological characterization of nanofibers: methods and application in practice. J Nanomater 327–369
41. Goonoo N, Bhaw-Luximon A, Jhurry D (2014) Drug loading and release from electrospun biodegradable nanofibers. J Biomed Nanotechnol 10:2173–2199

42. Hrib J, Sirc J, Hobzova R, Hampejsova Z, Bosakova Z, Munzarova M, Michalek J (2015) Nanofibers for drug delivery: incorporation and release of model molecules, influence of molecular weight and polymer structure. Beilstein J Nanotechnol 6:1939–1945
43. Jafari M, Doustdar F, Mehrnejad F (2019) Molecular self-assembly strategy for encapsulation of an amphipathic α-helical antimicrobial peptide into the different polymeric and copolymeric nanoparticles. J Chem Inf Model 59(1):550–563
44. Tallury SS, Pasquinelli MA (2010) Molecular dynamics simulations of flexible polymer chains wrapping single-walled carbon nanotubes. J Phys Chem B 114(12):4122–4129
45. Tallury SS, Pasquinelli MA (2010) Molecular dynamics simulations of polymers with stiff backbones interacting with single-walled carbon nanotubes. J Phys Chem B 114(29):9349–9355
46. Kyrychenko A, Pasko DA, Kalugin ON (2017) Poly(vinyl alcohol) as a water 4 protecting agent for silver nanoparticles: The role of polymer size and structure. Phys Chem Chem Phys 19(13):8742–8756
47. Weng L, Stott SL, Toner M (2018) Molecular dynamics at the interface between ice and poly(vinyl alcohol) and ice recrystallization inhibition. Langmuir 34(17):5116–5123
48. Chiu CW, Huang TK, Wang YC, Alamani BG, Lin JJ (2014) Intercalation strategies in clay/polymer hybrids. Prog Polym Sci 39(3):443–485
49. Gao D, Li R, Lv B, Ma J, Tian F, Zhang J (2015) Flammability, thermal and physical-mechanical properties of cationic polymer/montmorillonite composite on cotton fabric. Compos B Eng 77:329–337
50. Lowe DJ, Chapman AV, Cook S, Busfield JJ (2011) Micromechanical models of young's modulus of NR/organoclay nanocomposites. J Polym Sci, Part B: Polym Phys 49(22):1621–1627
51. Mattausch H, Laske S, Duretek I, Kreith J, Maier G, Holzer C (2013) Investigation of the influence of processing conditions on the thermal, rheological and mechanical behavior of polypropylene nanocomposites. Polym Eng Sci 53(5):1001–1010
52. Decker JJ, Meyers KP, Paul DR, Schiraldi DA, Hiltner A, Nazarenko S (2015) Polyethylene-based nanocomposites containing organoclay: a new approach to enhance gas barrier via multilayer coextrusion and interdiffusion. Polymer 61:42–54
53. Gooneie A, Nazockdast H, Shahsavan F (2015) Effect of selective localization of carbon nanotubes in PA6 dispersed phase of PP/PA6 blends on the morphology evolution with time, part 1: Droplet deformation under simple shear flows. Polym Eng Sci 55(7):1504–1519
54. Sepahvand R, Adeli M, Astinchap B, Kabiri R (2008) New nanocomposites containing metal nanoparticles, carbon nanotube and polymer. J Nanopart Res 10(8):1309–1318
55. Moniruzzaman M, Winey KI (2006) Polymer nanocomposites containing carbon nanotubes. Macromolecules 39(16):5194–5205
56. Elliott JA (2011) Novel approaches to multiscale modelling in materials science. Int Mater Rev 56(4):207–225
57. Zeng QH, Yu AB, Lu GQ (2008) Multiscale modeling and simulation of polymer nanocomposites. Progress Polym Sci (Oxford). 33(2):191–269
58. Ayton GS, Noid WG, Voth GA (2007) Multiscale modeling of biomolecular systems: in serial and in parallel. Curr Opin Struct Biol 17(2):192–198
59. Ayyaswamy PS, Muzykantov V, Eckmann DM, Radhakrishnan R (2013) Nanocarrier hydrodynamics and binding in targeted drug delivery: challenges in numerical modeling and experimental validation. J Nanotechnol Eng Med 4(1):011001
60. Raabe D (2002) Challenges in computational materials science. Adv Mater 14(9):639–650
61. Kremer K, Müller-Plathe F (2001) Multiscale problems in polymer science: simulation approaches. MRS Bull 26(3):205–210
62. Murtola T, Bunker A, Vattulainen I, Deserno M, Karttunen M (2009) Multiscale modeling of emergent materials: biological and soft matter. Phys Chem Chem Phys 11(12):1869–1892
63. Peter C, Kremer K (2009) Multiscale simulation of soft matter systems—from the atomistic to the coarse-grained level and back. Soft Matter 5(22):4357–4366

64. Sherwood P, Brooks BR, Sansom MS (2008) Multiscale methods for macro molecular simulations. Curr Opin Struct Biol 18(5):630–640
65. Arantes PR, Polêto MD, John EBO, Pedebos C, Grisci BI, Dorn M, Verli H (2019) Development of gromos-compatible parameter set for simulations of chalcones and flavonoids. J Phys Chem B 123(5):994–1008
66. Arantes PR, Sachett LG, Graebin CS, Verli H (2014) Conformational characterization of ipomotaosides and their recognition by Cox-1 and 2. Molecules 19(4):5421–5433
67. Pedebos C, Pol-Fachin L, Verli H (2012) Unrestrained conformational characterization of Stenocereus Eruca Saponins in aqueous and nonaqueous solvents. J Nat Prod 75(6):1196–1200
68. Pedebos C, Arantes PR, Giesel GM, Verli H (2015) In silico investigation of the Pglb active site reveals transient catalytic states and octahedral metal ion coordination. Glycobiology
69. Kern NR, Lee HS, Wu EL, Park S, Vanommeslaeghe K, Mackerell AD, Im W (2014) Lipid-linked oligosaccharides in membranes sample conformations that facilitate binding to oligosaccharyltransferase. Biophys J 107(8):1885–1895
70. Swet JH, Pacheco HJ, Iannitti DA, El-Ghanam A, McKillop AH (2014) J Biomed Mater Res B: Appl Biomater 102B:190–202
71. Bai J, Liu Y, Jiang X (2014) Multifunctional PEG-GO/CuS nanocomposites for near-infrared chemo-photothermal therapy. Biomaterials 35:5805–5813
72. Misak H, Zacharias N, Song Z, Hwang S, Man KP, Asmatulu R, Yang SYJ (2013) Skin cancer treatment by albumin/5-Fu loaded magnetic nanocomposite spheres in a mouse model. Biotechnol. 164:130–136
73. Liu C, Yang B, Chen X, Hu Z, Dai Z, Yang D, Zheng X, She X, Liu Q (2019) Capture and separation of circulating cells using functionalized magnetic nanocomposites with simultaneous in situ chemotherapy. Nanotech 30:285706
74. Carenza E, Jordan O, Segundo PMS, Jirik R, Starcuk Z, Borchard G et al (2015) Encapsulation of VEGF(165) into magnetic PLGA nanocapsules for potential local delivery and bioactivity in human brain endothelial cells. J Mater Chem B. 3:2538–2544
75. Dawson SJ, Tsui DW, Murtaza M, Biggs H, Rueda OM, Chin SF et al (2013) Analysis of circulating tumor DNA to monitor metastatic breast cancer. N Engl J Med 368:1199–1209
76. Leong SP, Tseng WW (2014) Micrometastatic cancer cells in lymph nodes, bone marrow, and blood: clinical significance and biologic implications. CA Cancer J Clin 64:195–206
77. Chambers AF, Groom AC, MacDonald IC (2002) Metastasis: dissemination and growth of cancer cells in metastatic sites. Nat Rev Cancer 2:563
78. Fidler IJ (2003) The pathogenesis of cancer metastasis: the 'seed and soil' hypothesis revisited. Nat Rev Cancer 3:453
79. Bhana S, Wang Y, Huang X (2015) Nanotechnology for enrichment and detection of circulating tumor cells. Nanomedicine 10:1973–1990
80. Yoon HJ, Kozminsky M, Nagrath S (2014) Emerging role of nanomaterials in circulating tumor cell isolation and analysis. ACS Nano 8:1995–2017
81. Olsson AK, Dimberg A, Kreuger J, Claesson-Welsh L (2006) VEGF receptor signalling? In control of vascular function. Nat Rev Mol Cell Biol 7:359–371
82. Simons M (2012) An inside view: VEGF receptor trafficking and signaling. Physiology 27:213–222
83. Taurone S, Galli F, Signore A, Agostinelli E, Dierckx R, Minni A et al (2016) VEGF in nuclear medicine: clinical application in cancer and future perspectives (Review). Int J Oncol 49:437–447
84. Irizarry LR, Hambardzumyan D, Nakano I, Gladson CL, Ahluwalia MS (2012) Therapeutic targeting of VEGF in the treatment of glioblastoma. Expert Opin Ther Targets 16:973–984
85. Bamburowicz-Klimkowska M, Poplawaska M, Grudzinski IP (2019) Nanocomposites as biomolecules delivery agentes in nanomedicine. J Nanobiotechnol 17:48
86. Castano AP, Mroz P, Hamblin MR (2006) Photodynamic therapy and anti-tumor immunity. Nat Rev Cancer 6:535–545
87. Luo L, Zhang Q, Luo Y, He Z, Tian X, Battaglia G (2019) Thermosensitive nanocomposite gel for intra-tumoral two-photon photodynamic therapy. J Control Release 298:99–109

88. Zhang N, Xu X, Zhang X, Qu D, Xue L, Mo R, Zhang C (2016) Nanocomposite hydrogel incorporating gold nanorods and paclitaxel-loaded chitosan micelles for combination photothermal-chemotherapy. Inter J Pharm 497:210–221
89. Templeton A, Hofer S, Topfer M, Sommacal A, Fretz C, Cerny T, Gillessen S (2008) Extraneural spread of glioblastoma: report of two cases. Onkologie 31:192–164 PMID:18418021
90. Lawson HC, Sampath P, Bohan E, Park MC, Hussain N, Olivi A, Weingart J, Kleinberg L, Brem H (2007) Interstitial chemotherapy for malignant gliomas: the Johns Hopkins experience. J Neuro-Oncol 83:61–70
91. Bodell WJ, Bodell AP, Giannini DD (2007) Levels and distribution of BCNU in GBM tumors following intratumoral injection of DTI-015 (BCNU-ethanol). Neuro-Oncology 9:12–19 PMID:17018699
92. Ding L, Wang Q, Shen M, Sun Y, Zhang X, Huang C, Chen J, Li R, Duan Y (2017) Thermoresponsive nanocomposite gel for local drug delivery to suppress the growth of glioma by inducing autophagy. Autophagy 13:1176–1190
93. Liu X, Chen XH, Chua MX, Li ZB, Loh XJ, Wu YL (2017) Injectable supramolecular hydrogels as delivery agents of Bcl-2 conversion gene for the effective shrinkage of therapeutic resistance tumors. Adv Healthc Mater. 6:11
94. Huang C, Soenen SJ, Rejman J, Trekker J, Chengxun L, Lagae L, Ceelen W, Wilhelm C, Demeester J, Smedt SCD (2012) Magnetic electrospun fibers for cancer therapy. Adv Funct Mater 22:2479–2486
95. Radmansouri M, Bahmani E, Sarikhani E, Rahmani K, Sharifianjazi F, Irani M (2018) Doxorubicinhydrochloride—Loaded electrospun chitosan/cobalt ferrite/titanium oxide nanofibers for hyperthermictumor cell treatment and controlled drug release. Int J Biol Macromol 116:378–384
96. Li W, Luo T, Yang Y, Tan X, Liu L (2015) Formation of controllable hydrophilic/hydrophobic drug deliverysystems by electrospinning of vesicles. Langmuir 31:5141–5146
97. Nair BP, Vaikkath D, Mohan DS, Nair PD (2014) Fabrication of a microvesicles-incorporated fibrous membrane for controlled delivery applications in tissue engineering. Biofabrication 6:045008
98. Yohe ST, Herrera VLM, Colson YL, Grinstaff MW (2012) 3D super hydrophobic electrospun meshes as reinforcement materials for sustained local drug delivery against colorectal cancer cells. J Control Release 162:92–101
99. Hu J, Zeng F, Wei J, Chen Y, Chen Y (2014) Novel controlled drug delivery system for multiple drugs based onelectrospun nanofibers containing nanomicelles. J Biomater Sci Polym Ed 25:257–268
100. Chen M, Feng W, Lin S, He C, Gao Y, Wang H (2014) Antitumor efficacy of a PLGA composite nanofiber embedded with doxorubicin@MSNs and hydroxycamptothecin@HANPs. RSC Adv 4:53344–53351
101. Qiu K, He C, Feng W, Wang W, Zhou X, Yin Z, Chen L, Wang H, Mo X (2013) Doxorubicin-loadedelectrospun poly(L-lactic acid)/mesoporous silica nanoparticles composite nanofibers for potential post surgical cancer treatment. J Mater Chem B 1:4601–4611
102. Irani M, Sadeghi GMM, Haririan I (2017) The sustained delivery of temozolomide from electrospunPCL-Diol-b-PU/gold nanocompsite nanofibers to treat glioblastoma tumors. Mater Sci Eng C Mater Biol Appl 75:165–174
103. Bruni L, Diaz M, Castellsague X, Ferrer E, Bosch FX, de Sanjose S (2010) Cervical human papillomavirus prevalence in 5 continents: meta-analysis of 1 million women with normal cytological findings. J Infect Dis 202:1789–1799
104. McLaughlin-Drubin ME, Munger K (2008) Viruses associated with human cancer. Biochim Biophys Acta 1782:127–150
105. Lallukka T, Millear A, Pain A, Cortinovis M, Giussani G (2017) Mortality and causes of death collaborators. Global, regional, and national life expectancy, all-cause mortality, and cause-specific mortality for 249 causes of death, 1980–2015: a systematic analysis for the global burden of disease study 2015. Lancet 389:1459–1544

106. Bodai BI, Tuso P (2015) Breast cancer survivorship: a comprehensive review of longterm medical issues and lifestyle recommendations. Perm J 19:48–79
107. Zhang Z, Wu Y, Kuang G, Liu S, Zhou D, Chen X, Jing X, Huang Y (2017) Pt(IV) prodrug-backboned micelle and DCA loaded nanofibers for enhanced local cancer treatment. J Mater Chem B 5:2115–2125
108. Zhang Z, Liu S, Xiong H, Jing X, Xie Z, Chen X, Huang Y (2015) Electrospun PLA/MWCNTs composite nanofibers for combined chemo- and photothermal therapy. Acta Biomater 26:115–123
109. Fitzmaurice C, Allen C, Barber RM, Barregard L, Bhutta ZA, Brenner H, Dicker DJ, Chimed-Orchir O, Dandona R, Dandona L, Al E (2017) Global, regional, and national cancer incidence, mortality, years of life lost, years lived with disability, and disability-adjusted life-years for 32 cancer groups, 1990 to 2015: A systematic analysis for the global burden of disease study. Jama Oncol 3:524–548
110. Ding Q, Li Z, Yang Y, Guo G, Luo F, Chen Z, Yang Y, Qian Z, Shi S (2016) Preparation and therapeutic application of docetaxel-loaded poly(D, L-lactide) nanofibers in preventing breast cancer recurrence. Drug Deliv 23:2677–2685
111. Yang G, Wang J, Wang Y, Li L, Guo X, Zhou S (2015) An implantable activetargeting micelle-in-nanofiber device for efficient and safe cancer therapy. ACS Nano 9:1161–1174
112. Wu H, Liu S, Xiao L, Dong X, Lu Q, Kaplan DL (2016) Injectable and pH-responsive silk nanofiber hydrogels for sustained anticancer drug delivery. ACS Appl Mater Interfaces 8:17118–17126

Tumor Microenvironment and Intracellular Signal-Activated Nanocomposites for Anticancer Drug Delivery

Yilan Huang, Yiheng Huang, Yuefei Zhu, Xiaowen Zhu, and Zhiqing Pang

1 Introduction

Malignant tumors have become a major disease threatening human health worldwide. The World Health Organization's statistical survey indicates that about 8.8 million people die of cancer each year worldwide. In 2010, the total cost of cancer on the global issue was about 1.16 trillion U.S. dollars, and this figure continues to grow. Research on cancer treatment has gradually attracted people's attention. The current clinical therapeutic methods of tumors mainly include surgery, chemotherapy, radiation therapy, and biological therapy, among which chemotherapy is very common and important [1]. However, as most chemotherapeutic drugs lack selectivity for normal cells and tumors, they not only affect the growth and proliferation of normal cells when killing tumor cells, but also are hardly accumulated in tumor tissues to achieve effective drug concentrations.

Y. Huang · Y. Huang · X. Zhu · Z. Pang (✉)
Key Laboratory of Smart Drug Delivery, School of Pharmacy, Ministry of Education, Fudan University, 826 Zhangheng Road, Shanghai 201203, China
e-mail: zqpang@fudan.edu.cn

Y. Huang
e-mail: 15301030080@fudan.edu.cn

Y. Huang
e-mail: 15301030005@fudan.edu.cn

X. Zhu
e-mail: 15301030063@fudan.edu.cn

Y. Zhu
Department of Biomedical Engineering, Columbia University Medical Center, 3960 Broadway Ave, New York, NY 10032, USA
e-mail: yz3713@columbia.edu

A. K. Nayak et al. (eds.), *Biomedical Composites*, Materials Horizons: From Nature to Nanomaterials, https://doi.org/10.1007/978-981-33-4753-3_8

In recent years, with the rapid development of nanotechnology, the unique advantages of nanocomposites in the field of targeted tumor therapy have become increasingly prominent. Due to their special size effect (1–100 nm), quantum effect and surface effect, nanocomposites could enhance aqueous solubility, stability, bioactivity, and bioavailability of drugs, and reduce their adverse toxicities [2]. Diverse classes of nanocomposites have been developed for improved anticancer drug delivery, including polymeric nanoparticles, liposomes, micelles, polymersomes, mesoporous silica nanoparticles, dendrimers, and inorganic nanoparticles made of iron oxide, quantum dots, gold, or metal oxide frameworks. Although much progress has been achieved in preclinical animal or human models, the majority of nanocomposites for anticancer drug delivery still suffers from their limitations, such as fast blood clearance, premature drug release in circulation, nontargeted accumulation in healthy tissues, poor tumor penetration capability, and insufficient drug release in tumor cells [3]. In order to resolve these limitations, maximizing the therapeutic efficacy and minimizing the side effects of loaded drugs, smart nanocomposites responding to environmental stimuli have been developed. These stimuli-activated nanocomposites are equipped with environmental sensitive modalities within their structures and can make responses to specific environmental stimuli (such as tumor microenvironment/intracellular signals) to change their chemical or physical properties, exhibiting greater potential in anticancer drug delivery than traditional nanocomposites [4].

The tumor microenvironment (TM) refers to the normal cells, molecules, and blood vessels that surround and feed tumor cells, including surrounding blood vessels, immune cells, fibroblastic cells, bone marrow-derived inflammatory cells, lymphocytes, signaling molecules and the extracellular matrix (ECM) composed of collagen and proteoglycans [5]. It is an integral part of cancer since it supplies a nurturing environment for the malignant process and contributes to cancer development and progression by affecting the complex interplay of genetic and epigenetic changes within the cells themselves [6]. Compared with normal tissues or cells, abnormal physiologic conditions exist in the TM such as acidic extracellular pH, high concentrations of specific enzymes (matrix metalloproteinases, hyaluronidase, cathepsin B, legumain, etc.), hypoxia, extra- and intracellular glutathione (GSH) and adenosine triphosphate (ATP) gradients, and excessive amounts of reactive oxygen species (ROS). Based on a well understanding of the TM, the prominent distinctions between tumor sites and normal tissues have been utilized as stimuli to trigger specific changes in stimuli-activated nanocomposites, thus enhancing the therapeutic efficacies and reducing the side effects of chemotherapeutics.

In this chapter, we will provide a comprehensive outline of the tumor microenvironment (Table 1) and intracellular signal-activated nanocomposites for anticancer drug delivery (Table 2) and try to provide an in-depth look at the design of these drug delivery systems to maximize their benefits for cancer treatment.

Table 1 A summary of tumor microenvironment and intracellular signals

Signals	Details	References
Low pH	Inflammation-associated acidic pH: 7.2–6.5 for extracellular pH in tumor	[7]
Over-expressed enzyme	Matrix metalloproteinases, hyaluronidase, cathepsin B, urokinase plasminogen activator, β-galactosidase, legumain, secretory phospholipase A_2	[8]
Glutathione	Intracellular environment: 0.5–10 mM; extracellular environment: 2–20 μM	[3]
Hypoxia	Hypoxia (low oxygen levels in tissues): regions with low oxygen concentration often exist in solid tumors	[9]
H_2O_2	Excessive amounts of H_2O_2 have been observed in various cancer cells	[10]
ATP	Intracellular environment: 1–10 mM; extracellular environment: <0.4 mM	[11]

2 Tumor Microenvironment and Intracellular Signal

2.1 *The Acidic pH*

It is well-known that pH values vary significantly in different tissues and cellular compartments, and in disease states, such as ischemia, infection, inflammation, and tumorigenesis [12] (Table 1). Studies over the last few decades have demonstrated that the pH value of the tumor extracellular environment is maintained with a range from 6.5 to 7.2 [13], whereas that in blood circulation and normal tissues is around 7.4 [7]. The pH decrease is mainly a consequence of enhanced glycolysis and plasma membrane proton-pump activity of tumor cells, both in aerobic and anaerobic conditions, making tumor cells produce more lactic acid than normal cells and then leach out the acid to the surrounding environment [14]. The high production of acidic metabolites, termed as the Warburg effect, provides a growth advantage for tumor cells in vivo [13]. Besides, much lower pH also exists in tumor intracellular microenvironment: the pH values decrease to the range 5–6, 4–5, and around 6.4 in endosomes, lysosomes, and Golgi apparatus, respectively [15].

2.2 *The Over-Expressed Enzyme*

Enzymes participate in a variety of physiological and pathophysiological processes and accelerate biochemical reactions by lowering their activation energy. Compared to normal tissues, a great number of enzymes are overexpressed in pathological conditions, such as cancer and inflammation (Table 1). The altered expression of specific enzymes in tumors is linked to several key events, including tumor progression, growth, neovascularization, intravasation, extravasation, and metastasis [16].

Matrix metalloproteinases (MMPs), especially MMP2 and MMP9, have been observed in almost every type of human tumors. The MMPs family is comprised of at least 23 structurally related, zinc-dependent extracellular matrix remodeling proteases [8a]. These endopeptidases play essential roles in a number of pathologic processes, including proteolytic degradation of ECM, alteration of the cell-cell and cell-ECM interactions, cell migration, and tumor angiogenesis [17]. Elevated cellular level of hyaluronidase (HAase), a kind of enzyme that can catalyze the degradation of hyaluronic acid through preferential cleavage of glycosidic linkages, has been particularly documented in the prostate, urinary bladder, breast, as well as laryngeal cancers [8b]. The effect of its overexpression is still not clear now, and a few reports suggest that it may be involved in the metastatic process [18]. Cathepsin B is a lysosomal cysteine protease that is involved in various pathologies and oncogenic processes in humans. Upregulation of cathepsin B is found in a wide variety of human cancers and often induces secretion into the ECM [8c]. Urokinase plasminogen activator (uPA), another important protease to degrade fibrin, is closely associated with tumor growth, invasion, and metastasis in a wide range of tumors, including glioma [19]. It has been demonstrated to be highly expressed in tumors (for instance, glioma neo-vascular cells and glioma cells, especially in the leading edge of glioma [20]), about 4–10 folds higher than that in normal tissues [21]. Other enzymes overexpressed in human malignant tumors include urokinase plasminogen activator, β-galactosidase, legumain, secretory phospholipase A_2 and so on [8d–g].

2.3 The Reductive Environment

Glutathione (γ-glutamyl-cysteinyl-glycine; GSH) is the most abundant low-molecular-weight biological thiol [22]. As an important antioxidant, GSH is involved in protecting cellular components from the noxious effect of reactive oxygen species, such as free radicals, peroxides, lipid peroxides, and heavy metals. It is also capable of forming disulfide bonds with cysteine residues of proteins, and participating in nonenzymatic conjugation with some chemicals [23]. Many evidence has shown that there is a large GSH concentration gradient between the intracellular environment and the external matrix of tumor tissues (Table 1). Generally, inside cells, the GSH concentration could be up to 0.5–10 mM, while only around 2–20 μM in extracellular environments [3]. In addition, studies have shown that the intracellular concentration of GSH in tumor tissues is at least four fold higher than that in the normal cells [24], and it is often elevated in multidrug resistance (MDR) cancer cells. GSH/glutathione disulfide (GSSG) is the major redox couple in human cells and determines the anti-oxidative capacity of human cells. GSH/GSSG is kept reduced by NADPH and glutathione reductase. The intracellular concentration of GSH is also dependent on other redox couples such as $NADH/NAD^+$, $NADPH/NADP^+$, and thioredoxinred/thioredoxinox [25]. Moreover, the endo/lysosome is also redox-active. It also possesses a high reducing potential controlled by a specific reducing enzyme gamma interferon-inducible lysosomal thiol reductase (GILT) in the co-presence of L-cysteine, not GSH [26]. The redox-active lysosome also contains low-mass iron. The iron is kept in a reduced state (Fe^{2+}) by the concentration of thiols.

2.4 The Hypoxia

Hypoxia, defined as a state of reduced tissue oxygen availability or decreased oxygen partial pressure, is a common characteristic of the tumor microenvironment [27] (Table 1). To grow beyond a diameter of approximately 1 mm, newly developing tumors must form their own vascular network and blood supply, which they accomplish either by incorporating preexisting host vessels or by forming new microvessels through the influence of tumor angiogenesis factors. However, the newly formed blood vessels in tumors are often highly abnormal, with dilations, incomplete or absent endothelial linings and basement membranes, leakiness, irregular and tortuous architecture [9]. Tumor hypoxia is generated when a tumor rapidly outgrows its blood supply, leaving portions of the tumor with regions where the oxygen concentration is significantly lower than in healthy tissues [28]. Emerging evidence has indicated that the hypoxic response is a series of hypoxia-induced proteomic and genomic changes, and primarily driven by the transcription factor hypoxia-inducible factor 1 (HIF-1) [29].

Many clinical studies have demonstrated that tissue oxygen partial pressures, which are measured from those cancer patients, are near zero mm Hg, which is significantly lower than that in normal tissue (about 30 mm Hg) [27]. Hypoxia is a negative factor for anticancer therapy. It has been shown that hypoxia is strongly associated with tumor propagation by enhancing MDR in cancer as well as malignant invasiveness and metastasis. Under hypoxia conditions, tumor cells divide more slowly than that in oxygen-rich environment, which makes these tumor cells less susceptible to traditional antiproliferative drugs that target rapidly dividing cells. In addition, tumor cells undergo genetic alterations that allow them to survive and grow in the hypoxic environment, such as local invasive growth, perifocal tumor cell spreading, and distant tumor cell spreading. However, hypoxia is rarely seen in normal tissue, indicating it is a primary target in the development of anticancer drug delivery.

2.5 Reactive Oxygen Species

Studies have shown that compared with normal cells, tumor cells can constantly produce high levels of reactive oxygen species (ROS), such as hydrogen peroxide

(H_2O_2), hydroxyl radicals (•OH), and superoxide anion ($O_2^{\bullet-}$), which are derived from by-products of aerobic metabolism due to oncogenic transformation (Table 1). Many evidence shows that tumor cells suffer from a higher level of ROS than normal cells due to the fast proliferation and enhanced metabolic rate of tumor cells. In addition, MDR cells show even higher ROS levels than their wild-type cells. It is because MDR cells need more energy to elevate mitochondrial activity for the over-expression of membrane transporters [30]. ROS is a double-edged sword. It can stimulate the growth and proliferation of tumor cells, and have genetic instability and senescence evasion [10]. ROS also play an important role in maintaining tumor phenotype. However, producing excessive ROS will cause cellular damage, such as lipid peroxidation, protein oxidation, DNA adduct formation, and so on. These damages can finally cause cell death by apoptosis or necrosis [31]. H_2O_2 is a major component of ROS and is a common marker of oxidative stress, which can induce apoptosis of many tumor cells. Hydroxyl radical is far more reactive than H_2O_2 and $O_2^{\bullet-}$, and can also mediate intracellular cytotoxicity [31].

2.6 *The ATP Gradient*

It is well-known that ATP, referred to as the "molecular unit of currency" of intracellular energy transfer, participates in various kinds of metabolic processes. Recently, researches revealed that ATP has a very high concentration (approximately 1–10 mM) in intracellular cytosol, much greater than that in the extracellular environment (below 0.4 mM) (Table 1) [11]. Such a prominent difference inspires researchers to exploit ATP as a new trigger for the controlled release of anticancer drugs both in vitro and in vivo [32].

3 Tumor Microenvironment and Intracellular Signal-Activated Nanocomposites for Anticancer Drug Delivery

3.1 *pH-Activated Nanocomposites for Anticancer Drug Delivery*

The acidic pH of the extracellular and intracellular environment of tumors is an appropriate internal trigger to control the release of anticancer drugs from pH-activated nanocomposites in tumor tissues and/or tumor cells (Table 2). Generally, pH-activated nanocomposites can be designed by incorporating pH-labile chemical bonds into the polymer components or the polymer-drug conjugates, such as acetal, hydrazone, cis-acotinyl, orthoester, b-carboxylic acid amide, and glycerol ester groups [3]. These nanocomposites are quite stable under neutral conditions,

Table 2 A summary of representative TM and intracellular signal-activated nanocomposites for anticancer drug delivery

Stimulus	Nanocomposites	Disease/Application	References
pH	Polydopamine-coated mesoporous silica nanoparticles	Colorectal cancer	[33]
	Nanoparticle composed of pH-responsive PEGylated polymer and cell-targeting and -penetrating peptide-amphiphile	Prostate cancer	[34]
	Nanoparticle constructed by assembly of platinum-prodrug conjugated polyamidoamine dendrimers	Pancreatic cancer	[35]
	Micelles composed of poly (lysine)-b-polycaprolactone and poly (glutamic acid)-g- methoxyl poly (ethylene glycol)	Non-small cell lung cancer	[36]
	Nanoparticle with poly(ethylene glycol)-b-poly(2-(hexamethyleneimino) ethyl methacrylate) diblock copolymer	Breast cancer	[37]
Enzyme	PbS/CdS/ZnS core/shell/shell quantum dots	Colon cancer	[38]
	Nanoparticle comprising MMP-cleavable octapeptide (Gly-Pro-Leu-Gly-Ile-Ala-Gly-Gln)	Breast cancer	[39]
	Nanoparticle developed from MMP2 and MMP9-sensitive copolymers and folate receptor targeted copolymers	Melanoma	[40]
	Hyaluronic acid-conjugated mesoporous silica nanoparticle	Colon cancer	[41]
	Polycation/DNA complexes coated with PEGylated hyaluronic acid	Liver cancer	[42]
	Rotaxane-modified MSNs based on alkoxysilane tether, α-cyclodextrin and multifunctional peptides	Cervical cancer	[43]
	Nanoparticle comprising human serum albumin and cathepsin B cleavable peptide	Pancreatic cancer	[44]
Reduction/GSH	Thiolated silica nanoparticle	Hepatocellular carcinoma	[45]
	Poly(amido amines) nanoconjugate	Liver cancer	[46]
H_2O_2	Nanoparticle comprised of semiconducting polymer core and manganese dioxide (MnO_2) shell	Photodynamic therapy	[47]
	Thioether-bridged dendritic mesoporous organosilica nanoparticle	Breast cancer	[48]

(continued)

Table 2 (continued)

Stimulus	Nanocomposites	Disease/Application	References
ATP	Nanocomplex based on ATP-responsive aptamer duplex	Lung adenocarcinoma	[49]
	Polypeptide-wrapped mesoporous-silica-coated nanoparticle	Real-time monitoring	[50]
pH and GSH	PEG-SS-PTMBPEC micelles	Cervical cancer	[51]
	Amphiphilic PCL-PDEA and MPEG-SS-PCL mixed micelles	Cervical cancer	[52]

but they can rupture at acidic pH to release their payload rapidly because of the hydrolysis of these bonds. Zheng et al. designed mussel-inspired polydopamine (PDA)-coated mesoporous silica nanoparticles (MSNs) as pH-responsive nanoparticles (MSNs@PDA) [53]. MSNs with a large pore diameter were synthesized by using 1,3,5-trimethylbenzene as a pore-expanding agent and were modified with a PDA coating by virtue of oxidative self-polymerization of dopamine in neutral pH. The tailored mesoporous structure, high surface area, and tunable pore size of MSNs gave them significant advantages in stimuli-activated controlled drug delivery [54]. Meanwhile, PDA coating worked as the "gatekeepers" to block the drugs inside MSNs at neutral pH and trigger their release at lower pH. In order to evaluate the drug release profiles, doxorubicin (DOX) was loaded in MSNs@PDA (MSNs-DOX@PDA). It was shown negligible release of DOX from MSNs-DOX@PDA was found in pH 7.4 PBS. Decreasing the pH value of the release medium could increase the release rate of DOX from MSNs-DOX@PDA. In addition, the color of the supernatant solution turned brown in acidic PBS but not in neutral one, which agreed with the PDA coating's peeling off from the surface of MSNs in an acidic environment.

RNA-based therapeutics, such as small-interfering RNAs (siRNAs), microRNAs (miRNAs), and antisense oligonucleotides (ASOs), have shown great potential in targeting a large part of the currently undruggable genes or gene products in cancer therapy [55]. However, existing challenges lie in the delivery of RNAi molecules such as siRNAs to diseased sites in vivo [56]. Recent years have witnessed developments in RNAi molecule delivery systems in nanooncology, especially nanoparticles encapsulating siRNA [57]. Xu et al. fabricated a pH-responsive multistage nanoparticle platform for effective and systemic targeted siRNA delivery to tumor cells [34]. This nanoparticle platform consisted of a PEG surface shell, a hydrophobic polymer poly(2-((hexamethyleneimino)ethyl methacrylate) (PHMEMA) core which was sensitive to tumor microenvironment extracellular (TME) low pH, and charge-mediated complexes of siRNA and cell-targeting and -penetrating peptide-amphiphile (TCPA) encapsulated in the nanoparticle core. As shown in Fig. 1, after intravenous injection, the pH-responsive multistage nanoparticles exhibited a long circulatory half-life because of PEG outer shell on these nanoparticles and thus enhanced their accumulation in the tumor tissue due to the enhanced permeability and retention (EPR) effect. When these nanoparticles entered tumors,

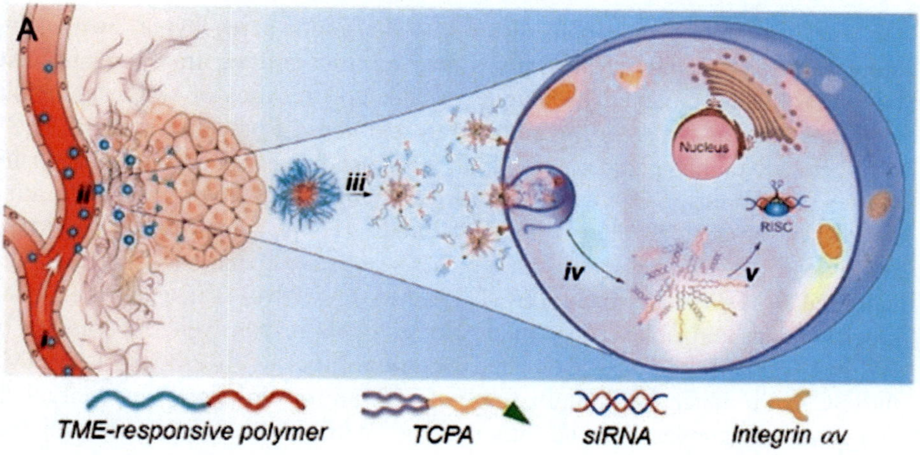

Fig. 1 Schematic illustration of the TME pH-responsive multistage nanoparticle platform for systemic targeted siRNA delivery and effective cancer therapy. Reprinted from Ref. [34] with permission, Copyright American Chemical Society, 2017

extracellular low pH-activated rapid nanoparticle disassembly led to the exposure of siRNA-TCPA complexes, which could then target and penetrate tumor cells via the RGD (Arg-Gly-Asp) ligand segment of TCPA. Finally, the cell-penetrating ability of TCPA attributable to its cationic polyarginine segment enhanced the cytosolic siRNA transport to achieve efficient gene silencing. In order to evaluate the pH-responsive behavior of these nanoparticles, transmission electron microscopy (TEM) imaging was employed. TEM images exhibited a spherical morphology of the nanoparticles at pH 7.4 PBS. After incubation at pH 6.8 PBS for 3 min, some large amorphous aggregates and small size particles were observed, which were confirmed as the ionized polymers and exposed siRNA-TCPA complexes. In addition, more than 80% of siRNA and less than 20% of siRNA was released from these nanoparticles when nanoparticles were incubated in pH 6.8 PBS and 7.4 PBS for 4 h, respectively. In vitro cellular uptake showed that HeLa cells took in more siRNA at pH 6.8, compared to cells incubated at pH 7.4. More importantly, the internalized siRNA molecules at pH 6.8 exhibited a much higher distribution in the cytoplasm than that of the internalized siRNA molecules at pH 7.4. Further flow cytometry demonstrated the siRNA uptake by cells incubated at pH 6.8 is >5-fold more than that by cells incubated at pH 7.4. Moreover, ex and in vivo results suggested the multistage siRNA delivery platform could silence the expression of bromodomain 4, a conserved member of the BET family of chromatin readers regulating signaling network in prostate cancer (PCa), and significantly inhibit PCa tumor growth [34].

Apart from low delivery efficiency to cancer, another challenge faced by cancer nanomedicine is poor tumor penetration. It is well-known that nanoscale drug delivery systems preferentially accumulate in solid tumors through the EPR effect due to their inherent size advantages [58]. However, the large size also limits the deep penetration of nanomedicine into the tumor parenchyma. Several literature reported that after nanosystems reached tumors, drug transport and distribution would be hindered by unique architectural features of solid tumors such as elevated interstitial fluid pressure (IFP) and dense extracellular matrix (ECM) [59]. To solve this paradox, Li et al. developed a class of ultra-pH-sensitive cluster nanobombs (SCNs) with instantaneous size switching in the acidic tumor microenvironment [35]. The platinum (Pt)-prodrug conjugated SCNs (SCNs/Pt) were self-assembled from poly(ethylene glycol)-b-poly(2-azepane ethyl methacrylate)-modified PAMAM dendrimers (PEG-b-PAEMA-PAMAM/Pt), in which the PAEMA block contains ionizable tertiary amine groups for pH-responsiveness. At neutral pH, PAEMA was hydrophobic and directed the assembly of PEG-b-PAEMA-PAMAM/Pt into SCNs/Pt, which had an initial size of ~80 nm for prolonged blood circulation and selective extravasation into tumor tissue. While in the acidic tumor microenvironment, PAEMA was rapidly protonated, leading to a hydrophobic-hydrophilic transformation and subsequently instantaneous disintegration of SCNs/Pt into small nanoparticles for deep penetration into tumor parenchyma. The Pt-prodrug was covalently attached to the polymer and specifically activated by intracellular redox, resulting in the release of active cisplatin [60]. The tumor-pH-triggered size transition of SCNs/Pt was testified by using dynamic light scattering (DLS) and TEM. At neutral pH, the size was 80 nm in diameter. Then the size showed a dramatic transition from ~80 nm at pH 6.8 to ~10 nm at pH 6.7, which was superfast and completed within seconds, indicating the rapid dissolution of the superstructure into small particles. The penetration capability of SCNs/Pt was investigated by employing BxPC-3 human pancreatic cancer cell-derived tumor spheres as an in vitro model. To monitor the penetration activity, SCNs and pH-insensitive cluster nanostructure (ICNs) were labeled with Cy5 (denoted as SCNs/Cy5 and ICNs/Cy5). Confocal laser scanning microscopy (CLSM) showed limited penetration of both SCNs/Cy5 and ICNs/Cy5 into tumor spheres when incubated at pH 7.4. However, a significant improvement of SCNs/Cy5 in penetration capability (at a depth of 85 μm into tumor spheres) was found at tumor extracellular pH (pH 6.7). Next, tumor growth inhibition study in nude mice bearing poorly permeable BxPC-3 tumor xenografts revealed tumor suppression rate of 22%, 24%, 41%, and 82%for PAMAM/Pt, free cisplatin, ICNs/Pt and SCNs/Pt (at a Pt dose of 2 mg/kg), respectively. All these results demonstrated that the pH-triggered size switching is a viable strategy for improving drug penetration and therapeutic efficacy of nanomedicine [35].

Surface charge also plays a determining role in the performance of nanoparticles. The high affinity of positively charged nanoparticles to biological interfaces makes them easily taken up by tumor cells via electrostatic attraction mediated targeting [61], but this strong interaction also limits their tumor permeation due to shortened blood circulation [62]. Thus, a class of novel delayed charge reversal nanoparticles were fabricated to investigate the influence of the charge reversal profile of nanoparticles on tumor penetration [36]. These nanoparticles were constructed by an anionic coating with poly (glutamic acid)-g-methoxyl poly (ethylene glycol) (PGlu-g-mPEG) and a drug-loading cationic inner core composed of poly (lysine)-b-polycaprolactone (PLys-b-PCL). Manipulation of charge reversal profile under lower pH was achieved by varying the shell cross-linking degree of the cationic core, prior to electrostatic interaction in the polyanion complex layer. The changes in ξ-potential of these nanoparticles with different cross-linking degrees (CTCL0, CTCL30, and CTCL70) incubated in pH 6.5 PBS revealed a relation between the charge reversal profile of nanoparticles and cross-linking degree. After 6-h incubation, no changes in the ξ-potential of CTCL0 nanoparticles were observed, indicating that it was non-charge reversible; CTCL70 nanoparticles accomplished charge reversal after 1-h incubation, while it took 2 h for the ξ-potential of CTCL30 to turn positive. However, the ξ-potential of the three nanoparticles remained constant after 24-h incubation in pH 7.4 PBS. The amount of remained PGlu-g-mPEG on nanoparticles during incubation was determined by gel permeation chromatography to study the charge reversal caused by decoating of PGlu-g-mPEG. Results showed the coating material remained on CTCL30 and CTCL70 nanoparticles were 34.7% ± 0.77% and 8.57% ± 6.1%, respectively, while the percentage was over 70% for CTCL0 nanoparticles. These results indicated that the lowered environmental pH was responsible for charge reversal, and controllable charge reversal could be achieved by changing the degree of cross-linking. The in vivo intratumoral penetration and cellular uptake of CTCL30 and CTCL70 nanoparticles were compared, and it was found that CTCL30 nanoparticles with a delayed charge reversal profile increased the number of cells exposed to the nanoparticles by the wider intratumoral nanoparticle distribution as well as improved nanoparticle–cell interaction due to tumor acidity induced time-dependent charge reversal.

PEGylation of nanoparticles is a good way to extend their blood circulation time, but researchers also found it could dramatically shield the targeting effect and suppress the cellular internalization. Thus, novel nanoparticles, including stimuli-activated nanoparticles, have been developed to address the above problem and improve tumor distribution of PEGylated nanoparticles. Wang et al. designed a tumor extracellular acidity-triggered ligand-presenting (ATLP) nanoparticle for tumor-specific penetration and cellular uptake [37]. The ATLP nanoparticles were composed of a Ce6-modified acid-responsive poly(ethylene glycol)-b-poly(2-(hexamethyleneimino) ethyl methacrylate) (PEG-b-PHMA, denoted as PHMA) diblock copolymer as a shield matrix and an iRGD-modified polymeric prodrug of doxorubicin (iPDOX) as an amphiphilic core. In the tumoral acidic microenvironment, PHMA diblock copolymer underwent amphiphilic to hydrophilic transition and dissociated from ATLP to expose the iRGD ligand for cellular uptake of the PDOX prodrug and restore the fluorescence imaging of Ce6 for real-time monitoring. The acid-sensitivity of the ATLP nanoparticles was evaluated by monitoring their size change by dynamic light scattering after incubation in PBS with different pH values. The average particle size and surface charge of ATLP at pH 7.4 were 34 ± 2.0 nm and 1.1 ± 0.4 mV, respectively. When pH was decreased to 6.6, the nanoparticles swelled dramatically to 155 ± 12 nm, with a positive surface charge of 4.6 ± 0.2 mV, presumably due to protonation-induced dissociation of the PHMA matrix and formation of iPDOX aggregates with iRGD presented on the surface. At the same time, no changes in particle size and surface charge of acid-insensitive nanoparticles at both pH values were observed. Next, the tumor penetration ability of ATLP was investigated using MCF-7/ADR-derived multicellular spheroid, which could mimic the acidic microenvironment of solid tumors. The DOX fluorescence of ATLP perfused throughout the tumor spheroid at pH 6.6, while the fluorescence distribution at pH 7.4 was limited, verifying acid-triggered exposure of iRGD ligand dramatically promoted tumor penetration of iPDOX.

3.2 *Enzyme-Activated Nanocomposites for Anticancer Drug Delivery*

3.2.1 Matrix Metalloproteinases

Fluorescence (FL) imaging into the second near-infrared region (NIR-II) at 1,000–1,700 nm have attracted much interest due to its high spatial resolution through deep tissues. However, bright and compact fluorophores are rare in this region, and sophisticated control over NIR-II probes has not been fully achieved yet. Thus, Jeong et al. fabricated a protease-activated NIR-II quantum dot probe (PA-NIRQD), which could exhibit FL upon matrix metalloprotease activity in the tumor microenvironment [38]. Bright and stable PbS/CdS/ZnS (core/shell/shell) NIRQDs that emitted at 1200 nm were synthesized, and activatable modulators, containing an MMP-cleavable peptide sequence and methylene blue (MB) which could quench FL via photoexcited electron transfer (PET), were attached to the surface ligands. The NIRQDs were quenched by PET from the MB when the QDs were in close proximity to MB, and FL was activated upon modulator cleavage. The peptide sequence was optimized to impart flexibility, facilitating enzyme access, and guaranteeing effective quenching upon activation. To evaluate the ability of PA-NIRQD to respond to the tumor microenvironment, fresh colon tissues were prepared from azoxymethane/dextran sulfate sodium-treated colorectal tumor-bearing mice due to the high-level expression of MMPs. The colon tissues were excited by a time-modulated 910 nm pulse laser, and then the FL intensities were recorded over time. A rapid increase in FL ranging from 150% to 300% was observed with 10 min at tumor site, while no changes occurred in the area of normal mucosa. Time-dependent contrast-to-noise ratio values at tumor site were also plotted, indicating the PA-NIRQD probe could distinguish tumors from the normal tissues. In contrast, administered PA-NIRQD in normal colon tissues showed minimal activation, and the FL intensity remained constant in another experiment using NIRQD instead of PA-NIRQD, attributed the low MMP activity and lack of activation event, respectively. In conclusion, PA-NIRQD enabled the simple and rapid visualization of MMP enzyme activity and successfully distinguished tumor sites in the colorectal tumor mouse model. Thus, it has the potential for the image-guided biopsy of cancers.

Zhu et al. designed a novel multifunctional liposomal nanocarrier, which could respond to the up-regulated extracellular MMP2 in tumors, resulting in the enhanced tumor targetability and internalization [39]. An MMP-cleavable octapeptide (Gly-Pro-Leu-Gly-Ile-Ala-Gly-Gln) was used as a sensitive linker between a liposomal lipid and its long-chain PEG block to serve as a steric shield for the nanoparticle and surface-attached cell-penetrating peptide TAT in the blood. The multifunctional liposomes could be retained in the tumor site due to the EPR effect and the active targeting effect by tumor cell-specific antinucleosome monoclonal antibody (mAb 2C5). Then the high expressed extracellular MMP2 cleaved the octapeptide linker and removed the protective long-chain PEG, providing the exposure of TATp for the enhanced intracellular uptake. Results showed that the MMP2-cleavable peptide did not lose its "cleavability" after the conjugation with both PEG and DOPE at its two ends. The MMP2-responsive liposomes were prepared using the film hydration method followed by extrusion and surface modification with mAb 2C5. The particle size and zeta-potential of the liposomes were 207.5 ± 75.5 nm and 10.23 ± 1.21 mV, respectively. To demonstrate the in vitro cellular uptake of the MMP2-responsive multifunctional liposomes, mouse breast cancer cells (4T1), and normal rat cardiomyocytes (H9C2) were treated with fluorescence-labeled liposomes at different conditions. Fluorescence-activated cell sorting suggested enhanced targetability and internalization of the MMP2-responsive multifunctional liposomes in 4T1 cancer cells, as expected. Besides, it was also shown that the MMP2-sensitive liposomes minimized nonspecific uptake in normal cells and maximized their penetration into cancer cells.

Folate receptor has been well documented to be over-expressed in some malignant tissues, which provides an effective target site for drug delivery systems [63]. To enhance programmable long circulation and tumor targeting ability, MMP-sensitive nanoparticles (SNPs) were constructed by mixing a MMP2/9-sensitive copolymers (mPEG-Pep-PCL) and folate receptor-targeted copolymers (FA-PEG-PCL) [40]. The SNPs exhibited a sandwich structure composed of a detachable PEG outmost layer, a targeting middle layer of folate ligands, and a hydrophobic innermost core loaded with camptothecin (CPT). In the tumor microenvironment, the MMP2/9-sensitive linker was expected to be cleaved, followed by the detachment of the PEG layer and then the exposure of folate, via which the selective binding to the tumor cells and the cellular uptake could be strengthened. The size of CPT-loaded SNPs was determined to be around 160 nm by DLS, in line with previous literature which claimed that the effective size for the drug delivery systems should be below 200 nm since the threshold vesicle size for the extravasation into a tumor's extracellular space was around 400 nm [64]. To demonstrate changes in SNPs in response to MMP, SNPs were incubated with gelatinases for 24 h. Serious aggregation and precipitation were observed in nanoparticle suspension, attributing to the detachment of the PEG layer induced by the breakup of mPEG-Pep-PCL. In contrast, MMP-insensitive nanoparticles (INPs) composed of mPEG-PCL and FA-PEG-PCL exhibited no significant size variation due to the absence of peptide. In vitro uptake by B16 melanoma cells was investigated using a fluorescence microscope to compare the cytotoxicity of CPT-loaded SNPs and INPs. After 4-h incubation, no fluorescence was found in the group of INPs, indicating no or negligible amount of CPT was inside the B16 cells. However, fluorescence was observed at the initial 2 h in the SNPs-treated group, and more fluorescence appeared when the incubation time was extended to 4 h, indicating that CPT was delivered into the B16 cells more efficiently by SNPs. In vivo antitumor studies in B16 melanoma xenograft models demonstrated that a significantly higher tumor group inhibition was achieved by CPT-loaded SNPs treatment compared with other treatments with saline, empty NPs, CPT, CPT-loaded INPs.

3.2.2 Hyaluronidase

Fluorouracil (5-FU) is a pyrimidine analog used to treat cancer of the breast, glioblastoma, colon, rectum, stomach, and pancreas [65]. It interferes with DNA and RNA synthesis by mimicking the building blocks necessary for synthesis. However, the growth of normal cells may also be affected by 5-FU. Side effects will also occur, such as fever, unusual bleeding, and trouble breathing. Thus, effective delivery strategies for 5-FU are needed. Jiang et al. prepared hyaluronic acid (HA)-conjugated 5-FU-loaded mesoporous silica nanoparticles (HA/FMSN) to improve its therapeutic efficacy in colon cancer [41]. The surface-conjugated HA with large molecular weight could prevent the drug release in normal conditions and target the CD44 receptors, which are overexpressed in tumor tissues. After entering the tumor site, HA would be degraded into lower molecules with the assistance of hyaluronidase (HAase), an overexpressed enzyme at the tumor microenvironment, leading to the release of the drug. HA/FMSN were perfectly spherical and well-ordered in appearance with particle size and zeta potential of around 190 nm and −18.5 mV, respectively. When incubated in colon cancer cells for 2 h, HA/FMSN exhibited superior targeting efficiency with a stronger fluorescence over FMSN, indicating a typical receptor-mediated endocytosis uptake of the former particles. In MTT assay, the IC50 value of free 5-FU, FMSN and HA/FMSN were 6.89 μg/ml, 4.52 μg/ml, and 1.08 μg/ml, respectively, which suggested a significantly higher anticancer effect of HA/FMSN compared with other formulations. As expected, higher apoptosis of cancer cells induced by HA/FMSN was also observed in colon cancer cell-bearing mice. Altogether, these results suggested that the hyaluronidase-responsive targeted nanoparticles could hold enormous potential in colon cancer targeting.

PEGylation can not only improve the biocompatibility and long-circulation of nanoparticles in vivo, but also inhibits cellular uptake and endosomal escape and leads to low transfection efficiency [66]. To overcome the "PEG dilemma," PEGylated HA, i.e., HA-g-PEG (HgP), was synthesized to coat the binary complexes of core-shell cationic polycaprolactone-graft-poly (N, N-dimethylaminoethyl methacrylate) nanoparticles/DNA (NP-D) for constructing a novel enzyme-activated polycation/DNA complexes (NP-D-HgP) for gene therapy [42]. The coating of HgP on NP-D could inhibit the release of the inner layer of nucleic acid due to strong electrostatic interaction between the polyelectrolytes. DNA release could be achieved by the high level of HAase in the tumor microenvironment, which promoted HA degradation, resulting in enhanced endosome escape and transfection efficacy. To confirm the detachment of HgP coating, zeta potential and size of NP-D-HgP were measured in the presence of HAase at different times. Results showed a dramatic increase in the diameter of NP-D-HgP from 60 to 8591 nm within the prior 1 h and a subsequent decrease to 200 nm. The zeta potential of NP-D-HgP went through a charge reverse from −10 mV to +5 mV, which could be attributed to the detachment of the HA layer. Compared with the control nanoparticles composed of PCL-g-PDMAEMA NPs/DNA/PGA-g-PEG (NP-D-PGgP), the nanoparticle uptake by HepG2 cells was enhanced by HgP coating, partly caused by the CD44-mediated endocytosis. The effect of HgP and PGgP coating on cellular uptake was further evaluated using confocal microscopy. After incubation for 9 h, NP-D nanoparticles were found to aggregate outside the cell and stay with the surface of cells, while NP-D-HgP were stable and the cell uptake efficiency was remarkably improved. In vivo RFP gene transfection of NP-D-HgP was performed at a dose of 1.5 mg/kg of RFP plasmid DNA via intravenous injection. It was shown that mice treated with NP-D died in several hours after injection, while no death occurred in NP-D-HgP and NP-D-PGgP groups. At 24 h after injection, higher RFP expression was observed in mice administered with NP-D-HgP than other groups. Taken together, NP-D-HgP showed high potential in gene delivery to cancer cells.

3.2.3 Cathepsin B

Among various kinds of nanocarriers, functional mesoporous silica nanoparticles (MSNs) have exhibited unique advantageous properties, such as chemically inert, easy functionalization, and large loading capacity, attracting research interest for drug storage and delivery. In order to achieve efficient drug delivery and reduce adverse effects, functionalized MSNs are usually constructed by covering the surface of MSNs with nanovalves, including rotaxanes, polypseudorotaxanes, cucurbiturils, and so on, which serve as a "gatekeeper" to retain drugs until external stimuli. Cheng et al. designed a drug delivery system based on rotaxane-modified MSNs [43]. The rotaxane structure was formed with alkoxysilane tether, α-cyclodextrin, and multifunctional peptides, which were composed of a cell-penetrating peptide of seven arginine (R7) sequence, a cathepsin B-cleavable peptide of GFLG, and a tumor-targeting peptide of RGDS. When incubated with tumor and normal cells, the DOX-loaded MSNs could target tumor cells via the specific interaction between RGDS and integrins receptor overexpressed on tumor cells, followed by penetrating cell membrane with the aid of R7 sequence. After endocytosis into tumor cells, drug release was triggered due to the breakage of GFLG peptide induced by cathepsin B, resulting in enhanced antitumor activity. In vitro drug release behavior of the DOX-loaded MSNs was investigated in PBS buffer at pH 7.4 or 5.0, simulating physiological environment and lysosome environment, respectively. In the presence of cathepsin B, about 60% of DOX has been released from the nanoparticles at pH 7.4 PBS for 24 h. While in the absence of cathepsin B, less than 10% DOX has been released within the same time scale, attributed to the protection of the gatekeeper on the surface of the MSNs. The DOX release behavior at pH 5.0 also confirmed that cathepsin B could cleave the GFLG sequence and thus remove the gatekeeper from the surface of the MSNs. These results showed MSNs had great potential in controlling drug release in cancer cells.

Gemcitabine (GEM) is a chemotherapy medication used in a number of types of carcinomas, including breast cancer, ovarian cancer, non-small cell lung cancer, pancreatic cancer, and bladder cancer. However, the therapeutic efficacy of GEM is limited because it could be rapidly delaminated in the blood to the inactive metabolite 2′,2′-difluorodeoxyuridine, causing a short plasma half-life [67]. To increase the stability and bioavailability of GEM in vivo, new GEM delivery strategy should be designed. Han et al. developed enzyme-sensitive gemcitabine-conjugated albumin nanoparticles as a versatile theranostic nano platform for pancreatic cancer treatment [44]. The novel nano platform was developed by covalent conjugation of GEM to human serum albumin (HSA) via cathepsin B cleavable peptide GFLG and then bounding with near-infrared (NIR) cyanine dye IR780 (HAS-GEM/IR780). It was shown the release of GEM from HAS-GEM/IR780 was inhibited until the breakage of GFLG triggered by cathepsin B, which is overexpressed in pancreatic cancer cells, leading to superior tumor inhibition activity with minimal side effects. As expected, improved cellular proliferation inhibition of HSA-GEM/IR780 was observed in MTT assays, which suggested efficient drug release and cytotoxicity toward pancreatic cancer cells. Furthermore, the in vivo antitumor efficacy revealed that HSA-GEM/IR780 exhibited significantly stronger tumor inhibition compared with other GEM formulations, indicating a distinct improvement of the therapeutic effect of GEM.

3.3 Reduction (or GSH)-Activated Nanocomposites for Anticancer Drug Delivery

Reduction-sensitive nanocomposites can be applied for the development of anticancer drug delivery systems (Table 2). These nanocomposites usually contain characteristic disulfide bonds (S-S) in the main chain, in the side chain, or in the cross-linker [68]. The disulfide bonds are sufficiently stable in the extracellular environment, but under reductive conditions in the cytoplasm, they tend to be rapidly cleaved after minutes or hours through thiol–disulfide exchange reactions [69]. Thiol-disulfide exchange reactions are rapid and readily reversible. GSH is the most abundant thiol in mammalian cells [23], and the significant difference in GSH concentration between tumor cells and extracellular fluid renders GSH-responsive nanocomposites most appealing for targeted anticancer drug delivery.

Due to the reducing potential in tumor cells, reduction-responsive nanocomposites targeting tumor cells can significantly enhance the release of intracellular drugs and achieve the rapid accumulation of drugs (Fig. 2). In addition, compared with the pH-sensitive nanocomposites, reduction-sensitive nanocomposites have several unique advantages, such as high stability against hydrolytic degradation, rapid response to the intracellular reducing environment, and the selective and rapid release of anticancer drugs, at the same time, reducing side effects at normal tissues. Reduction-responsive nanocomposites (also called redox-responsive nanocomposites) have

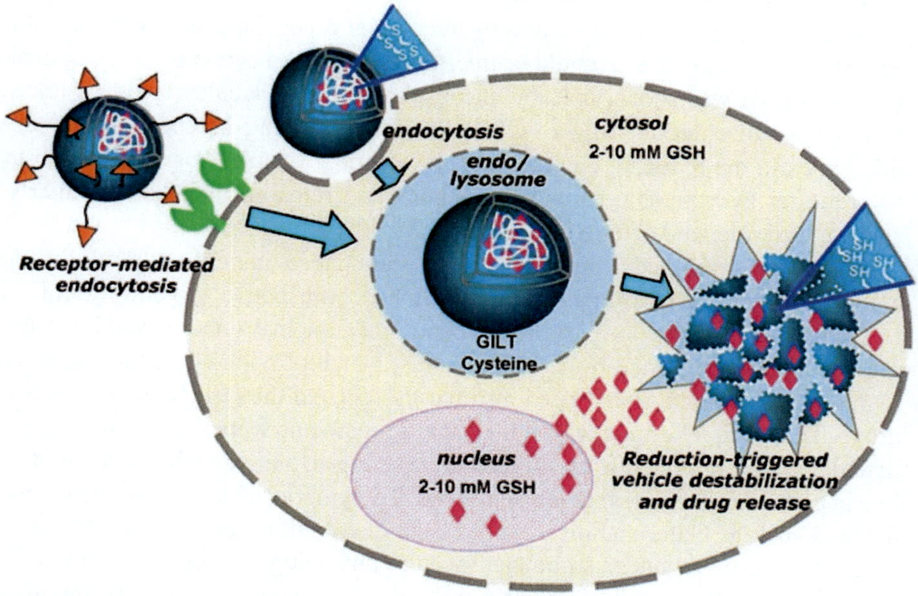

Fig. 2 The illustration of the intracellular trafficking pathway of GSH-responsive nanocomposites including steps of cellular internalization, endosomal escape, reduction-triggered vehicle degradation, and drug release. The redox potential of the cytosol is mainly determined by GSH/GSSG, while that of the endo/lysosome is controlled by GILT and co-factor cysteine. GSH-responsive nanocomposites may also be partially degraded in the endo/lysosomal compartments. Reprinted from Ref. [25] with permission, Copyright Elsevier, 2011

been used to confer sensitivity to the intracellular reducing potential for the release of anticancer drugs into tumor cells, such as liposomes [70], polymersomes, polymeric nanogels [71], micelles [72], dendrimers, and protein nanocapsules [73]. The disulfide bond has been widely used as the cleavable/reversible linker in nanocomposites to render redox potential sensitivity to nanocomposites [26]. In addition, disulfide bonds can be introduced in the repeated units of the polymers, and cause the rapid degradation of the polymer nanocomposites under reductive conditions. In general, reduction-activated nanocomposites contain characteristic disulfide bonds which can be rapidly cut in highly reducing intracellular environments by a GSH-mediated thiol-disulfide exchange reaction to release anticancer drugs [3], showing great potential in targeted anticancer drug delivery.

Ong et al. developed reduction-activated liposomes comprised of quinone-dioleoyl phosphatidylethanolamine (Q-DOPE) lipids for site-specific release [70]. In these liposomes, a "trimethyl-locked" quinone redox switch was attached to the N-terminus of DOPE lipids, and underwent a cleavage (liposome opening) by two-electron reduction to liberate the payload. Ryu et al. synthesized highly stable polymeric nanogels with lots of advantages, such as high encapsulation capability, tunable release kinetics, and no loss of drugs before entering tumor cells [71]. It was a self-cross-linked polymeric nanogel composed of pyridyl disulfide-grafted oligoethyleneglycol-based polymers by intra-/intermolecular disulfide cross-linking. The kinetics of drug release could be fine-tuned by control over cross-linking density in nanogels. It was shown that the encapsulated dye molecules could be released in response to a redox trigger (GSH) via the degrading of cross-linking disulfide bonds in the reducing environment. In addition, these nanogels could be used to sequester hydrophobic chemotherapeutics and released them intracellularly to achieve drug-induced cytotoxicity [71].

Shi et al. developed a multifunctional star-shaped micellar system based on a four-arm amphiphilic copolymer, which had both active targeting ability and redox-responsive behavior [72]. In this copolymer, the hydrophobic and hydrophilic segments were connected by disulfide bonds, which provided the redox-sensitivity to the multifunctional star-shaped micellar system and then triggered the release of drugs. The terminal groups of the hydrophilic segments were decorated with folate ligands, which provided a highly active targeting to tumor cells. Compared with linear polymeric micelles, these star-shaped polymeric micelles exhibited unique features such as better stability, smaller size distribution, and higher drug loading capacity. Moreover, under stimulated normal physiological conditions, the redox-responsive star-shaped micelles could maintain great stability to avoid the premature drug release, while under reducing and acid conditions, drugs would be rapidly released. In vivo fluorescence imaging showed that these folic acid-functionalized star-shaped polymeric micelles had much better specificity to target solid tumors. In vivo antitumor efficacy study demonstrated that the active targeting and redox-responsive star-shaped micelles had better therapeutic effects to solid tumors and higher safety to the normal tissues in contrast to the redox-insensitive micelles [72].

Direct delivery of proteins to the cytosol of cells has tremendous potential in biomedical applications, and biodegradable polymeric nanocapsules have attracted much attention for anticancer therapy. Zhao et al. prepared redox-responsive single-protein nanocapsules for intracellular protein delivery. The target protein could be encapsulated into a positively charged polymeric shell interconnected by disulfide cross-linking. Under reducing conditions, the polymeric shell would dissociate, and the target protein would be released. Studies showed that the active caspase 3 could be efficiently internalized and induced apoptosis in human tumor cells (such as HeLa, MCF-7, and U-87 MG) by the nanocapsules [73a].

Drug resistance in tumor cells and unexpected cytotoxicity in healthy cells are major problems in tumor chemotherapy. To overcome these obstacles, Chang et al. constructed a redox-responsive DOX/siRNA (small interfering RNA) co-delivery system (cationic vesicle) consisting of a novel ferrocenium capped amphiphilic pillar[5]-arene (FCAP) [74]. Due to the high GSH level in tumor cells, this cationic vesicle could disassemble after entering cancer cells, leading to a rapid release of encapsulated siRNA and drug. SiRNA could knock down drug resistance genes and restore the drug sensitivity of cells to enable effective chemotherapy. Meanwhile, due to the absence of the specific stimulus, drug release did not occur or occurred slowly in healthy cells, thus efficiently avoiding severe side effects. It was shown that the co-delivery of siRNA and DOX into cancer cells by cationic vesicles resulted in a clear synergistic inhibition of SKOV-3 cells.

In addition to the disulfide bonds, more reactive diselenide (Se–Se) bonds have been integrated into nanocomposites to achieve redox responsiveness. Ma et al. synthesized a redox-sensitive micelle that was self-assembled by an amphiphilic diselenide-containing triblock copolymer (PEG–PUSeSe–PEG) [75]. The results revealed that the redox responsive micelles were quite stable and showed a high sensitivity to the external redox stimulus.

In conclusion, the reduction-sensitive nanocomposites meet the requirements of an ideal anticancer drug delivery system, such as high stability in circulation, rapid degradation in tumor cells, enhanced therapeutic effect, and minimal carrier-related in vitro and in vivo cytotoxicity. Reduction-sensitive nanocomposites are in particular suited for the efficient delivery of active nucleic acid drugs like siRNA to the cytosol of tumor cells. However, the reduction-sensitive nanocomposites still need further improvement, especially in aspects such as reduction-sensitivity, biocompatibility, long circulation time, and cellular or intracellular targeting ability for better clinical applications.

3.4 Hypoxia-Activated Nanocomposites for Anticancer Drug Delivery

The tumor microenvironment is commonly hypoxic. It is different from normal tissues due to the imbalance between the supply and the consumption of oxygen. The vascular networks within solid tumors are very irregular and are unable to deliver sufficient oxygen and nutrients to all regions. That situation then results in the lack of oxygen in tumor cells [76]. Hypoxia in tumors is mainly a pathophysiologic consequence of structurally and functionally disturbed microcirculation.

Studies have reported that low oxygen pressure can cause conformational changes in hypoxia-responsive nanocomposites via the cleavage or reduction of hypoxia-responsive functional groups (Table 2) [77]. There are three representative classes of hypoxia-responsive moieties, including nitroimidazoles, nitrobenzyl alcohols, and azo linkers, which are useful for the design of nanocomposites for cancer therapy. These hypoxia-responsive moieties contain reducible functional groups (such as nitro, azo, and quinoa), which can accept electrons under hypoxic conditions. When the hypoxia-responsive nanocomposites are exposed to a hypoxia environment, the hydrophilicity of the nanocomposites can be altered as the reduction of hypoxia-responsive moieties produces hydrophilic functional groups [77].

Thambi et al. reported self-assembled nanoparticles (HR-NPs) that could selectively release the hydrophobic agents under hypoxic conditions (Fig. 3) [78]. The HR-NPs were composed of carboxymethyl dextran integrated with a hydrophobic 2-nitroimidazole derivative for the hypoxia-triggered drug release. HR-NPs were stable under physiological conditions and were capable of selectively releasing hydrophobic drugs under hypoxic conditions. Studies showed that DOX-loaded HR-NPs had higher cytotoxicity to hypoxic tumor cells than to normal cells, and

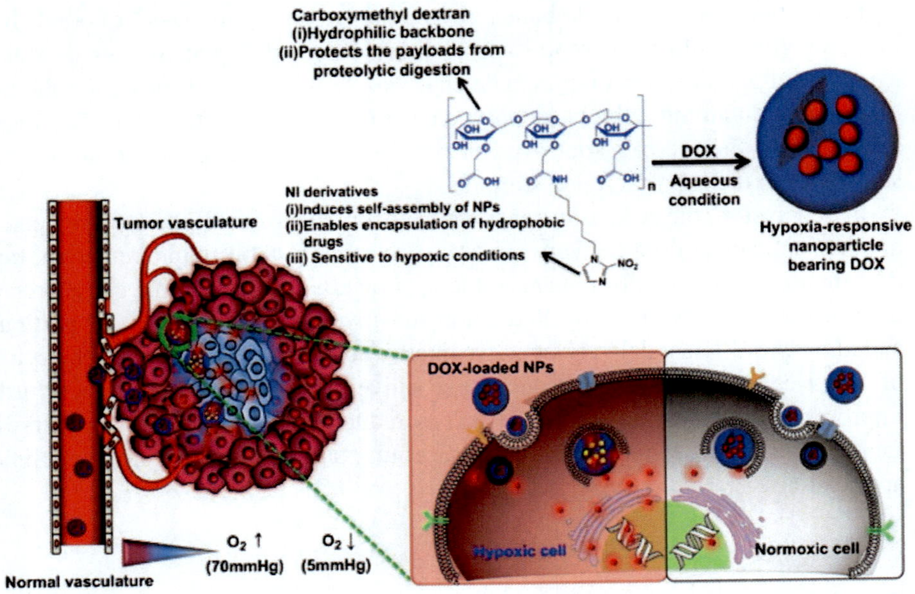

Fig. 3 Schematic illustration of the formation of drug-loaded HR-NPs and in vivo tumor-targeting pathways. The HR-NPs can reach the tumor site via the EPR effect, followed by intracellular drug release at hypoxic tissues. Reprinted from Ref. [78] with permission, Copyright Elsevier, 2014

HR-NPs could effectively accumulate at the tumor site, resulting in significantly higher anticancer efficacy than free DOX. To enhance anticancer therapy, Thambi et al. also developed a novel type of hypoxia-sensitive polymeric micelles (HS-PMs) [79] exhibiting significant physicochemical changes at the early stage of tumor progression and responsiveness to tumor cells [80]. It was shown that HS-PMs composed of an amphiphilic block copolymer poly(ethylene glycol)-poly(e-(4-nitro)benzyloxycarbonyl-L-lysine) had higher cytotoxicity to SCC7 (squamous cell carcinoma) cells under hypoxic conditions than under normal conditions.

Since hypoxic regions are distant from blood vessels and have increased efflux transporters for siRNA delivery to tumor cells with nanocomposites, it is necessary to protect siRNA from degradation and to promote its cellular internalization and endosomal escape. Perche et al. developed hypoxia-activated nanoparticles composed of polyethyleneglycol 2000, azobenzene, polyethyleneimine (PEI), and 1,2-dioleyl-sn-glycero-3-phosphoethanolamine (DOPE) units for siRNA delivery to tumor cells [81]. It was shown that these nanoparticles exhibit hypoxia-activated gene-silencing capacity both in vitro and in vivo.

Qian et al. reported on light-activated hypoxia-responsive nanocomposites for enhanced anticancer therapy [82]. The nanoparticle was composed of three parts: ROS-generating and hypoxia-sensitive 2-nitroimidazole-grafted conjugated polymer (CP-NI), polyvinyl alcohol (PVA)-based surface coatings, and the encapsulated DOX. It was a novel drug delivery system for the first time using a light-activated hypoxia-responsive pathway, which was able to produce reactive oxygen species and induce hypoxia to release drugs into tumor cells. This drug delivery system combined both photodynamic therapy and chemotherapy and could enhance the efficacy of anticancer therapy. Photodynamic therapy depended on the ability of photosensitizers to transfer energy from light to the surrounding molecular oxygen ($^3O_2$) to produce cytotoxic ROS (such as singlet oxygen $^1O_2$). During the conversion from molecular oxygen ($^3O_2$) to singlet oxygen ($^1O_2$), oxygen was consumed and blood vessels constricted, which produced a hypoxic condition. Therefore, these nanoparticles combining a photodynamic system with a hypoxia-responsive drug delivery system efficiently enhanced the anticancer therapy.

In conclusion, tumors are complex and exhibit molecular and cellular heterogeneity. Hypoxia is the hallmark of solid tumors. The properties of the hypoxic tumor environment can be used in diagnostic and therapeutic strategies [77]. Targeting the tumor hypoxia microenvironment provides an important strategy to overcome the problem of tumor heterogeneity, and the development of hypoxia-activated nanocomposites is important for enhancing anticancer therapy. However, there still are some challenges in hypoxia-activated nanocomposites for anticancer drug delivery. For example, the ability of nanocomposites to deliver anticancer drugs to tumors depends on their physicochemical and biological properties, such as the circulation time, targeting ability, and the uptake of tumor cells. A small variation in these parameters may lead to a dramatic alteration in their circulation and tumor targeting efficiency. Although there are many limitations in hypoxia-activated nanocomposites due to the lack of biodegradability and biocompatibility, which limit the clinical application of nanocomposites, they are still promising in anticancer drug delivery for targeted tumor therapy.

3.5 *ROS-Activated Nanocomposites for Anticancer Drug Delivery*

The increased cellular levels of ROS are the biochemical property of tumor cells. ROS-responsive drug delivery systems targeting oxidative microenvironment in tumor cells can selectively deliver anticancer drugs to tumor cells and can effectively reverse their drug resistance (Table 2) [31]. de Gracia Lux et al. developed new biocompatible, self-degradable, and H_2O_2-triggered nanoparticles that could release drugs upon exposure to the low concentrations of H_2O_2 [83]. These nanoparticles were prepared from a bio-responsive polyester bearing aryl boronic ester groups that had a high sensitivity to H_2O_2. Every cleavage of the boronic ester led to polymer backbone degradation and could be degraded easily. At the inflamed or cancerous cells, the boronic ester cap was removed to produce phenols and led to a quinonemethide rearrangement, thus disrupting the nanoparticles and releasing the drugs [3]. The nanoparticles could be degraded into small molecules, which were easily cleared by human body. Since cytokines and chemokines produced by neutrophils could affect the recruitment and activation of inflammatory cells in the tumor microenvironment, and provide an immunosuppressive microenvironment for tumor growth, metastasis, and angiogenesis, using these nanoparticles to deliver drugs that inhibited neutrophils recruitment in tumor tissues could limit tissue damage and improve anticancer therapy.

Shim and Xia synthesized a cationic polymer poly(amino thioketal) (PATK) with ROS-cleavable thioketal linkages for mediating intracellular gene delivery [84]. In the study, PATK and DNA could self-assemble into nanoparticles, and these nanoparticles could be easily cleaved under ROS-abundant conditions due to the biodegradable thioketal linkages. It was the probable mechanism that the ROS-cleavable thioketal-based nanoparticles could enhance the efficiency of the intracellular gene delivery in tumor cells by facilitating the intracellular release of the encapsulated nucleic acids in response to the high levels of intracellular ROS. The results showed that the DNA/PATK nanoparticles exhibited efficient gene transfection in prostate cancer cells.

Liu et al. developed H_2O_2-responsive nanoparticles based on an amphiphilic hyperbranched polymer with hydrophobic selenide groups and hydrophilic phosphate segments in the dendritic backbone framework [85]. The nanoparticles could be disassembled under an exclusive oxidative tumor microenvironment that contained a high level of H_2O_2 (the level of H_2O_2 in tumor cells was increased at the speed of 0.5 nmol/10^4 cells/h [86], compared with normal cells). Because hydrophobic selenide groups could be easily oxidized into hydrophilic selenone groups, the amphiphilic hyperbranched polymers became hydrophilic hyperbranched polymers after oxidation and then nanoparticles could rapidly and selectively release drugs in tumor cells. In this study, the DOX-loaded H_2O_2-responsive nanoparticles could be rapidly disassembled in the target tumor cells under intracellular oxidative conditions, thus rapidly releasing DOX for inducing cell apoptosis. In addition, selenium compounds enhanced the immune response, produced anticancer metabolites that could efficiently disturb the metabolism of tumor cells, inhibited angiogenesis and finally induced the apoptosis of tumor cells [85]. Therefore, these nanoparticles that contained selenium compounds showed potential in cancer therapy.

High ROS level in tumor cells is essential for ROS-activated anticancer drug release in tumor cells. Su et al. developed novel ROS-activated and positive feedback micelles that could replenish the ROS upon their disassembly to maintain high ROS levels [87]. These micelles not only responded to original ROS signals in tumor cells for drug release, but also responded to self-regenerating ROS signals in the cytoplasm by amplifying ROS signal. These micelles were prepared by a new amphiphilic polymer (TBH) that was composed of D-α-tocopherol polyethylene glycol 1000 succinate (TPGS), hyaluronic acid (HA), and arylboronic ester. TPGS was an effective P-gp efflux inhibitor, selective tumor cell killing agent and, most importantly, a powerful ROS inducer upon its interaction with mitochondrial respiratory complex II [88]. The arylboronic ester was as a ROS sensing linker for imaging and targeting of tumor cells. TBH was synthesized by linking HA as hydrophilic chains to the hydrophobic TPGS with arylboronic esters [87]. Therefore, TBH micelles could efficiently degrade in response to ROS in tumor cells and disassembled to release

TPGS rapidly. Because of the ROS-induced activity of TPGS, ROS regenerated and increased. These results could be observed both in MCF-7/ADR cells and MCF-7/ADR tumor xenografts after TBH micelles treatment. In addition, HA was an active target ligand for CD44 positive malignant cells and it could increase the uptake of the TBH nanoparticles [89]. DOX, as a model anticancer agent, was loaded in TBH micelles (DOX-TBH). The results showed that compared with free DOX, DOX-TBH could significantly reduce the tumor volume, increase the percentage of apoptotic or necrotic tumor cells, and prolong the survival time of tumor-bearing nude mice. In addition, after DOX-TBH treatment, no obvious pathological abnormalities were observed in normal tissues such as heart, lung, spleen, liver, and kidney.

In conclusion, many studies have observed an increased generation of ROS and altered redox status in tumor cells. Targeting ROS signal in tumor cells provides the possibility for selectively eliminating tumor cells and some studies have achieved ROS-activated anticancer drug delivery by using ROS-sensitive nanocomposites. ROS has a mixed effect on tumor cells, both good and bad. Evidence suggests that transformed cells use ROS signals to promote proliferation. However, this confers a state of increased basal oxidative stress for tumors that can make them more vulnerable to chemotherapeutic agents and can further augment the generation of ROS or weaken antioxidant defenses of cells [90]. Developing ROS-activated nanocomposites to amplify ROS level in tumors, keeping the intracellular ROS at a high level, increasing the oxidant stress and sensitivity to chemotherapeutic agents, and finally causing cell death represents a new way for tumor treatment.

3.6 ATP-Activated Nanocomposites for Anticancer Drug Delivery

ATP is the basic energy currency in the physiology environment. It has a sharp concentration contrast between extracellular (<0.4 mM) and intracellular (1–10 mM) environments, especially in the tumor sites, which makes ATP a feasible cue for regulating drug release (Table 2) [91]. ATP-activated drug release is always achieved by competitive binding of ATP with other DNA sequences to ATP aptamer. Generally, drugs (for example, DOX) are loaded inside the double-chain structure composed of ATP aptamer (DNA sequence) and its corresponding cDNA. When it enters tumor cells that are full of ATP, ATP will compete with cDNA to bind with the ATP aptamer. The binding of ATP and ATP aptamer is much stronger than the interaction between ATP aptamer and cDNA [92] and thus the previous double-chain structure is destroyed, which results in drug release.

Yang et al. firstly demonstrated that ATP aptamer-ATP interaction could be used as a stimuli-responsive model for controlling drug release. They prepared fluorescein isothiocyanate (FITC) dye-loaded mesoporous silica nanoparticles (MSN), which were capped with ATP aptamer-based molecular gates to control the drug release from MSN [93]. First, they covered MSN with a derivative of ATP through amidation reaction. Then ATP aptamer-modified Au nanoparticles (AuNPs) acted as molecular gates to close the pores of MSN through adenosine-aptamer interaction (MSA). When MSA met ATP in tumor cells, ATP would compete to bind the ATP aptamer, which replaced the previous interaction of aptamer and adenosine, causing the pores to open and drug releasing from MSA. It was shown that AuNPs could cap MSN and dissociate from MSA once exposed to ATP, confirmed by TEM microscopy. In vitro experiments showed that the drug release of MSA was nearly 0% without ATP and the drug release of uncapped MSN was very rapid. In the presence of 8 mM ATP, drug release from MSA reached around 40% after 10 h. The results showed MSA could effectively control drug release in response to ATP.

As AuNPs still have some cytotoxicity and may accumulate in the liver, Ozalp et al. simplified these molecular gates by extending 7 more bases to 3′ end of ATP aptamer (5′-CACCTGGGGGAGTATTGCGGAGGAAGGTTCCAGGTG-NH2-3′) where the first 7 bases in 5′-end could be hybridized with the extended 7 bases to close the pores of MSN [94]. It was demonstrated that the release of drug from MSN increased dramatically in the presence of 1 mM ATP and it reached around 80% release after 1-h incubation. To further confirm the specificity of ATP aptamer to ATP, 1 mM GTP was added to aptamer-capped MSN and it was found that drug release from MSN was very slow. In addition, drug release from MSN increased with increasing concentration of ATP, indicating that drug release from MSN could be controlled in response to ATP.

Changing ATP aptamer sequence may impact its affinity with ATP. To avoid this problem, He et al. designed a “sandwich structure” composed of ATP aptamer, ssDNA1 and ssDNA2 to serve as molecular gates of MSN (Fig. 4) [95]. The sandwich structure would close the pores of MSN by the interaction of ssDNA1 and ssDNA2 with ATP aptamer until it met ATP, which could bind with ATP aptamer, resulting in the opening of pores. To test the ATP-responsive drug release system, $Ru(bipy)_3^{2+}$, as the model drug, was loaded in MSN. It was shown that $Ru(bipy)_3^{2+}$ release from MSN after 48-h incubation correlated with the increasing ATP concentration and it reached a platform when ATP concentration was above 10 mM. These results indicated that the more ATP existed in cells, the more drugs were released from MSN. To test the stability of the aptamer-MSN system, He et al. added it into mouse serum to test their precipitation at 37 °C. After incubating mouse serum with MSN for about 3 h, the supernatant was still transparent, which indicated the nanoparticles were quite stable. Moreover, to demonstrate the ATP-response mechanism, the drug release of MSN was tested in Ramos cells as a large amount of ATP appeared to be in Ramo cells. Once MSN was incubated with broken Ramo cells, drug release from MSN increased dramatically, suggesting great potential in cancer drug delivery.

Mo et al. used ATP aptamer as the connecting arm, which joined DNA-graphene oxide (GO) complex pieces together to develop ATP-activated graphene nanosheets (Fig. 5) [96]. In this system, DNA1 and DNA2 had sequences of “head and tail,” which would bind to aptamer (with head) and GO (with tail). ATP aptamer was added to the DNA-GO complex to cause the DNA-GO complex to self-assemble into ATP-activated graphene nanosheets. DOX, as a model drug, was loaded into DNA-GO nanosheets with high loading efficiency because of π–π stacking interaction

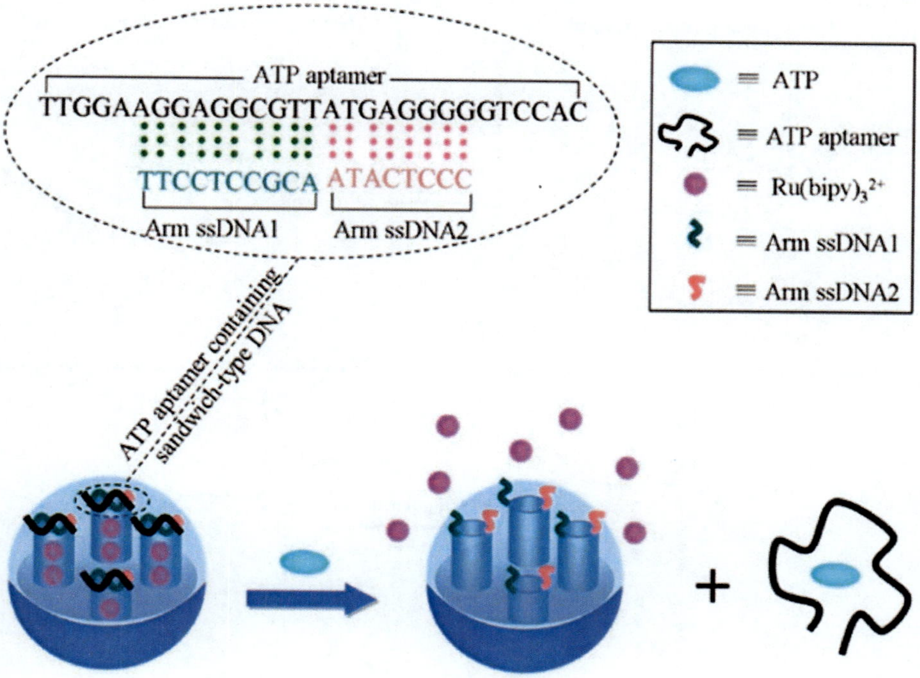

Fig. 4 Schematic illustration of aptamer-based ATP-responsive MSN system. ATP aptamer combined with ssDNA1 and ssDNA2 forming sandwich-type DNA which could block the drug release in the absence of ATP. When it was exposed to ATP, the sandwich structure was destroyed and triggering drug release. Reprinted from Ref. [95] with permission, Copyright American Chemical Society, 2012

between DOX and GO. Once ATP appeared, ATP aptamer would bind with ATP and thus dissociated from the DNA-GO nanosheets causing the release of DOX. The results demonstrated that in the presence of 3 mM ATP, the particle size of DNA-GO nanosheets reduced from 550 to 270 nm. To test the intracellular ATP-responsive DOX release, the uptake of DNA-GA nanosheets on HeLa cells was tested by measuring the fluorescence intensity of tumor cells. It was shown the fluorescence intensity of DOX in tumor cells increased significantly after incubating DNA-GA nanosheets with HeLa cells at 37 °C. To confirm the triggering of drug release was caused by ATP, the temperature was reduced to 4 °C to stop ATP synthesis of tumor cells and almost no increase in DOX release after incubating HeLa cells with DNA-GA nanosheets, indicating DNA-GA nanosheets had a good response to ATP.

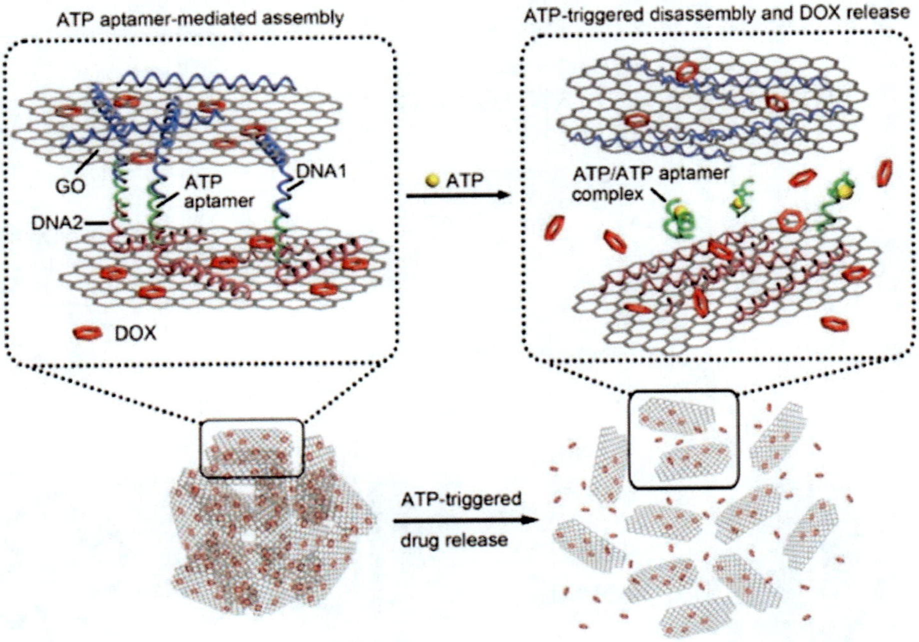

Fig. 5 Schematic illustration of ATP-responsive DNA-GA nanosheets for controlled drug delivery. Reprinted from Ref. [96] with permission, Copyright Elsevier, 2015

3.7 Multi-stimuli Activated Nanocomposites for Anticancer Drug Delivery

Single stimuli-responsive nanocomposites have achieved some progress in anticancer drug delivery. Smart nanocomposites responding to multiple (two or more than two) stimulus can take advantages of different responsive mechanisms and make up their disadvantages, which may fulfill the different requirements in anticancer drug delivery and has become an important direction in tumor drug delivery (Table 2). Notably, these combined stimulus activations take place either simultaneously at the same pathological site or in a sequential manner in different settings. These multi-stimuli activated nanocomposites with programmed site-specific drug delivery features have shown superior in vitro and/or in vivo anticancer efficacy.

DOX is an important anticancer drug, but it has high cytotoxicity, such as cardiotoxicity. Zhang et al. designed enzyme/pH-responsive nanoparticles to control the DOX release in tumor cells and thus reduce the cytotoxicity (Fig. 6) [97]. In these nanoparticles, DOX was inserted into the negative double strand of DNA, which was complexed with cationic gelatin to form compact nanoscale complexes (CPX1). Another pH-sensitive material histamine (His)-modified alginate (pKa 6.9) was used to cover CPX1 to form enzyme/pH-responsive nanoparticles (CPX2). Under a physiological environment (pH 7.2), these nanoparticles were stable. Once these nanoparticles went into the tumor sites (pH 6.2–6.7), His-alginate would become

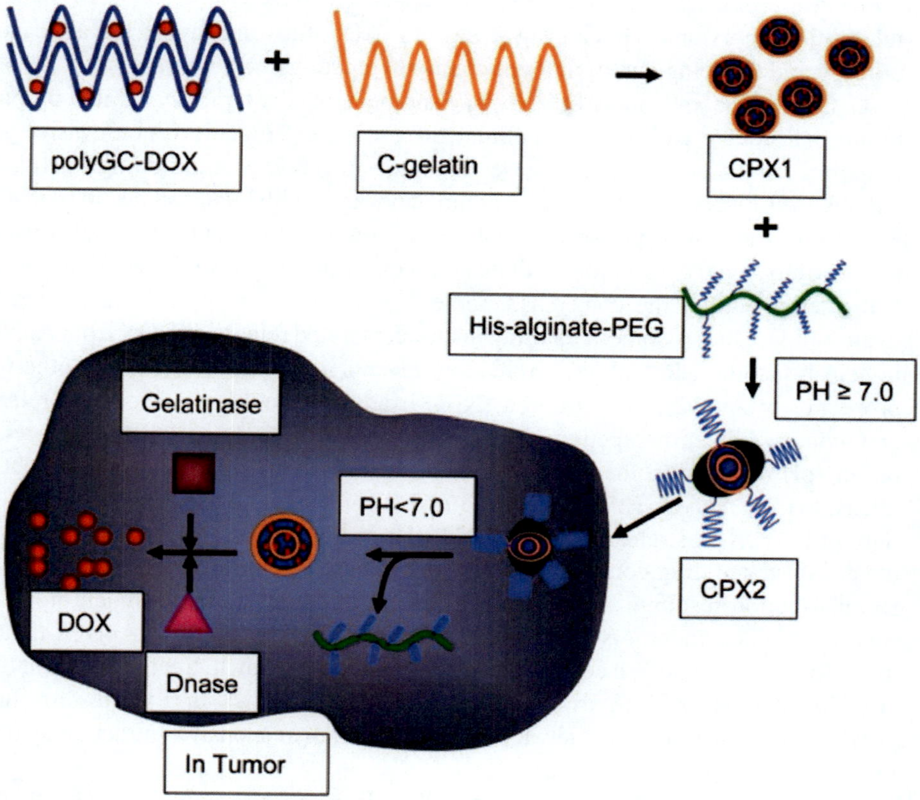

Fig. 6 Fabrication of CXP1 and CPX2 and the mechanism of enzyme/pH-triggered drug release. First, polyGC-DOX and C-gelatin formed into CPX1, and then CPX1 was covered by His-alginate-PEG to form CPX2. When CPX2 was under physiological condition (pH > 7.0), it was stable. When it was exposed to lower pH and enzyme in tumor tissues, it would be broken and release DOX. Reprinted from Ref. [97] with permission, Copyright Elsevier, 2010

cationic and dissociate from CPX2. Then, the cationic gelatin could be digested by gelatinase, which was highly expressed in tumors, and DOX was released after DNA was digested with DNase. It was shown that CPX2 and CPX1 both had a high DOX release (nearly 80%) within 5 h after incubation with the tumor tissue homogenate supernatant, indicating that these enzyme/pH-responsive nanoparticles might work well in the tumor environment. In addition, as CPX2 was covered by His-alginate, which responded to low pH, it showed low DOX release (<30%) in liver tissue homogenate supernatant (LHS) while CPX1 has a relatively high DOX release in LHS. In vivo bio-distribution results revealed that CPX2 has the highest DOX release in tumor cells and the lowest release in the liver at 48 h after an intravenous injection of CPX2 at the dose of 20 mg DOX/kg body weight. Anticancer experiments in sarcoma tumor-bearing mice demonstrated that only 30% of the mice survived after free DOX treatment, while 90% of the mice survived more than 4 weeks after CPX2 treatment. The results indicated that CPX2 could effectively increase the drug release in tumors and reduce drug release in livers, thus improving the anticancer efficacy and reducing cytotoxicity of chemotherapeutics.

Generally, the anticancer efficacy of nanomedicines is typically limited by poor tumor cell uptake and low intracellular drug release. To achieve noninvasive and pinpointed intracellular drug release, Wang et al. reported multi-stimuli-activated nanohybrids by coating DOX-loaded gold nanocages with hyaluronic acid (AuNCs-HA) [98]. These well-prepared nanohybrids could be efficiently internalized into cancer cells by the receptor-mediated process via HA-CD44 interactions. After cell internalization, the coated HA molecules on AuNCs could be degraded by hyaluronidase in lysosomes, resulting in the accelerated release of DOX from AuNCs in acid lysosomes. In addition, by taking advantage of the excellent photothermal properties of AuNCs, near-infrared (NIR) irradiation could enhance DOX release and induce a higher therapeutic efficacy. In vitro results confirmed that in the acetate buffer (pH 4.5) and in the presence of hyaluronidase, there was subsequent DOX releasing from AuNCs-HA. While in pH 7.4 PBS and in the absence of hyaluronidase, only little DOX was released from AuNCs-HA within 12 h. These results indicated the encapsulated drug could be pinpointedly released in intracellular environments. Biodistribution results in breast cancer-bearing mice revealed that much more Au was accumulated in tumors of AuNCs-HA-treated mice than those of AuNCs-treated mice. More importantly, a complete inhibition of tumor growth treated with AuNCs-HA with the help of NIR irradiation was observed in vivo. In a word, this study provided new insights into developing multi-stimuli responsive intracellular drug release systems for cancer therapy.

Breast cancer appears to have higher ATP concentration in tumor cells (1–10 mM) than that in tumor extracellular space (<0.4 mM) and pH inside tumor cells is 5.0–5.5. According to these pathology properties, Yu Zhang et al. designed ATP/pH-activated nanoparticles for breast cancer therapy, in order to improve tumor cell uptake and intracellular drug release of nanomedicine [99]. First, DOX was conjugated to PEG-poly(aspartic acid) to form PA-DOX, which could be hydrolyzed in low pH. Then PA-DOX was connected with ATP aptamer by intercalating DOX into the double-strand DNA to form nanoparticles. In this case, positive-charged polymers were not needed in this nanoparticle structure, which implied lower cytotoxicity to the body. Moreover, Des-Arg analog D-[des-Arg10] kallidin was conjugated to nanoparticles (D-PA-DOX) for active targeting to bradykinin receptor1 upregulated in the breast cancer cells. When the nanoparticles came to the tumor tissue, [des-Arg10] kallidin facilitated the endocytosis by tumor cells quickly. In the tumor cells, the nanoparticles could be destroyed by the binding of ATP to the aptamer and the breakdown of the acid liable bond, and release DOX dramatically to improve the killing ability of DOX. It was shown that nearly 70% of DOX was released from D-PA-DOX at pH 5.0 in the presence of 4 mM ATP within 24 h. At a lower ATP concentration (0.4 mM) and pH 5.0, DOX release from D-PA-DOX reached only about 40%. In addition, at pH 7.4 and 4 mM ATP, DOX release from D-PA-DOX was even lower than 30%. These results proved D-PA-DOX had a dual response to ATP and pH in vitro. Orthodox tumor-bearing nude mice were used as the animal model to test the in vivo biodistribution and tumor accumulation of D-PA-DOX. It was shown that

DOX concentration in tumors for the D-PA-DOX group was approximately 100-fold that for the free DOX group, suggesting good accumulation of D-PA-DOX in tumors. After 3 cycles of treatment at the DOX dose of 5 mg/kg, breast cancer cell-bearing mice in the D-PA-DOX group exhibited significantly smaller tumor volume than any other group, indicating these ATP/pH-activated nanoparticles could effectively control the drug release and treat tumors.

Redox and low pH are the two most appealing stimuli because they exist naturally in all cancer cells, and they have been well identified as ideal stimuli for triggering rapid destabilization of nanoparticle drug delivery systems inside cancer cells. To provide effective therapeutic and diagnostic strategies against ovarian cancers, Lin et al. developed a redox and pH-sensitive self-assembling hyaluronic acid nanoparticle with active targeting peptide LHRH [100]. Anticancer drug DOX is grafted onto hyaluronic acid via cis-aconityl linkage and disulfide bond to possess pH sensitivity and redox property, respectively. With the amphiphilic property, this conjugate could self assemble to nanoparticle and load NIR for in vivo tracking. The results showed that the nanoparticles have a drug loading amount of about 6.2% and their particle size was approximately 229.0 nm. Synergistic controlled DOX release from these nanoparticles was triggered by acidic and reductive environment, indicating these nanoparticles could release DOX rapidly under the trigger of reductive reagents in the cytoplasm and low pH value in lysosomes. In orthotopic ovarian tumor models, these nanoparticles possessed a preferable tumor imaging capability and an excellent antitumor ability to almost 30% of original size in 20 days.

Apart from smart nanocomposites described above, which were all activated simultaneously in tumor cells, nanocomposites experiencing distinct two-step delivery kinetics under the stimuli of the tumor microenvironment and intracellular microenvironments, have been developed for deep tumor tissue penetration, cellular uptake, and controlled drug release. Ruan et al. developed a kind of novel nanocarrier, G-AuNPs-DOX-PEG, which was constructed with shrinkable gelatin nanoparticles, DOX tethered gold nanoparticles, and long-chain polyethylene glycol (PEG) [101]. GAuNPs-DOX-PEG could be degraded by MMP-2 proteins, an overexpressed extracellular enzyme, with a size shrink from 186.5 to 59.3 nm. The release of DOX from G-AuNPs-DOX-PEG was in a pH-dependent manner. Tumor spheroid penetration and collagen gel diffusion showed G-AuNPs-DOX-PEG with pre-incubation with MMP-2 could significantly enhance its penetrating efficiency. In vivo and ex vivo imaging revealed that G-AuNPs-DOX-PEG could distribute into 4T1 and B16F10 tumors at the highest intensity. Correspondingly, 4T1 and B16F10 tumor-bearing mice treated with G-AuNPs-DOX-PEG displayed the lowest tumor growth rate. These results suggested that G-AuNPs-DOX-PEG could enter the deep tumor region and then release DOX, resulting in a powerful antitumor effect.

4 Conclusion and Perspectives

With the up-to-date advancements in nanotechnology, biopharmaceutics, materials science, and chemistry, the design and application of stimuli-triggered nanodevices for drug delivery have drawn our increasing attention and sprung up one after another. Based upon the in-depth study of pathophysiological characteristics of the endogenous tumor microenvironment and the discrepancies in the physiological and biological characteristics between the tumor and healthy tissue, cancer-associated stimuli-responsive nanosystems have been increasingly considered for the delivery of anticancer drugs. This review has demonstrated various applications for which nanoparticles are being employed in the fight against cancer. In accordance with the deeper understanding of the realistic conditions and interactions involved in the tumor microenvironment, all kinds of strategies have been carried out to treat and control the evolution and neoplastic nature of cancer. Nanoparticles drug delivery systems with elaborately designed synthetic nanocomposites have attested to be a promising alternative for the targeted delivery of anticancer drugs. The unique attributes of nanocomposites like pH-activated, enzyme-activated, GSH-activated, and hypoxia-activated peculiarities may have allowed clinicians to furnish them either as new treatments (monotherapy) or as adjuncts to existing treatments (combined therapy) to enhance therapeutic efficacy.

Regardless of the current remarkable progress, only a minority of these nanoparticles have been examined in preclinical animal models in vivo, and few have been applied clinically. For example, nanovehicles with multi-stimuli-activated features or thermosensitive liposomes are still in their infancy. The efficacy of those stimulus-responsive nanosystems depends mainly on their high specificity and sensitivity of tumor-targeting so as to reduce systemic toxic side effects at normal tissues and increase drug accumulation at the tumor site or at intracellular cytosolic compartments in tumor cells. However, this remains many challenges, which incorporate the off-tumor targeting caused by the exaggerated EPR effect in humans, non-specific stimuli, a lack of understanding of the distribution and activity of some internal stimuli, the low levels of some triggering signals that cannot meet the signal response needs and so on. With regard to these, those newly developed nanocarriers need to be thoroughly characterized pharmacologically, physiochemically, and immunologically before they can be approved for use in humans. Clinicians may also need to tightly regulate the distribution of the nanoparticle size, consistency, and uniformity between batches. Additionally, owing to the lack of adequate evidence and in-depth researches on the safety, biocompatibility, and degradability of these novel materials, transition of their applications from animal to human models has been considerably hindered. In view of that, even though the potential of intracellular signal-activated nanocomposites for anticancer drug delivery to enhance drug delivery cannot be denied, their transition from bench to bedside still strongly rests upon the emerging region of nanotoxicology to assess the toxic effects and elucidation of the mechanism of reversing chemoresistance. Nevertheless, with the deeper study of tumor acidity, enzymes, endo/lysosomal pH, GSH, ROS, etc., we firmly believe that these tumor microenvironment-targeting strategies will offer a good chance for the practical translation of nanoparticle formulations into the clinic. Moreover, these further well-established nanosystems may control the drug-releasing pattern spatiotemporally and will improve the accumulation of chemotherapeutic agents in MDR tumors or cells.

In general, this review discussed the exploitation of the tumor microenvironment signal-activated nanocomposites for cancer therapy. The utility of tumor microenvironment-sensitive nanoparticles holds promise in oncotherapy. We are convinced that with rational design and continuing researches, a rosy future of these smart nanocomposites can be foreseen for the effective treatment of multifarious cancers.

Acknowledgements This work was supported by the National Natural Science Foundation of China (81773283, 81472757, 81361140344).

References

1. Manchun S, Dass CR, Sriamornsak P (2012) Life Sci 90:381–387
2. Glisoni RJ, Sosnik A (2014) Macromol Biosci 14:1639–1651
3. Mo R, Gu Z (2016) Mater Today 19:274–283
4. Liu MR, Du HL, Zhang WJ, Zhai GX (2017) Mater Sci Eng C-Mater Biol Appl 71:1267–1280
5. Joyce JA, Fearon DT (2015) Science 348:74–80
6. (a) Mbeunkui F, Johann DJ (2009) Cancer Chemother Pharmacol 63:571–582; (b) Spill F, Reynolds DS, Kamm RD, Zaman MH (2016) Curr Opin Biotechnol 40:41–48
7. Mura S, Nicolas J, Couvreur P (2013) Nat Mater 12:991–1003
8. (a) Kessenbrock K, Plaks V, Werb Z (2010) Cell 141:52–67; (b) Stern R (2008) Semin Cancer Biol 18:275–280; (c) Aggarwal N, Sloane BF (2014) Proteom Clin Appl 8:427–437; (d) Basel MT, Shrestha TB, Troyer DL, Bossmann SH (2011) Acs Nano 5:2162–2175; (e) Candel I, Aznar E, Mondragon L, de la Torre C, Martinez-Manez R, Sancenon F, Marcos MD, Amoros P, Guillem C, Perez-Paya E, Costero A, Gil S, Parra M (2012) Nanoscale 4:7237–7245; (f) He XY, Cao HQ, Wang H, Tan T, Yu HJ, Zhang PC, Yin Q, Zhang ZW, Li YP (2017) Nano Lett 17:5546–5554; (g) Jin Y, Yang F, Du L (2013) Colloids Surf B Biointerfaces 112:421–428
9. Vaupel P (2004) Oncologist 9:10–17
10. Liou GY, Storz P (2010) Free Radic Res 44:479–496
11. Mo R, Jiang T, DiSanto R, Tai W, Gu Z (2014) Nat Commun 5:3364
12. Liu J, Huang YR, Kumar A, Tan A, Jin SB, Mozhi A, Liang XJ (2014) Biotechnol Adv 32:693–710
13. Du JZ, Mao CQ, Yuan YY, Yang XZ, Wang J (2014) Biotechnol Adv 32:789–803
14. Heiden MGV, Cantley LC, Thompson CB (2009) Science 324:1029–1033
15. (a) Fleige E, Quadir MA, Haag R (2012) Adv Drug Deliv Rev 64:866–884; (b) Karimi M, Ghasemi A, Zangabad PS, Rahighi R, Basri SMM, Mirshekari H, Amiri M, Pishabad ZS, Aslani A, Bozorgomid M, Ghosh D, Beyzavi A, Vaseghi A, Aref AR, Haghani L, Bahrami S, Hamblin MR (2016) Chem Soc Rev 45:1457–1501
16. Liu XM, Zhang XF, Lv Y, Xiang JX, Shi JH (2014) Oncol Rep 32:2810–2816
17. Gialeli C, Theocharis AD, Karamanos NK (2011) FEBS J 278:16–27
18. Bertrand P, Girard N, Duval C, DAnjou J, Chauzy C, Menard JF, Delpech B (1997) Int J Cancer 73:327–331
19. (a) Rao JS (2003) Nat Rev Cancer 3:489–501; (b) Mohanam S, Go Y, Sawaya R, Venkaiah B, Mohan PM, Kouraklis GP, Gokaslan ZL, Lagos GK, Rao JS (1999) Int J Oncol 14:169–174

20. Zhang X, Bu XY, Zhen HN, Fei Z, Zhang JN, Fu LA (2000) J Clin Neurosci 7:116–119
21. (a) Kim TD, Song KS, Li G, Choi H, Park HD, Lim K, Hwang BD, Yoon WH (2006) BMC Cancer 6:211; (b) Miyake K, Shimada M, Nishioka M, Sugimoto K, Batmunkh E, Uto Y, Nagasawa H, Hori H (2009) Cancer Chemother Pharmacol 64:885–892; (c) Roomi MW, Kalinovsky T, Niedzwiecki A, Rath M (2013) Int J Oncol 43:1027–1035
22. Wu G, Fang YZ, Yang S, Lupton JR, Turner ND (2004) J Nutr 134:489–492
23. Pompella A, Visvikis A, Paolicchi A, De Tata V, Casini AF (2003) Biochem Pharmacol 66:1499–1503
24. Kuppusamy P, Li H, Ilangovan G, Cardounel AJ, Zweier JL, Yamada K, Krishna MC, Mitchell JB (2002) Can Res 62:307–312
25. Cheng R, Feng F, Meng F, Deng C, Feijen J, Zhong Z (2011) J Control Release 152:2–12
26. Li R, Xie Y (2017) J Control Release 251:49–67
27. Hockel M, Vaupel P (2001) J Natl Cancer Inst 93:266–276
28. Gilkes DM, Semenza GL, Wirtz D (2014) Nat Rev Cancer 14:430–439
29. Nurwidya F, Takahashi F, Minakata K, Murakami A, Takahashi K (2012) Anat Cell Biol 45:73–78
30. Pelicano H, Carney D, Huang P (2004) Drug Resist Updates 7:97–110
31. Liu Q, Shuhendler A, Cheng J, Rauth AM, O'Brien P, Wu XY (2010) Breast Cancer Res Treat 121:323–333
32. Gorman MW, Feigl EO, Buffington CW (2007) Clin Chem 53:318–325
33. Cui X, Guan X, Zhong S, Chen J, Zhu H, Li Z, Xu F, Chen P, Wang H (2017) Ultrason Sonochem 38:145–153
34. Xu X, Saw PE, Tao W, Li Y, Ji X, Yu M, Mahmoudi M, Rasmussen J, Ayyash D, Zhou Y, Farokhzad OC, Shi J (2017) Nano Lett 17:4427–4435
35. Li HJ, Du JZ, Liu J, Du XJ, Shen S, Zhu YH, Wang X, Ye X, Nie S, Wang J (2016) ACS Nano 10:6753–6761
36. Gou J, Liang Y, Miao L, Guo W, Chao Y, He H, Zhang Y, Yang J, Wu C, Yin T, Wang Y, Tang X (2017) Acta Biomater 62:157–166
37. Wang T, Wang D, Liu J, Feng B, Zhou F, Zhang H, Zhou L, Yin Q, Zhang Z, Cao Z, Yu H, Li Y (2017) Nano Lett 17:5429–5436
38. Jeong S, Song J, Lee W, Ryu YM, Jung Y, Kim SY, Kim K, Hong SC, Myung SJ, Kim S (2017) Nano Lett 17:1378–1386
39. Zhu L, Kate P, Torchilin VP (2012) ACS Nano 6:3491–3498
40. Yu H, Chen J, Liu S, Lu Q, He J, Zhou Z, Hu Y (2015) J Control Release 216:111–120
41. Jiang H, Shi X, Yu X, He X, An Y, Lu H (2018) Pharm Res 35:73
42. Zhang W, Cheng Q, Guo S, Lin D, Huang P, Liu J, Wei T, Deng L, Liang Z, Liang XJ, Dong A (2013) Biomaterials 34:6495–6503
43. Cheng YJ, Luo GF, Zhu JY, Xu XD, Zeng X, Cheng DB, Li YM, Wu Y, Zhang XZ, Zhuo RX, He F (2015) ACS Appl Mater Interfaces 7:9078–9087
44. Han H, Wang J, Chen T, Yin L, Jin Q, Ji J (2017) J Colloid Interface Sci 507:217–224
45. Wang JK, Zhou YY, Guo SJ, Wang YY, Nie CJ, Wang HL, Wang JL, Zhao Y, Li XY, Chen XJ (2017) Mater Sci Eng C Mater Biol Appl 76:944–950
46. Ma P, Sun Y, Chen J, Li H, Zhu H, Gao X, Bi X, Zhang Y (2018) Drug Deliv 25:153–165
47. Zhu H, Li J, Qi X, Chen P, Pu K (2018) Nano Lett 18:586–594
48. Liu T, Zhang N, Wang Z, Wu M, Chen Y, Ma M, Chen H, Shi J (2017) ACS Nano 11:9093–9102
49. Yu X, Mu Y, Xu M, Xia G, Wang J, Liu Y, Chen X (2017) Carbohydr Polym 173:600–609
50. Lai J, Shah BP, Zhang Y, Yang L, Lee KB (2015) ACS Nano 9:5234–5245
51. Chen W, Zhong P, Meng F, Cheng R, Deng C, Feijen J, Zhong Z (2013) J Control Release 169:171–179
52. Cai M, Zhu K, Qiu Y, Liu X, Chen Y, Luo X (2014) Colloids Surf B Biointerfaces 116:424–431
53. Zheng QS, Lin TR, Wu HY, Guo LQ, Ye PR, Hao YL, Guo QQ, Jiang JZ, Fu FF, Chen GN (2014) Int J Pharm 463:22–26
54. Tang F, Li L, Chen D (2012) Adv Mater 24:1504–1534
55. Dowdy SF (2017) Nat Biotechnol 35:222–229

56. Xin Y, Huang M, Guo WW, Huang Q, Zhang LZ, Jiang G (2017) Mol Cancer 16:134
57. Young SW, Stenzel M, Yang JL (2016) Crit Rev Oncol Hematol 98:159–169
58. Fang J, Nakamura H, Maeda H (2011) Adv Drug Deliv Rev 63:136–151
59. Waite CL, Roth CM (2012) Crit Rev Biomed Eng 40:21–41
60. Kim J, Pramanick S, Lee D, Park H, Kim WJ (2015) Biomater Sci 3:1002–1017
61. Zhang H, Kong X, Tang Y, Lin W (2016) ACS Appl Mater Interfaces 8:16227–16239
62. Alexis F, Pridgen E, Molnar LK, Farokhzad OC (2008) Mol Pharm 5:505–515
63. Yoo HS, Park TG (2004) J Control Release 96:273–283
64. (a) Awasthi VD, Garcia D, Goins BA, Phillips WT (2003) Int J Pharm 253:121–132; (b) Klibanov AL, Maruyama K, Torchilin VP, Huang L (1990) FEBS Lett 268:235–237
65. Longley DB, Harkin DP, Johnston PG (2003) Nat Rev Cancer 3:330–338
66. Li SD, Huang L (2010) J Control Release 145:178–181
67. Moysan E, Bastiat G, Benoit JP (2013) Mol Pharm 10:430–444
68. Meng F, Hennink WE, Zhong Z (2009) Biomaterials 30:2180–2198
69. Gilbert HF (1995) Methods Enzymol 251:8–28
70. Ong W, Yang Y, Cruciano AC, McCarley RL (2008) J Am Chem Soc 130:14739–14744
71. Ryu J-H, Chacko RT, Jiwpanich S, Bickerton S, Babu RP, Thayumanavan S (2010) J Am Chem Soc 132:17227–17235
72. Shi C, Guo X, Qu Q, Tang Z, Wang Y, Zhou S (2014) Biomaterials 35:8711–8722
73. (a) Zhao M, Biswas A, Hu B, Joo K-I, Wang P, Gu Z, Tang Y (2011) Biomaterials 32:5223–5230; (b) Zhao M, Liu Y, Hsieh RS, Wang N, Tai W, Joo K-I, Wang P, Gu Z, Tang Y (2014) J Am Chem Soc 136:15319–15325
74. Chang Y, Yang K, Wei P, Huang S, Pei Y, Zhao W, Pei Z (2014) Angew Chem Int Ed 53:13126–13130
75. Ma N, Li Y, Xu H, Wang Z, Zhang X (2009) J Am Chem Soc 132:442–443
76. Du J, Lane LA, Nie S (2015) J Control Release 219:205–214
77. Thambi T, Park JH, Lee DS (2016) Chem Commun 52:8492–8500
78. Thambi T, Deepagan V, Yoon HY, Han HS, Kim S-H, Son S, Jo D-G, Ahn C-H, Suh YD, Kim K (2014) Biomaterials 35:1735–1743
79. Thambi T, Son S, Lee DS, Park JH (2016) Acta Biomater 29:261–270
80. Zheng X, Wang X, Mao H, Wu W, Liu B, Jiang X (2015) Nat Commun 6:5834
81. Perche F, Biswas S, Wang T, Zhu L, Torchilin V (2014) Angew Chem 126:3430–3434
82. Qian C, Yu J, Chen Y, Hu Q, Xiao X, Sun W, Wang C, Feng P, Shen QD, Gu Z (2016) Adv Mater 28:3313–3320
83. de Gracia Lux C, Joshi-Barr S, Nguyen T, Mahmoud E, Schopf E, Fomina N, Almutairi A (2012) J Am Chem Soc 134:15758–15764
84. Shim MS, Xia Y (2013) Angew Chem Int Ed 52:6926–6929
85. Liu J, Pang Y, Zhu Z, Wang D, Li C, Huang W, Zhu X, Yan D (2013) Biomacromol 14:1627–1636
86. Lim SD, Sun C, Lambeth JD, Marshall F, Amin M, Chung L, Petros JA, Arnold RS (2005) Prostate 62:200–207
87. Su Z, Chen M, Xiao Y, Sun M, Zong L, Asghar S, Dong M, Li H, Ping Q, Zhang C (2014) J Control Release 196:370–383
88. Youk H-J, Lee E, Choi M-K, Lee Y-J, Chung JH, Kim S-H, Lee C-H, Lim S-J (2005) J Control Release 107:43–52
89. Han M, Lv Q, Tang X-J, Hu Y-L, Xu D-H, Li F-Z, Liang W-Q, Gao J-Q (2012) J Controlled Release 163:136–144
90. Schumacker PT (2006) Cancer Cell 10:175–176
91. Sun W, Gu Z (2016) Expert Opin Drug Deliv 13:311–314
92. Mo R, Jiang T, Gu Z (2014) Angew Chem Int Ed Engl 53:5815–5820
93. Zhu CL, Lu CH, Song XY, Yang HH, Wang XR (2011) J Am Chem Soc 133:1278–1281
94. Ozalp VC, Schafer T (2011) Chemistry 17:9893–9896
95. He X, Zhao Y, He D, Wang K, Xu F, Tang J (2012) Langmuir 28:12909–12915
96. Mo R, Jiang T, Sun W, Gu Z (2015) Biomaterials 50:67–74

97. Dong L, Xia S, Wu K, Huang Z, Chen H, Chen J, Zhang J (2010) Biomaterials 31:6309–6316
98. Wang Z, Chen Z, Liu Z, Shi P, Dong K, Ju E, Ren J, Qu X (2014) Biomaterials 35:9678–9688
99. Zhang Y, Lu Y, Wang F, An S, Zhang Y, Sun T, Zhu J, Jiang C (2017) Small 13
100. Lin CJ, Kuan CH, Wang LW, Wu HC, Chen Y, Chang CW, Huang RY, Wang TW (2016) Biomaterials 90:12–26
101. Ruan S, Cao X, Cun X, Hu G, Zhou Y, Zhang Y, Lu L, He Q, Gao H (2015) Biomaterials 60:100–110

Nanocomposites for Cancer Targeted Drug Delivery Therapeutics

Francisco N. Figueroa, Dariana Aristizabal Bedoya, Miriam C. Strumia, and Micaela A. Macchione

1 Introduction

Cancer is a major public health problem and is the second leading cause of death worldwide [1–3]. Globally, about 1 in 6 deaths is due to cancer. The 5-year relative survival rate for all cancers combined has increased substantially since the early 1960s, but there are still many improvements to make advances in treatment, as well as in earlier diagnosis [4].

Chemotherapy is the most common cancer treatment and plays a vital role in treating undetectable cancer microfocuses and free cancer cells [2]. This treatment uses highly potent drugs causing toxicity and subsequently, cell death [5]. Fast-growing cells are the main targets of chemotherapeutics. However, traditional chemotherapeutic agents do not discriminate between normal cells that divide rapidly and tumor cells causing severe side effects [2]. This lack of specificity of anticancer drugs is the main drawback associated with conventional chemotherapy which makes indispensable the development of novel formulations that significantly increase the survival of these patients. In this sense, the superior anticancer ability of nanosized carriers, usually called "nanocarriers," has attracted the attention of scientists [6]. Nanocarriers can enhance the intracellular concentration of drugs in cancer cells by using passive or/and active targeting strategies. "Passive targeting" is given by the property of nanoparticles (NPs) of suitable sizes to accumulate in tumor tissues

F. N. Figueroa · D. Aristizabal Bedoya · M. C. Strumia · M. A. Macchione (✉)
Departamento de Química Orgánica, Facultad de Ciencias Químicas, Universidad Nacional de Córdoba, Av. Haya de la Torre & Av. Medina Allende, Córdoba X5000HUA, Argentina
e-mail: micaela.macchione@unc.edu.ar; mmacchione@fcq.unc.edu.ar

CONICET, Instituto de Investigación y Desarrollo en Ingeniería de Procesos y Química Aplicada (IPQA), Av. Velez Sárfield 1611, Córdoba X5000HUA, Argentina

M. A. Macchione
Instituto Académico Pedagógico de Ciencias Humanas, Universidad Nacional de Villa María, Arturo Jauretche 1555, Villa María, Córdoba X5900KBJ, Argentina

A. K. Nayak et al. (eds.), *Biomedical Composites*, Materials Horizons: From Nature to Nanomaterials, https://doi.org/10.1007/978-981-33-4753-3_9

much more than in normal tissues [7]. This effect is called the enhanced permeability and retention (EPR) effect [8]. The reason for this phenomenon is that the abnormal tumor blood vessels have larger fenestrations and consequently, promote vascular permeability. On the contrary, free drugs can freely pass through the vasculature of any tissue. On the other hand, "active targeting" is used to describe drug accumulation within a target zone by, for example, specific interactions between drug/carrier and the target cells, usually through specific ligand–receptor interactions [9].

In this context, pharmaceutical industry is evolving toward the development of targeted drug delivery nanocarriers with the ability to transport a high amount of the therapeutic agent specifically to cancer cells, decreasing damage to healthy tissues. Furthermore, drug delivery systems should allow the administration of a drug to the body with an improvement of its pharmacokinetic and biodistribution profiles [10].

Despite the fact that significant advances have been made in this research area, substantial improvements are still needed to reach the next level of chemotherapeutic formulations. Accordingly, it is imperative to understand the processes which involve transport of drug to a target site after intravenous administration as well as relevant issues for specific target diseases and the response of the body towards a drug delivery system [9]. Rational planning of the nanocarrier structure by choosing the components and combining them into an appropriate design should be done considering structure/properties relationships. Finally, identifying a successful targeted drug delivery strategy provides insights to enhance the development of the next nanocarriers.

Among nanocarriers, nanocomposites are systems integrated by more than one material which have at least one physical dimension in the nanometer range [6]. This multicomponent system should have different phase domains in which at least one type of the phase domains is a continuous phase [11]. In particular, nanostructures that incorporate nanosized particles into a matrix of another material are usually called nanocomposites [12].

Nanocomposites for cancer targeted drug delivery should be able to facilitate that the therapeutic cargo reaches the site of action from the administration site with minimal nonspecific accumulations, release the therapeutic payload in/around the target site achieving effective therapeutic levels at the site of action, protect the therapeutic molecules from the detrimental effects of environmental factors and, finally, avoid toxicity or adverse reactions of the chemotherapeutic molecule on healthy tissues [10]. Furthermore, it is also important to mention that the interaction of the drug with the nanocarrier should not alter its therapeutic action.

The interest in applying nanocomposites as drug delivery carriers is based on their outstanding properties. Firstly, the multicomponent nature of nanocomposites makes the resulting physicochemical properties at least an additive combination of the properties of the individual constituents but, but even a synergistic effect can take place giving unique properties to the composite. Additionally, nanocomposites can be functionalized to selectively deliver drugs to specific cancer cells, thereby improving therapeutic efficacy. Moreover, the wide variety of synthetic strategies and nanomaterials which can be handle by the research community opens the possibility of designing nanocomposites with multiresponsiveness to achieve better targeting on a specific site of action.

Among nanocomposites, those which combine hard and soft nanomaterials which are usually described as organic/inorganic nanohybrids have attracted intense attention for biomedical application due to their ability to respond to external stimuli such as light, electric or magnetic field, etc., and their biological properties resulting from the synergy of combining the advantageous physicochemical properties of both the inorganic and organic component. In general, this kind of nanocomposites is formed by inorganic NPs covered by an organic brush or shell. There are many strategies for the construction of organic/inorganic nanohybrids which can be categorized into three main kinds of methods, including surface functionalization, one-pot synthesis, and wrapping (Fig. 1) [13].

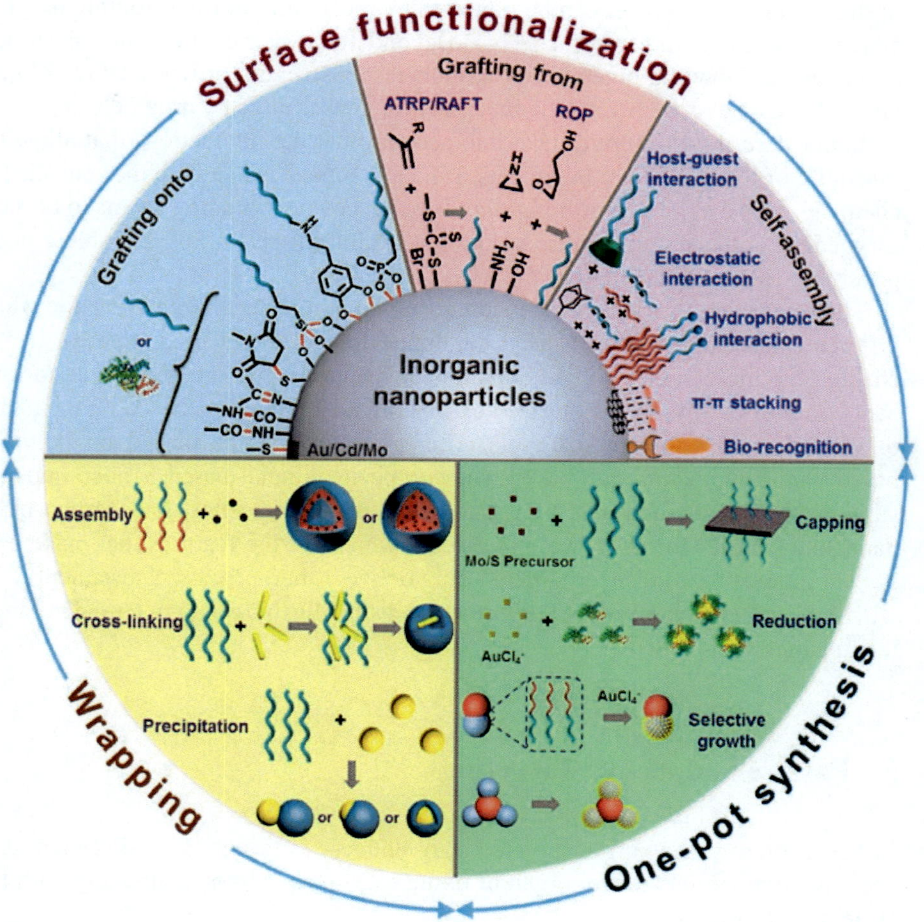

Fig. 1 Construction of organic/inorganic nanohybrids through surface functionalization, wrapping, and one-pot synthesis. Reprinted with permission from [13]. Copyright (2019) American Chemical Society

Surface functionalization is a powerful tool to provide nanohybrids with biocompatible hydrophilic surfaces and functional groups capable of interacting with biomolecules for targeting or delivery. Otherwise, one-pot synthesis offers a simple and efficient approach in which one of the components is formed directly during a one-step reaction in the presence of the other component. Wrapping means the encapsulation of one of the components with the other through noncovalent interactions [13].

Furthermore, those nanocomposites formed by polymers are a particularly interesting type of nanocarriers. Polymers exhibit therapeutic and diagnostic advantages over conventional medicines which lead them to be prevalent materials within those approved by the FDA [14]. Polymeric materials modulate the toxicity and improve cellular interactions, stability, and drug-release kinetics [15]. The polymer-drug conjugates have an increased drug half-life and bioavailability [14]. The composition of the protein corona layer can be modified by the addition of low-fouling polymer coatings on the particle surface [14]. Beyond this, their facile synthesis procedures which allow the obtention of stimuli-responsiveness or degradable polymer architectures make them versatile materials for cancer targeted drug delivery.

In our description, we will take into account that the interactions that allow the assembly of the different components of nanocomposites can be either physical or chemical. However, it is worth highlighting that covalent binding seems to be safer for medical applications because it does not exhibit the risk of disassembling in the physiological environment [16].

There is a huge variety of nanocomposites reported in the literature for which different materials have been used combined into different architectures. In this chapter, the most important combinations of materials that can be merged to form nanocomposites will be summarized by making a classification based on the nature of one of the constituent materials. The emphasis will be placed on polymer-based nanocomposites, clay-based nanocomposites, metal-based nanocomposites, silica-based nanocomposites, magnetic-based nanocomposites, and carbon-based nanocomposites. Moreover, we will discuss which are the features that make each system interesting for cancer targeted drug delivery therapeutics. Subsequently, we will highlight some examples of recent research that illustrate their potential in cancer therapeutics.

2 Passive and Active Targeting

In general terms, the aim of targeted drug delivery is to obtain predominant drug accumulation within a target organ or tissue selectively and quantitatively with low systemic exposure [17].

Passive targeting means the extravasation of nanosized particles through leaky tumor vasculature and a dysfunctional lymphatic drainage system which is characteristic for solid tumors and inflamed tissues, leading to preferential accumulation through the EPR effect [17, 18]. In the ideal case, the drug release occurs when NPs translocate across the capillary endothelium and enter the interstitial space of the tumor, producing an accumulation of pharmaceutical agents in the tumor interstitium. In order to attain effective passive targeting, it is necessary to have enough plasma concentrations of the drug/carrier conjugate for prolonged blood circulation time. However, the majority (>95%) of administered NPs are known to accumulate in other organs, in particular the liver, spleen, and lungs [9]. Consequently, nowadays the focus is placed on developing drug delivery systems that can show passive and active targeting simultaneously.

As it was mentioned in the introduction, active targeting means the functionalization of drug carriers with the purpose of giving them affinity toward specific receptors/markers on cells, tissues, or organs. The more quantity of available targetable components on the target organ/cell (e.g., transferrin receptors in tumor) than in normal cells is the basic principle of active targeting [10]. In addition, the employment of external stimuli, such as magnetic fields or ultrasound, acting on nanocarriers to target and release drugs in the target tissue is also another kind of active targeting [10].

3 Polymer-Based Nanocomposites

In polymer-based nanocomposites, the filler or the matrix are polymeric compounds, the most common systems of this type are, generally, made up of a polymeric nanoparticle core and a lipid or another polymer around. One of the major advantages of these type of nanocomposites is the capability to employ natural polymers or polymers like aliphatic polyesters such as polylactide, polyglycolides (PGA), and poly (ε-caprolactone) (PCL) that are biodegradable and biocompatible with the human body [19, 20].

Although natural polymers have good biocompatibility and biodegradability for biomedical applications, they present low solubility and poor mechanical and thermal properties [20]. An efficient alternative to overcome these disadvantages lies in the possibility of combining natural and synthetic polymers, to achieve new materials with a desired combination of properties which are inherent to the polymers employed [21]. Polymeric-based nanocomposites employed as drug delivery systems show stable and controlled release, as well as, enhanced encapsulation efficiency and loading capacity [22].

These features make polymer-based nanocomposites suitable for cancer therapy, involving drug delivery, photothermal therapy, and photo-dynamic therapy, among others. A large variety of properties can be found in polymer-based nanocomposites, such as their nontoxic nature, the capability to target nucleus or another subcellular compartment, passive targeting, on demand drug release and stimuli-responsive systems, etc., which convert these systems into potential cancer treatment devices.

One particular characteristic that has to be controlled over the synthesis of nanocomposites, in particular polymeric ones, is their size and polydispersity. The strict control over these qualities remains a major challenge to solve in upcoming years. Different properties of the final material are related to the size, such as kinetic stability, cellular uptake, biodistribution, and drug release kinetics.

Nanocomposites present several advantages over conventional cancer treatments, including targeted drug delivery, strict control over the release rate of the drug, enhanced drug absorption into tumors, and reduced side effects of the free drug by avoiding healthy cells interactions [23, 24].

One of the most widely used polymers in the formulation of a drug delivery nanocomposite is chitosan [25]. Chitosan-based nanocomposites have a very important role in targeted drug delivery providing a nontoxic, biocompatible, stable, target specific, and biodegradable delivery device. Chitosan possesses unique properties including low cytotoxicity and inmunogenicity, biodegradability, and biocompatibility, making this polymer a good material for cancer therapy and other biomedical applications [26]. Luo et al. prepared a chitosan-based nanocomposite using charge-driven self-assembly in aqueous buffer, preventing the side effects of employing organic solvent in synthetic methodologies. Doxorubicin (DOX) was covalently attached to chitosan, which interacted with methoxyl poly (ethylene glycol)-block-poly (glutamic acid) copolymer (PEG-PGA), to form nanocomposites with loaded anticancer drug and legumain substrate peptide to confer targeted drug delivery. These systems show high cytotoxicity against choroidal melanoma cell line (Mum-2C) and reduced cytotoxicity on normal human corneal epithelial cells (HCEC). Drug release remains a major problem in these systems, reaching a maximum release of 42% of encapsulated drugs [27].

Xu et al. have described a monomethoxy (polyethylene glycol)-poly(D, L-lactide-co-glycolide) (mPEGPLGA) nanocomposite, for paclitaxel (PTX) and temozolomide (TMZ) drug delivery. The thermo-sensitive gels exhibit a passive targeting through an average size of 110 nm. This system has the potential to become a drug delivery device for the interstitial chemotherapy of glioblastoma [28].

4 Inorganic-Based Nanocomposites

Inorganic nanomaterials have unique physicochemical properties which make them extremely interesting for biomedical uses. Besides, synthesis and modification of inorganic nanomaterials are usually easier procedures compared with those involved in organic nanostructures. The available synthetic methods allow one access to the control of the nanoparticle size, shape, and surface functionalization [29]. Inorganic NPs can be made of robust materials that allow encapsulation and protection of sufficient amounts of cargos for cancer treatment [30]. Furthermore, certain inorganic nanomaterials have intrinsic functionalities that are not normally seen in organic materials such as the ability to respond to external stimuli such as magnetic fields or near-infrared (NIR) light [31]. They also exhibit thermal stability which is often an advantage over organic nanocarriers.

Some inorganic nanocomposite can display targeting mediated by external stimuli by themselves, but organic functionalization is commonly used to achieve active targeting. Hydrophilic groups are also used to increase the stability of the nanocomposite dispersion in aqueous solution and consequently, prevent aggregation [32] and improve compatibility [33]. Organic coatings are also used to give the chance to encapsulate therapeutic drugs in hard NPs.

Furthermore, there are inorganic NPs which can load therapeutic agents inside their pores, mesoporous inorganic NPs. However, in pure inorganic mesoporous nanosystems the drug would be released immediately after administration because it is simply physically adsorbed in the channels [34]. Once more, smart polymeric frameworks are used as gatekeeper to ensure a controlled drug release in these cases.

The strategy of functionalization of the surface of inorganic nanocomposites with organic molecules is very often used to low systemic toxicity [35]. It is well known that pure inorganic nanocomposites become surrounded by a nonspecific protein coat after being exposed to serum or extracellular environments, a process described as the protein corona effect. This kind of biological coating masks the chemical nature of this nanomaterials and provides them with biological properties [36]. Functionalization with polyethylene glycol (PEG), a strategy also known as PEGylation, is the most widely used coating to shield the surface of NPs from aggregation, opsonization, and phagocytosis, thereby prolonging circulation time [37].

In the next sections, different types of inorganic-based nanocomposites will be described.

4.1 Clay-Based Nanocomposites

These nanocomposites are multicomponent systems where one of its components consists of clay NPs, whether the clay NPs could be dispersed/covered by a polymer matrix or clay NPs could be decorated with NPs of different nature. As it has been mentioned throughout this chapter, nanocomposites have unique properties that cannot be obtained by the materials separately [38]. The combination of these two materials causes a synergistic effect in the improvement of physical properties, including mechanical strength, thermal stability, among others [39]. But it is noteworthy that clay-based nanocomposite has unique properties given by this kind of inorganic nanoparticle, that make this nanocomposite has been reported as the most used in various fields [40]. For example, most clays present notably biocompatibility and biodegradation, they have lower toxicity compared to another kind of inorganic NPs as those based on metal, oxides, silica, or carbon nanotubes [41]. Also, they can be obtained in nature as clay minerals or by an easy and low-cost synthetic route [6]. Besides, their two-dimensional (2D) structure gives them a high ions reservation capacity, they can incorporate ions both between the sheets and on their surface [42]. And in the same way, clays can carry out ion exchange with ions from the surrounding environment [41].

In general, clay structure is characterized by layered platelets stacked through electrostatic interactions. Depending on the type of clay, it presents different kinds of isomorphic cation substitution in their structure that result in excess charge (positive or negative) on layers, this charge is balanced with the presence of water and ions between the layered [43]. Clays can be classified into two large groups as natural (clay minerals) or synthetic. On the one hand, clay minerals are composed of alternating negatively charged aluminosilicate layers, with exchangeable interlaminar cations [42]. In this group are kaolinite, illite, smectite (montmorillonite), chlorite, halloysite, and vermiculite. The most used clays for biomedical applications are kaolinite ($Si_2Al_2O_5(OH)_4$) and montmorillonite ($AlSi_2O_5(OH).xH_2O$) [44]. On the other hand, among the synthetic clays, the lamellar double hydroxides (LDHs), also known as anionic clays, stand out. They present stacks of brucite ($Mg(OH)_2$)-like layers that are positively charged, as a result of the isomorphic substitution of divalent by trivalent cations. In this case, there are anions and water between interlaminar spaces to balance the charge [45].

Features exposed above have allowed clays to be widely used as excipient or active substances in pharmaceutical technology and dermopharmacy; for example LDHs can be found in commercial antacid formulations (Almax© and Talcid©) [46]. Currently, the possibility of reducing its size to nanoscale and incorporating ionic drugs or biomolecules between clay sheets stabilized through electrostatic interaction, makes them very attractive for the development of nanocarriers with controlled release ability, by exchanging the drug with other ions present in biological environment [42]. Though, clay-based nanocarriers have the drawback that under physiological conditions, clay particles could have cytotoxic properties linked to their dispersion pattern, they could form agglomerates due to the high ionic strength and presence of different electrolytes in biological fluid [42]. However, the formation of nanocomposites with clay and organic materials, like polymer or proteins, offers an improvement for this inconvenience. For example, hydrophilic polymers offer colloidal stability for inorganic NPs, that result in prolonged blood circulation and low uptake by the reticuloendothelial system (RES) organs [13]. Likewise, another advantage of this nanocomposite is that drug can be linked to organic material as well as in the clay, increasing its storage drug capacity [6].

Depending on the organic material and clay employed, nanocomposites could have three different morphologies: (**a**) tactoid, (**b**) intercalated, or (**c**) exfoliated [42], as illustrated in Fig. 2.

Tactoid structures result when the polymer closes around to clay agglomerates. Their X-ray diffraction (XRD) shows that the basal spacing of the clay does not expand, due to poor affinity of the polymer for the clay [39].

Intercalated: in this case, the polymer presents moderate affinity for clay sheets and their chains penetrate between of them. This causes an increase in basal space of clay but the shape of the layered stack remains [47].

Exfoliated or delaminated structures are obtained when the polymer presents high affinity for clay sheets, their chains penetrate between the sheets widening the interlayer space until stacks of clay layers become disordered. Thereby, X-ray cannot detect any regular structure and the diffractogram does not show any peaks [48].

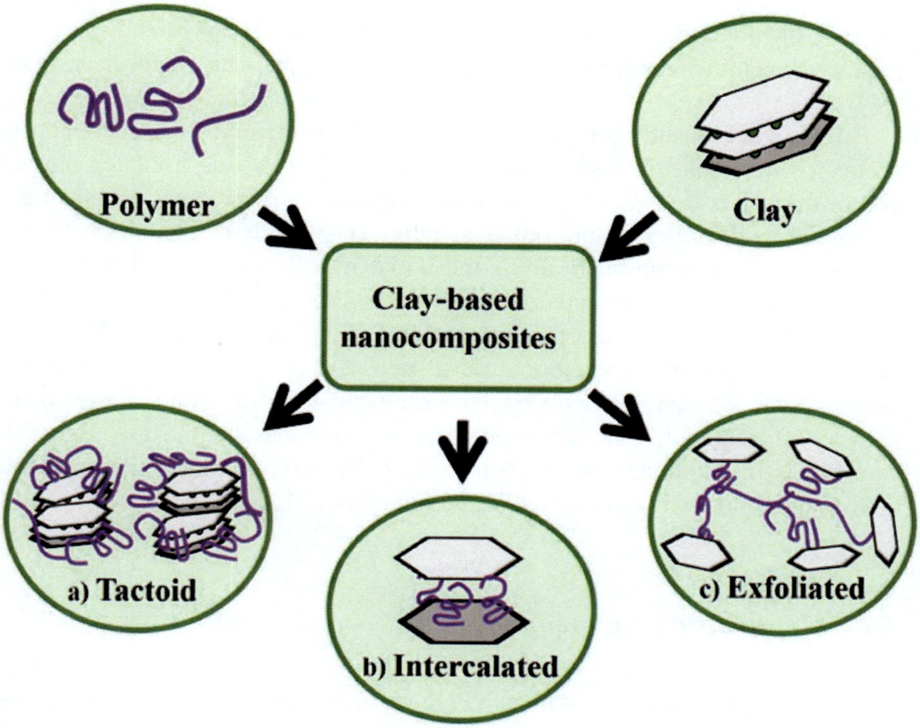

Fig. 2 Possible morphologies obtained for clay-polymer nanocomposites

Specifically, the use of both natural and synthetic clay-based nanocomposites has been reported for cancer targeted drug delivery therapeutics, some examples are mentioned below.

Yang et al. have developed a nanocomposite based on Halloysite nanotubes (HNTs) with chitosan oligosaccharide-grafted as a DOX carrier for treating breast cancer. Through the grafting of these two materials (DOX@HNTs-g-COS), it was possible to increase the biocompatibility of HNTs, as well as, the cellular uptake ability of DOX. The studies in vitro and in vivo conditions showed in both cases an improved chemotherapy effect of nanocomposite compared with the free drug. In vitro study in MCF-7 cells showed that IC50 value was 1.17 μg/mL, while it was 2.43 μg/for free DOX. Similarly, in vivo study in 4T1-bearing mice, the tumor-inhibition ratio of DOX@HNTs-g-COS was 83.5%, while it was 46.1% for free DOX [49].

Another kind of natural layered clay used in nanocomposites is montmorillonite (MMT), Phan et al. synthesized a nanocomposite hydrogel for controlled release of gemcitabine (GEM). They first formed a complex MMT-GEM through intercalation and adsorption of the drug in clay NPs, then this complex was dispersed in a temperature-sensitive poly(ϵ-caprolactone-co-lactide)-b-poly(ethylene glycol)-b-poly(ϵ-caprolactone-co-lactide) hydrogel. In vitro GEM release studies were performed between pristine hydrogel and the nanocomposite, the latter exhibited longer GEM release time and a much lower initial burst. Likewise, nanocomposite produced a significant inhibition of tumor growth on pancreatic tumor-bearing mice [50].

On the other hand, Ray et al. designed a synthetic clay-based nanocomposite for treatment of osteosarcoma. An LDHs matrix was used for intercalation of methotrexate (MTX), at the same time this matrix was encapsulated using a nontoxic and biodegradable polymer, poly (D,L-lactide-co-glycolide) (PLGA). In this way, they achieved to arrest the initial release and dose-dumping-related toxicity. The authors conducted comparative studies between the nanocomposite (PLGA-LDH-MTX), and its counterparts LDH, PLGA-MTX, and free MTX. Pharmacokinetics, tissue distribution, survival rate, and in vivo antitumor efficacy in osteosarcoma induced Balb/c nude mice were evaluated. The results showed that the delivery efficiency of MTX using the nanocomposite was significantly higher than bare MTX. In addition, this nanocomposite presented remarkably more effective in vivo antitumor activity than free MTX and PLGA-MTX [51].

4.2 Metal-Based Nanocomposites

Metal NPs, especially gold and silver NPs, have been exploited as a potential platform in controlled drug delivery for cancer therapy.

Gold nanoparticles (AuNPs) have promising applications in the fields of drug delivery and photothermal therapy because of their unique optical and photothermal properties. In addition, the size and shape of the particles can be easily tuned by the appropriate choice of synthetic method, which allows tunable optical absorption properties which make them especially suitable candidates for drug delivery applications. It is also worth highlighting that colloidal dispersions of AuNPs are one of the most stable among metal NPs [18]. Moreover, their surface can be modified in order to conjugate them with organic molecules.

The combination of AuNPs with other materials such as biomolecules, polymers, and proteins can result in nanoplatforms with improved therapeutic properties. As an example, it is worth mentioning that naked gold nanostructures are extremely toxic, however, surface modification can decrease their toxicity [52]. PEGylation renders the nanoparticle perfectly soluble in water and minimizes any protein adsorption which can postpone or prevent the rapid clearance by the RES [52]. By PEGylation, it is also possible to increase the tumor accumulation of gold-based nanocomposites through the EPR effect. In particular, thiolated poly(ethylene) glycol is very often used to functionalize the surface of AuNPs. Recently, Halas and collaborators have reported the initial results of a clinical trial in which laser-excited gold-silica nanoshells functionalized with thiolated poly(ethylene) glycol were used for localized photothermal ablation of prostate tumors [53]. Although it is not a drug delivery itself, cell death and tumor remission can be achieved in this case by photothermal cancer therapy taking advantage of the ability of nanosized gold to convert absorbed NIR light into heat and induce highly localized hyperthermia. At this point, it is important to note that light penetration depth in tissue dramatically increases with the increasing wavelength.

Furthermore, PEGylated magneto-plasmonic NPs with a hollow or semi-hollow interior have been successfully synthesized and evaluated as drug carrier using Rhodamine B isothiocyanate as a model compound [54].

Targeted DOX-loaded rattle-type gold nanorods/porous-SiO_2 nanocomposites were prepared as NIR-activated drug delivery systems for combined cancer chemo–photothermal therapy. PEG molecules and cyclic Arg-Gly-Asp peptides were covalently conjugated to the surface of the nanocomposites, providing them with biocompatibility and targeting capabilities. The multicomponent nanostructures exhibited combined cancer chemo–photothermal therapy, which was much more efficient in inhibiting cancer cell proliferation than chemotherapy or photothermal therapy alone. They also exhibited higher drug loading capacity than conventional core–shell nanostructures because of their hollow interior [55].

AuNPs coated with serum albumin have also improved properties such as greater compatibility, bioavailability, longer circulation times, lower toxicity, and selective bioaccumulation [56].

Many cancer cell types have abundant expression of transferrin receptors, consequently, by decorating AuNPs with transferrin, the intracellular delivery of these carriers is significantly increased compared with their non-targeted counterparts [57].

Polymers are also fascinating materials for the combination with gold in nanocomposites with defined architecture. Recently, soft poly(N-vinylcaprolactam) nanogels surface-decorated with AuNPs were prepared and tested as potential functional carriers. Small AuNPs on the surface of poly(N-vinylcaprolactam) nanogels are heated via light or radiofrequency radiation triggering the nanogels to shrink [58].

On the other hand, it is well known that silver-based nanocomposites exhibit remarkable antimicrobial effects but conversely to gold, they are not extensively applied in cancer drug delivery applications. However, they are a promising tool as anticancer agents in diagnostics and probing with strong effects against different cancer cell lines. Similar to gold, silver can be synthesized by reproducible routes, its surface can be easily functionalized and its optical properties can be tuned to have an absorption at specific wavelengths that is useful for imaging and photothermal applications in tissue [59]. AgNPs are reported to have anticancer property themselves because they can damage structures of cells, which finally induce cytotoxicity, genotoxicity, immunological responses, and even cell death. Additionally, silver-based nanocomposites loaded with anticancer drugs exhibit synergistic antitumor effect of both drug and AgNPs. For example, DOX loaded nanocomposites based on AgNPs with a polyvinylpyrrolidone (PVP) coating, in combination with iodine-125 isotope to track them in vivo after intravenous injection in normal and solid tumor bearing mice, provided new radiolabeled Ag-based nanocomposites as tumor-specific agents for both diagnostic and therapeutic applications [60].

4.3 Silica-Based Nanocomposites

Silica-based materials have unique properties that justify their use in cancer targeted drug delivery nanocomposites as for example hydrophilic surface, versatility for surface functionalization by simple exploitation of the well-established silicate chemistry, ability to easily obtain different shapes and sizes, easy synthetic protocols on a large scale with low production costs, lack of toxicity, biocompatibility, biodegradability, and controllable superficial charge [61, 62]. Due to their exceptional features, these materials have been recognized as one of the most promising biomedical platforms for therapeutic, diagnostic, prognostic, and combined applications [5]. In addition, silica-based materials have been approved by the FDA for use in diagnostic imaging and silica NPs have been approved for human clinical trials. Therefore, it is expected that in a short time they can be approved as a drug [63].

Silica/stimuli responsive polymer nanocomposites have interesting properties for drug/siRNA delivery. For example, nanostructured silica cores wrapped within a tuneable pH-responsive coating of poly(2-diethylaminoethyl methacrylate) hydrogel have been synthesized in a one-pot step obtaining uniformly dispersed nanocarriers in the aqueous phase [64]. In this case, thermogravimetric analysis shows that the formation of covalent chemical bonds between the silica and the polymer increases the stability of the organic phase around the inorganic core. These nanocomposites were loaded with small interfering RNA (siRNA) which was efficiently delivered into the cytoplasm of MDA-MB-231 breast cancer cells and inhibited the protein expression of CXCR4 with an efficacy comparable to that of a commercial transfection reagent. In vivo studies show that these nanocomposites preferentially accumulated in the tumors of human breast cancer-bearing mice and mediated CXCR4 silencing in these tumors.

Among silica-based nanomaterials, ordered mesoporous silica materials (OMMs) have received increasing popularity in drug delivery. These materials have a highly mesoporous structure which provides a large surface area and high pore volume which allow large quantities of drug molecules to be loaded into their structure, which is one of the major challenges in the development of carriers for anticancer drugs [5]. The size of the mesoporous framework can be tuned during synthesis, making these structures adaptable hosts for load agents of various sizes [65]. Pores and the surface of this kind of NPs can be modified with a huge number of molecules including drugs and fluorescent dyes (Fig. 3).

Their unique architecture of mesoporous silica allows the development of prolonged release systems for hydrophobic drugs. As an example, nanocomposites made of mesoporous silica NPs within a hyperbranched polymer network made from poly(vinyl alcohol) (PVA) and poly(acrylic acid) (PAA) were employed as ideal host materials with high adsorption capacity for a model hydrophobic drug, anti-inflammatory agent dexamethasone, exhibiting prolonged, sustained release profiles for as long as two months [66]. Biological characterization shows that these nanocomposites can, thus, be highly appealing biomaterials for sustained and prolonged drug release, such as wound dressing systems.

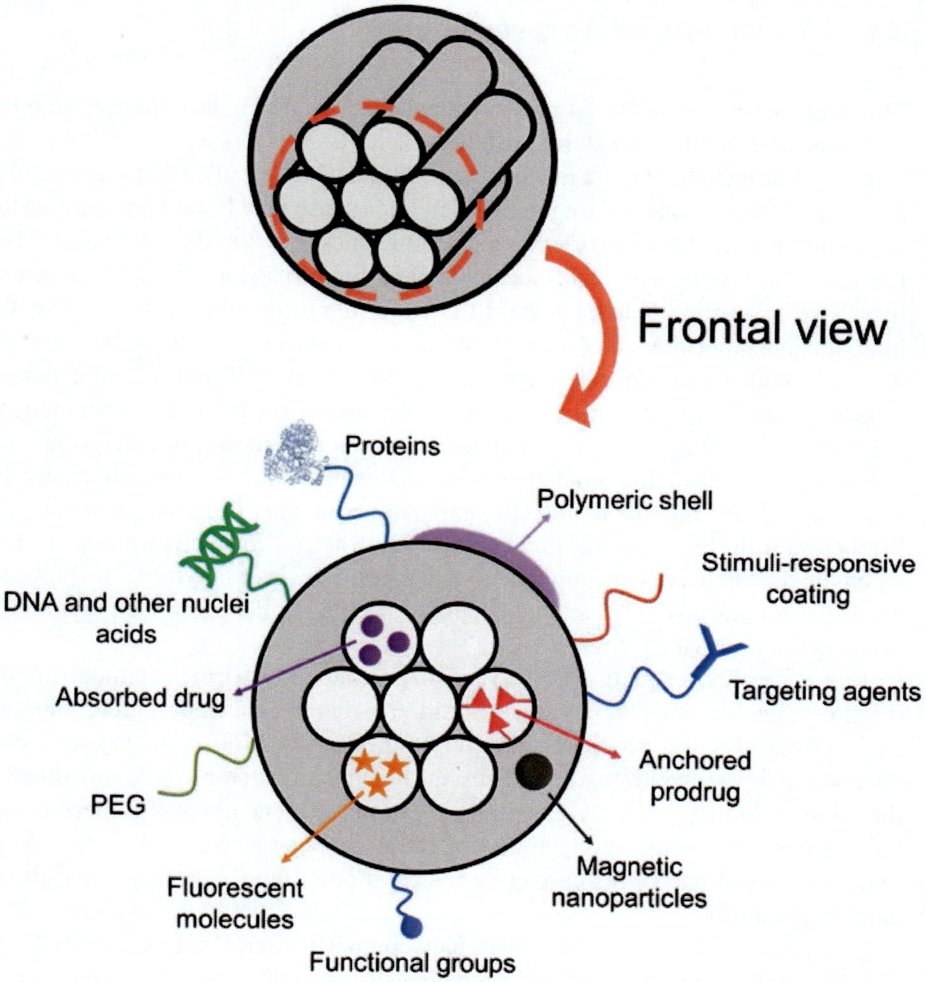

Fig. 3 Mesoporous silica nanoparticles and their functionalization alternatives

As it was mentioned in the introduction, drug delivery vehicles without side effects to normal physiological tissues are the actual challenge for safety and effective nanomedicine. Recently, a multifunctional drug delivery vehicle containing an ordered mesoporous resin as a polymer core and homogeneous Fe nanodots-doped silica as the biodegradable shell was developed [67]. The Fe-doped silica shell acts as a compact inorganic cap to seal DOX into the mesoporous polymer cores, but also serves as a superparamagnetic agent for magnetic targeting and magnetic resonance imaging (MRI). Via Fe extraction-induced degradation, the loaded drug can be slowly released under the acidic tumor environment, achieving ultralow drug leakage under normal in vivo blood circulation (physiological environment).

4.4 Magnetic-Based Nanocomposites

This multicomponent material presents magnetic NPs that induce the response to a noninvasive stimulus, the external magnetic field [68].

Currently, there has been a notable increase in the interest of developing magnetic nanocomposites, mainly in drug delivery field, because they have three features that highlight them among over of other nanocomposites. First of all, the magnetic NPs present in these nanocomposites mean that they can be used in MRI [13]. Secondly, they have magnetic guidance toward the target tumor or organ [40]. Finally, this kind of nanocomposites can be used as theragnostic systems by themselves, namely, they have both applications in diagnostic and therapy [69]. Therapy applications of magnetic nanocomposites involve a hyperthermia process which consists in inducing local heat when the magnetic nanoparticle is subjected to an oscillating magnetic field [70]. Nowadays, this is an approved treatment as a therapy against cancer, but it is necessary to combine it with chemotherapy or another different treatment [71]. Furthermore, the hyperthermia process can be used as a mechanism to release the drug when the magnetic NPs are conjugated to a thermosensitive polymer. The magnetic heating can induce conformational changes (swelling/collapse) in the polymer that produce drug release [72].

Iron oxide NPs, magnetite (Fe_3O_4) and maghemite (Fe_2O_3), are commonly used to develop magnetic nanocomposites. In addition to their magnetic properties, these are easily obtained through a coprecipitation synthesis [13]. Moreover, they have good chemical stability and low toxicity. Thus, the combination of these NPs with different materials including polymers, clays, silica, carbon, or metal organic frameworks has been extensively studied [68]. Some of these nanocomposites, whether they are magnetic NPs as part of the core or as a decoration of the mentioned materials, are described below.

Delavari et al. designed a theranostic nanocomposite for DOX release using Fe_3O_4 NPs as a MRI contrast agent and a coating of α-lactalbumin which was precipitated and cross-linked with poly(ethylene glycol) and glutaraldehyde. In addition, the resulting nanocomposite was pH-sensitive due to the formation of imine bonds between glutaraldehyde and amine groups on α-lactalbumin and polyethylenimine that resulted in the higher release of DOX at acidic pHs. The studies both in vitro and in vivo reflected greater action by the DOX-loaded nanocomposite compared to the free drug. Likewise, nanocomposite was less immunogenic than free DOX [73].

Another example of magnetic nanocomposite with dual capacity was developed by Meenach et al. Their system had the ability to deliver paclitaxel (PTX), as an anticancer drug, and produce heat by an alternating magnetic force (AMF) for the potential treatment of cancer. This consists of a hydrogel of poly(ethylene glycol) loaded physically with iron oxide NPs entrapped and PTX as chemotherapeutic agent. Their swelling, heating, and mechanical characteristics were studied according to the cross-linking density. Also, cytotoxicity assays on three different cell lines including the M059K (glioblastoma), MDA-MB-231 (breast carcinoma), and A549 (lung denocarcinoma) were conducted by exposing them to PTX only, hyperthermia only, and both PTX and hyperthermia. The drug release experiments showed that the lower cross-linked hydrogels released PTX more quickly, and in vitro cell studies showed that combined PTX and hyperthermia treatment increases the antitumor efficacy for A549 lung carcinoma cells [74].

4.5 Carbon-Based Nanocomposites

Since the discovery of graphene, the studies of carbon allotropes have gained significant attention because of their versatile applications. Graphene and its derivatives, and single or multiple walled carbon nanotubes (SWCNTs or MWCNTs) are the most employed systems in carbon allotropes family [10, 75]. They possess unique properties that make them potential anticancer drug delivery devices, such as good chemical and thermal stability, flexibility, large surface area with a high density of easily modifiable functional groups, good biocompatibility, and high capability of drug loading by π-π physical interaction [76–78].

Graphene oxide (GO) and reduced graphene oxide (rGO) nanocomposites have drawn extensive attention in biomedical applications like drug delivery, bioimaging, and sensing, owing to their biocompatibility, aqueous solubility, and ease of functionalization [79]. GO and rGO toxicity has been widely studied, due to its physicochemical characteristics like purity and 2-dimensional size may influence nanocomposites cellular uptake [75]. It has been proved that GO exhibits cell toxicity depending on time and dose of exposure [80]. GO can also destroy the cell membrane, their unique structure enables it to develop strong electrostatic interactions with cellular membrane, given by π-π stacking with biomolecules, proteins, and/or nucleic acids [81]. Interactions with cellular membrane may be also given by negatively charged oxygen in GO and the positively charged lipids in the membrane [82, 83].

Li et al. studied the interaction of graphene with three different cell type by molecular dynamics simulations and fluorescence and microscopic imaging. They discovered that the edge of graphene could penetrate through cellular membrane in human skin and immune cells [84]. Taking this into consideration, greater efforts are being made to further modify graphene in order to cause less toxicity [85].

Graphene-based nanocomposites are being widely used in biomedical applications, for all the previously mentioned qualities. This kind of nanocomposites can be classified into different groups for cancer drug delivery, graphene-based metallic nanocomposites, graphene-based quantum dots, graphene-based iron oxide magnetic NPs and graphene-based polymeric nanocomposites.

Graphene-based metallic composites are synthesized by incorporation of metallic NPs into graphene sheets. As stated previously, graphene sheets present strong Van der Wall forces causing restacking issues in synthesis methods, which limits GO and rGO applications. This problem is solved by the incorporation of metallic NPs (metal and metal oxides NPs) into graphene; metallic NPs reduces significantly the π-π stacking interactions. Gold (Au), silver (Ag), platinum (Pt), and cadmium (Cd) NPs are the most employed in the formation of graphene-metal nanocomposites [75].

The combination of carbon allotropes with metals or metallic oxides has become an attractive alternative for developing novel nanocomposites with unique properties to achieve efficient drug delivery.

Despite the inherent toxicity of GO, which can generate reactive oxygen species (ROS) or destroy de cell membrane through strong electrostatic interaction, different graphene-based nanocomposites with reduced adverse effects were synthesized and employed as drug carriers. Kakran et al. have covalently functionalized GO with hydrophilic and biocompatible Pluronic F38, Tween 80, and maltodextrin for loading and delivery of a poorly water-soluble antioxidant and anticancer drug, ellagic acid (EA). Studies have shown that loaded EA presents higher cytotoxicity rather than free EA. Therefore, all three functionalized GO composites are suitable nanocarriers for drug delivery because of their nontoxicity and high drug loading capacity [86].

Shi et al. synthesized a GO@Ag nanocomposite by chemical deposition of AgNPs onto GO. DOX, one of the most effective drugs against a wide range of cancers, was employed as drug with a high drug loading efficiency. GO@Ag-DOX nanocomposite was further functionalized with PEG2000-co-(Asn-Gly-Arg) peptide, to achieve active tumor-targeting capacity and excellent stability in physiological solutions. Compared with free DOX, these nanocomposites afforded much higher antitumor efficacy without obvious toxicity in a tumor model in vivo. This system can also be employed as an agent for photothermal ablation of tumor. The ability of the GO-NPs to combine chemotherapy with external photothermal therapy significantly improved the therapeutic efficacy [87].

A new class of nanomaterial emerged when 2D graphene folds on itself, to form a triangular 0D structure. Graphene quantum dots (GQD) exhibit strong quantum confinement and edge effects with a size smaller than 10 nm [88], making these systems potential drug delivery and theragnostic devices. In addition, they present great stability and can be obtained by easy synthetic methods. GQD possess some interesting properties, such as, great spin, optical and electrical properties. They can be easily modified by carboxylic acid making them more hydrophilic resulting less cytotoxic and more biocompatible [89].

GQD can be employed as targeted drug delivery systems because they can facilitate the transportation of the loaded molecule to the cell membrane and nucleus. Moreover, QGD can efficiently improve the nuclear accumulation rate of drugs, also enhanced the activity and cytotoxicity of these drugs against cancer cells [90]. Javanbakht et al. designed a hydrogel nanocomposite film by incorporation of GQD as NPs into a carboxymethyl cellulose hydrogel. DOX was efficiently loaded into the nanocomposite, as a broad-spectrum anticancer drug. GQD-carboxymethyl cellulose nanocomposites showed an improvement in degradation, pH-sensitive drug delivery, and significant toxicity against blood cancer cells [91].

An interesting alternative of graphene materials relies on combining GO with iron oxide NPs. Fe_3O_4 is less toxic, superparamagnetic, biodegradable, and biocompatible than other magnetic NPs resulting in a promising graphene-based composite [75]. GO acts as an oxidizing agent converting Fe_2 into Fe_3, generating Fe_3O_4 and turning itself into rGO, thus in situ deposition of the iron oxide on the rGO surfaces is achieved. The final material shows properties from both components [92].

Carbon-based polymer nanocomposites are considered unique candidates to be employed as drug delivery carriers. Inorganic carbon provides excellent thermal, mechanical, and optical properties [93], and polymers are capable of encapsulating and releasing cargo molecules with the ability to achieve active and passive targeting. The physicochemical properties of the nanocomposites will be determined by the dispersion of the graphene sheets inside the polymer matrix. Graphene sheets modifications with polymers result in increased biocompatibility and higher drug loading efficiency. Goncalves et al. grafted poly(methyl metacrylate) from the GO surface via atom transfer radical polymerization (ATRP) to generate a novel nanocomposite with a homogenous dispersion of GO in the polymer matrix [22].

The most employed polymer in carbon-based composites is chitosan, its biocompatibility and cationic nature enables to target cancer cell based receptors and negatively charged surfaces of the cell [94]. Wang et al. investigated a novel drug delivery system, composed of NPs of galactosylated chitosan/graphene oxide/DOX for cancer treatment. The synthesized nanocomposite exhibited a high drug loading capacity and presented a pH-responsive behavior to release the loaded molecule in a low pH environment. Additionally, nanocomposite showed higher cytotoxicity compared to the non-glycosylated nanocomposite (chitosan/graphene oxide/doxorubicin). Furthermore, Wang et al. synthesized a novel drug delivery system by encapsulation of superparamagnetic GO and DOX with folic acid conjugated chitosan. Nanocomposites exhibited a high loading efficiency and a prolonged release rate, which can be controlled by varying the pH. This system could be applied as dual-targeted drug nanocarrier, biological and magnetic targeting capabilities [81, 95]. These results showed that different components, such as polymers and magnetic NPs, can be combined with GO/rGO to obtain an improved material with the properties of all components.

Carbon nanotubes (CNTs) are a carbon allotrope which is cylindrical. Carbon nanotubes are hollow cylinders formed by graphene sheets rolled up on itself. These systems can be single walled carbon nanotubes (SWCNT), in which a single graphene sheet is rolled, or multiwalled carbon nanotubes (MWCNT), consisting of several SWCNT of different diameter inside each other. Their aromatic structures enable them to absorb several anticancer drugs, as well as, polymeric anticancer molecules by π-π stacking interactions [10]. Taking this into account, CNTs have emerged as a potential carrier for anticancer drug delivery and diagnosis. Chen et al. developed a CdTe quantum dots with Fe_3O_4-filled CNTs. This specific nanoplatform achieved magnetically guided anticancer drug delivery. The combination of iron oxide inside CNTs not only provides the magnetic property but also protects the CNTs from agglomeration, enhancing their stability and drug loading capacity. By coating the CNTs surfaces with transferrin, dual responsiveness is achieved. These properties make CNTs interesting nanocarriers for drug delivery [96].

5 Conclusions and Perspectives

Becoming quite popular in the last years, nanocomposites have gained a lot of attention in the field of biomedicine due to their improved physicochemical properties. The use of targeted nanocomposites in drug delivery ensures a controlled release rate in the desired site avoiding the side effects of highly specifics drugs. The possibility of combining two or more components into a single material enables them to possess unique properties, such as high loading capacity, enhanced biocompatibility and biodegradability, decreased toxicity, improved mechanical properties, and the capacity to image guidance.

In this chapter, we have summarized the most important concepts and recent advances in the use of nanocomposites in targeted cancer drug delivery. From the discussion made, it is possible to conclude that the multifunctional characteristics of nanocomposites make them highly suitable for this biomedical application. Nonetheless, overcoming toxicity drawbacks related to the used of nanocomposites in medicine remains a challenge to be solved in the upcoming years. Several issues as possible side effects, the long-term effects of biodistribution, the accumulation in nontarget tissues, as well as, the toxicity effects are imperative to be addressed in depth by the research community in order to move along to clinical trials and increase the number of FDA-approved nanomedicines. The need of interdisciplinary points of view in order to achieved better understanding of the structure/property relationships, biological performance and pharmaceutical effects of nanocomposites should not be overlooked. Therefore, both the development of novel nanocomposites and the in vivo evaluation of the safety and efficacy of those already reported ones are necessary for continuous progress.

Furthermore, multicomponent systems can offer the opportunity of developing image-guided cancer treatments or improving the antitumoral effect. Finally, all the considerations made let us to conclude that inorganic materials will display better targeted drug delivery properties when they are interacting with soft organic systems as polymeric materials.

All in all, much work is still needed for the development of improved nanocomposites which can become the next generation of efficient chemotherapeutic formulations.

References

1. Siegel RL, Miller KD, Jemal A (2019) Cancer statistics, 2019. CA Cancer J Clin 69(1):7–34
2. Hossen S, Hossain MK, Basher MK, Mia MNH, Rahman MT, Uddin MJ (2019) Smart nanocarrier-based drug delivery systems for cancer therapy and toxicity studies: a review. J Adv Res 15:1–18
3. www.who.int/news-room/fact-sheets/detail/cancer
4. Facts & Figures (2019) Am Cancer Soc 76
5. Baek S et al (2015) Smart multifunctional drug delivery towards anticancer therapy harmonized in mesoporous nanoparticles. Nanoscale 7:14191–14216

6. Mishra DK, Yadav KS, Prabhakar B, Gaud RS (2018) Nanocomposite for cancer targeted drug delivery. In: Applications of nanocomposite materials in drug delivery. Elsevier Inc., pp 323–337
7. Dai Y, Xu C, Chen X (2017) Nanoparticle design strategies for enhanced anticancer therapy by exploiting the tumour microenvironment. Chem Soc Rev 46:3830–3852
8. Wolfram J et al (2014) Safety of nanoparticles in medicine. Curr Drug Targets 16:1671–1681
9. Bae YH, Park K (2011) Targeted drug delivery to tumors: myths, reality and possibility. J Control Release 153(3):198–205
10. Pattni BS, Torchilin VP (2015) Targeted drug delivery: concepts and design. In: Targeted drug delivery: concepts and design, pp 3–38
11. Alemán J et al (2007) Definitions of terms relating to the structure and processing of sols, gels, networks, and inorganic-organic hybrid materials (IUPAC recommendations 2007). Pure Appl Chem 79(10):1801–1829
12. Merino S, Martín C, Kostarelos K, Prato M, Vázquez E (2015) Nanocomposite hydrogels: 3D polymer-nanoparticle synergies for on-demand drug delivery. ACS Nano 9(5):4686–4697
13. Zhao N et al (2019) Versatile types of organic/inorganic nanohybrids: from strategic design to biomedical applications. Chem Rev 119:1666
14. Bobo D, Robinson KJ, Islam J, Thurecht KJ, Corrie SR (2016) Nanoparticle-based medicines: a review of FDA-approved materials and clinical trials to date. Pharm Res 33:2373–2387
15. Jayalekshmi AC (2018) Nanocomposites used for drug delivery applications. In: Drug delivery nanosystems for biomedical applications. Elsevier Inc., pp 181–199
16. Conzatti G, Cavalie S, Combes C, Torrisani J, Carrere N, Tourrette A (2017) Biointerfaces PNIPAM grafted surfaces through ATRP and RAFT polymerization: chemistry and bioadhesion. Colloids Surf. B Biointerfaces 151:143–155
17. Pelissari FM, Neri-Numa IA, Molina G, Ferreira DC, Pastore G (2018) Potential of nanoparticles as drug delivery system for cancer treatment. In: Applications of nanocomposite materials in drug delivery. Elsevier Inc., pp 431–468
18. Santos HA, Bimbo LM, Peltonen L, Hirvonen J (2015) Inorganic nanoparticles in targeted drug delivery and imaging. In: Targeted drug delivery: concepts and design, pp 571–613
19. Feldman D (2016) Polymer nanocomposites in medicine. J Macromol Sci Part A Pure Appl Chem 53(1):55–62
20. Goonoo N, Bhaw-Luximon A, Bowlin GL, Jhurry D (2013) An assessment of biopolymer- and synthetic polymer-based scaffolds for bone and vascular tissue engineering. Polym Int 62(4):523–533
21. Sahoo S, Sasmal A, Nanda R, Phani AR, Nayak PL (2010) Synthesis of chitosan-polycaprolactone blend for control delivery of ofloxacin drug. Carbohydr Polym 79(1):106–113
22. Paredes M, Pulcinelli SH, Peniche C, Gonçalves V, Santilli CV (2014) Chitosan/(ureasil–PEO hybrid) blend for drug delivery. J Sol-Gel Sci Technol 72(2):233–238
23. Duncan R, Vicent MJ (2013) Polymer therapeutics-prospects for 21st century: the end of the beginning. Adv Drug Deliv Rev 65(1):60–70
24. Tebaldi ML, Belardi RM, Montoro SR (2016) Polymers with nano-encapsulated functional polymers: encapsulated nanoparticles for treatment of cancer cells. Encapsulated nanoparticles for treatment of cancer cells. Elsevier Inc.
25. Ali A, Ahmed S (2018) A review on chitosan and its nanocomposites in drug delivery. Int J Biol Macromol 109:273–286
26. Jayakumar R, Menon D, Manzoor K, Nair SV, Tamura H (2010) Biomedical applications of chitin and chitosan based nanomaterials—a short review. Carbohydr Polym 82(2):227–232
27. Luo M et al (2018) Fabrication of chitosan based nanocomposite with legumain sensitive properties using charge driven self-assembly strategy. J Mater Sci Mater Med 29(9)
28. Xu Y, Shen M, Sun Y, Gao P, Duan Y (2015) Polymer nanocomposites based thermo-sensitive gel for paclitaxel and temozolomide co-delivery to glioblastoma cells. J Nanosci Nanotechnol 15(12):9777–9787
29. Liang R, Wei M, Evans DG, Duan X (2014) Inorganic nanomaterials for bioimaging, targeted drug delivery and therapeutics. Chem Commun 50(91):14071–14081

30. Zhang L, Li Y, Yu JC (2014) Chemical modification of inorganic nanostructures for targeted and controlled drug delivery in cancer treatment. J Mater Chem B 2:452–470
31. Anselmo AC, Mitragotri S (2015) A review of clinical translation of inorganic nanoparticles. AAPS J 17(5):1041–1054
32. Liong M et al (2008) Multifunctional inorganic nanoparticles for imaging, targeting, and drug delivery. ACS Nano 2(5):889–896
33. Nalone LA et al (2018) Applications of nanocomposite materials in the delivery of anticancer drugs. In: Applications of nanocomposite materials in drug delivery, pp 339–352
34. Yuan L, Tang Q, Yang D, Zhang JZ, Zhang F, Hu J (2011) Preparation of pH-responsive mesoporous silica nanoparticles and their application in controlled drug delivery. J Phys Chem C 115:9926–9932
35. Ma P et al (2015) Inorganic nanocarriers for platinum drug delivery. Mater Today 18(10):554–564
36. Kim CS, Tonga GY, Solfiell D, Rotello VM (2013) Inorganic nanosystems for therapeutic delivery: Status and prospects. Adv Drug Deliv Rev 65(1):93–99
37. Suk JS, Xu Q, Kim N, Hanes J, Ensign LM (2016) PEGylation as a strategy for improving nanoparticle-based drug and gene delivery. Adv Drug Deliv Rev 99:28–51
38. Bogdal D, Bednarz S, Matras-Postolek K (2014) Microwave-assisted synthesis of hybrid polymer materials and composites. Biol Res 5(1):241–294
39. Arora A, Padua GW (2010) Review: Nanocomposites in food packaging. J Food Sci 75(1):43–49
40. Castro PM, Sarmento B, Madureira AR, Pintado ME (2019) Organic nanocomposites for the delivery of bioactive molecules. In: Natural polysaccharides in drug delivery and biomedical applications. Elsevier Inc., pp 471–493
41. Mousavi SM, Hashemi SA, Salahi S, Hosseini M, Amani AM, Babapoor A (2018) Development of clay nanoparticles toward bio and medical applications. In: Current topics in the utilization of clay in industrial and medical applications, vol i, no tourism. InTech, p 13
42. Jafarbeglou M, Abdouss M, Shoushtari AM, Jafarbeglou M (2016) Clay nanocomposites as engineered drug delivery systems. RSC Adv 6(55):50002–50016
43. Sposito G, Skipper NT, Sutton R, Park SH, Soper AK, Greathouse JA (1999) Surface geochemistry of the clay minerals. Proc Natl Acad Sci USA 96(7):3358–3364
44. Zhang Y et al (2017) Intercalated 2D nanoclay for emerging drug delivery in cancer therapy. Nano Res. 10(8):2633–2643
45. Bedoya DA, Vasti C, Rojas R, Giacomelli CE (2017) Risedronate functionalized layered double hydroxides nanoparticles with bone targeting capabilities. Appl Clay Sci 141:257–264
46. Alcântara ACS, Darder M (2018) Building up functional bionanocomposites from the assembly of clays and biopolymers. Chem Rec 18(7):696–712
47. Mittal V (2009) Polymer layered silicate nanocomposites: a review. Materials (Basel) 2(3):992–1057
48. Pavlidou S, Papaspyrides CD (2008) A review on polymer-layered silicate nanocomposites. Prog Polym Sci 33(12):1119–1198
49. Yang J et al (2016) Enhanced therapeutic efficacy of doxorubicin for breast cancer using chitosan oligosaccharide-modified halloysite nanotubes. ACS Appl Mater Interfaces 8(40):26578–26590
50. Phan VHG et al (2016) Pancreatic cancer therapy using an injectable nanobiohybrid hydrogel. RSC Adv 6(47):41644–41655
51. Ray S, Saha S, Sa B, Chakraborty J (2017) In vivo pharmacological evaluation and efficacy study of methotrexate-encapsulated polymer-coated layered double hydroxide nanoparticles for possible application in the treatment of osteosarcoma. Drug Deliv Transl Res 7(2):259–275
52. Connor DM, Broome AM (2018) Gold nanoparticles for the delivery of cancer therapeutics. In: Advances in cancer research, 1st ed., vol 139. Elsevier Inc., pp 163–184
53. Rastinehad AR et al (2019) Gold nanoshell-localized photothermal ablation of prostate tumors in a clinical pilot device study. Proc Natl Acad Sci USA 116(37):18590–18596

54. Urries I et al (2014) Magneto-plasmonic nanoparticles as theranostic platforms for magnetic resonance imaging, drug delivery and NIR hyperthermia applications. Nanoscale 6(15):9230
55. Yu Y et al (2019) Rattle-type gold nanorods/porous-SiO2 nanocomposites as near-infrared light-activated drug delivery systems for cancer combined chemo-photothermal therapy. Mol Pharm 16(5):1929–1938
56. Bolaños K, Kogan MJ, Araya E (2019) Capping gold nanoparticles with albumin to improve their biomedical properties. Int J Nanomed 14:6387–6406
57. Cai C et al (2018) Transferrin adsorbed on PEGylated gold nanoparticles and its relevance to targeting specificity. J Nanosci Nanotechnol 18(8):5306–5313
58. Siirilä J, Karesoja M, Pulkkinen P, Malho JM, Tenhu H (2019) Soft poly(N-vinylcaprolactam) nanogels surface-decorated with AuNPs. Response to temperature, light, and RF-field. Eur Polym J 115(January):59–69
59. De Matteis V, Cascione M, Toma CC, Leporatti S (2018) Silver nanoparticles: synthetic routes, in vitro toxicity and theranostic applications for cancer disease. Nanomaterials 8(5)
60. Farrag NS, El-Sabagh HA, Al-mahallawi AM, Amin AM, AbdEl-Bary A, Mamdouh W (2017) Comparative study on radiolabeling and biodistribution of core-shell silver/polymeric nanoparticles-based theranostics for tumor targeting. Int J Pharm 529(1–2):123–133
61. Bagheri E et al (2018) Silica based hybrid materials for drug delivery and bioimaging. J Control Release 277:57–76
62. Arriagada F, Nonell S, Morales J (2019) Silica-based nanosystems for therapeutic applications in the skin. Nanomedicine 14(16):2243–2267
63. Lindén JMRVMCSM (2012) Nanoparticles in targeted cancer therapy mesoporous silica nanoparticles entering preclinical development stage. Nanomedicine 7(1):111–120
64. Khaled SZ et al (2016) One-pot synthesis of pH-responsive hybrid nanogel particles for the intracellular delivery of small interfering RNA. Biomaterials 87:57–68
65. Maggini L, Cabrera I, Ruiz-Carretero A, Prasetyanto EA, Robinet E, De Cola L (2016) Breakable mesoporous silica nanoparticles for targeted drug delivery. Nanoscale 8(13):7240–7247
66. Follmann HDM et al (2018) Multifunctional hybrid aerogels: hyperbranched polymer-trapped mesoporous silica nanoparticles for sustained and prolonged drug release. Nanoscale 10(4):1704–1715
67. Li C et al (2018) Side effects-avoided theranostics achieved by biodegradable magnetic silica-sealed mesoporous polymer-drug with ultralow leakage. Biomaterials 186:1–7
68. Behrens S, Appel I (2016) Magnetic nanocomposites. Curr Opin Biotechnol 39:89–96
69. Karimi M et al (2016) Smart micro/nanoparticles in stimulus-responsive drug/gene delivery systems, vol 45, no 5. Royal Society of Chemistry
70. Bañobre-López M, Teijeiro A, Rivas J (2013) Magnetic nanoparticle-based hyperthermia for cancer treatment. Reports Pract Oncol Radiother 18(6):397–400
71. Etheridge ML, Campbell SA, Erdman AG, Haynes CL, Wolf SM, McCullough J (2013) "The big picture on nanomedicine: the state of investigational and approved nanomedicine products. Nanomed Nanotechnol Biol Med 9(1):1–14
72. Louguet S et al (2012) Thermoresponsive polymer brush-functionalized magnetic manganite nanoparticles for remotely triggered drug release. Polym Chem 3(6):1408–1417
73. Delavari B et al (2019) Theranostic α-lactalbumin-polymer-based nanocomposite as a drug delivery carrier for cancer therapy. ACS Biomater Sci Eng 5(10):5189–5208
74. Meenach SA, Shapiro JM, Hilt JZ, Anderson KW (2013) Characterization of PEG-iron oxide hydrogel nanocomposites for dual hyperthermia and paclitaxel delivery. J Biomater Sci Polym Ed 24(9):1112–1126
75. Madni A et al (2018) Graphene-based nanocomposites: synthesis and their theranostic applications. J Drug Target 26(10):858–883
76. Yuan XY, Zou LL, Liao CC, Dai JW (2012) Improved properties of chemically modified graphene/poly(methyl methacrylate) nanocomposites via a facile in-situ bulk polymerization. Express Polym Lett 6(10):847–858
77. Liu Z et al (2013) Alginic acid-coated chitosan nanoparticles loaded with legumain DNA vaccine: effect against breast cancer in mice. PLoS One 8(4)

78. Zhang L, Xia J, Zhao Q, Liu L, Zhang Z (2010) Functional graphene oxide as a nanocarrier for controlled loading and targeted delivery of mixed anticancer drugs. Small 6(4):537–544
79. Rui L et al (2015) Reduced graphene oxide directed self-assembly of phospholipid monolayers in liquid and gel phases. Biochim Biophys Acta -Biomembr 1848(5):1203–1211
80. Chang Y et al (2011) In vitro toxicity evaluation of graphene oxide on A549 cells. Toxicol Lett 200(3):201–210
81. Wang C, Zhang Z, Chen B, Gu L, Li Y, Yu S (2018) Design and evaluation of galactosylated chitosan/graphene oxide nanoparticles as a drug delivery system. J Colloid Interface Sci 516:332–341
82. Mendes RG et al (2015) A size dependent evaluation of the cytotoxicity and uptake of nanographene oxide. J Mater Chem B 3(12):2522–2529
83. Yue H et al (2012) The role of the lateral dimension of graphene oxide in the regulation of cellular responses. Biomaterials 33(16):4013–4021
84. Li Y et al (2013) Graphene microsheets enter cells through spontaneous membrane penetration at edge asperities and corner sites. Proc Natl Acad Sci USA 110(30):12295–12300
85. Dhas N, Parekh K, Pandey A, Kudarha R, Mutalik S, Mehta T (2019) Two dimensional carbon based nanocomposites as multimodal therapeutic and diagnostic platform: a biomedical and toxicological perspective. J Control Release 308(July):130–161
86. Kakran M, Sahoo NG, Bao H, Pan Y, Li L (2011) Functionalized graphene oxide as nanocarrier for loading and delivery of ellagic acid. Curr Med Chem 18(29):4503–4512
87. Shi J et al (2014) A tumor-targeting near-infrared laser-triggered drug delivery system based on GO@Ag nanoparticles for chemo-photothermal therapy and X-ray imaging. Biomaterials 35(22):5847–5861
88. Zhao J et al (2014) Chlorine doped graphene quantum dots: Preparation, properties, and photovoltaic detectors. Appl Phys Lett 105(11)
89. Shen J, Zhu Y, Yang X, Li C (2012) Graphene quantum dots: emergent nanolights for bioimaging, sensors, catalysis and photovoltaic devices. Chem Commun 48(31):3686–3699
90. Wang C et al (2013) Enhancing cell nucleus accumulation and DNA cleavage activity of anti-cancer drug via graphene quantum dots. Sci Rep 3:1–8
91. Javanbakht S, Namazi H (2018) Doxorubicin loaded carboxymethyl cellulose/graphene quantum dot nanocomposite hydrogel films as a potential anticancer drug delivery system, vol 87. Elsevier B.V
92. Xue Y et al (2011) Oxidizing metal ions with graphene oxide: The in situ formation of magnetic nanoparticles on self-reduced graphene sheets for multifunctional applications. Chem Commun 47(42):11689–11691
93. Santos CM, Mangadlao J, Ahmed F, Leon A, Advincula RC, Rodrigues DF (2012) Graphene nanocomposite for biomedical applications: fabrication, antimicrobial and cytotoxic investigations. Nanotechnology 23(39)
94. Rao W et al (2015) Chitosan-decorated doxorubicin-encapsulated nanoparticle targets and eliminates tumor reinitiating cancer stem-like cells. ACS Nano 9(6):5725–5740
95. Wang Z et al (2013) Fabrication and characterization of a triple functionalization of graphene oxide with Fe3O4, folic acid and doxorubicin as dual-targeted drug nanocarrier. Colloids Surf B Biointerfaces 106:60–65
96. Chen ML, He YJ, Chen XW, Wang JH (2012) Quantum dots conjugated with Fe3O4-filled carbon nanotubes for cancer-targeted imaging and magnetically guided drug delivery. Langmuir 28(47):16469–16476